Signals and Communication Technology

For further volumes:
http://www.springer.com/series/4748

Apurba Das

Signal Conditioning

An Introduction to Continuous Wave Communication and Signal Processing

Springer

Apurba Das
Senior Technical Lead
Imaging Tech Lab
Engineering and R&D
HCL Technologies Ltd.
Chennai
India

Formerly

Scientist
Image Processing Lab
CDAC, Kolkatta
Ministry of Communication and IT
Kolkatta
West Bengal
India
e-mail: apurba_das1@hotmail.com; apurba.das@hcl.com

Additional material to this book can be downloaded from http://extras.springer.com/

ISSN 1860-4862
ISBN 978-3-642-44188-2 ISBN 978-3-642-28818-0 (eBook)
DOI 10.1007/978-3-642-28818-0
Springer Heidelberg New York Dordrecht London

Printed on acid-free paper

Springer is part of Springer Science+Business Media (www.springer.com)

To
My Wife
Mrs. Angana Das Ghosh

Preface

For the past few years there was a requirement for a complete book volume which presents the fundamental techniques of signal conditioning along with their state of art applications both in the domain of continuous wave communication, single and two dimensional signal processing. In September 2010, just after the successful publication of my book "Digital communication—Principles and System Modelling" from Springer, the senior editor of Springer's "Signals and Communication Technology" series Dr. Christoph Baumann discussed about that requirement with me. He proposed me to write a book to meet the requirement. He opined that the present book along with the aforementioned book in digital communication can be an impressive two-volume archive for understanding the flavor of the subjects like communication, signal processing and image processing and their interdependencies. With this thought and motivation we both have prepared the draft table of content for the present book. This sincere effort to fulfill the expectations of all the readers including undergraduate and post graduate students, teachers, illustrators, domain professionals and researchers, started just one year back.

The generic chapter on preview and introduction starts with discussing about the fundamental properties and operations in signals and systems. The entire content is intentionally segregated in two parts namely "Continuous wave communication and analog signal conditioning" and "Discrete signal conditioning: 1D and 2D". The 1st part comprises of the continuous time Fourier series and transform, and the basic analog modulations like amplitude, frequency and phase modulation. The 2nd part starts with basic operations in discrete time signals and systems and ends with detailed discussion on different transformations like DTFS, DTFT, DFT, FFT, ZT and DWT.

I found a common problem among the students of undergraduate and even post graduate levels of incomplete and insufficient understanding of the transformed domain interpretation of signals and systems, physically. Most of them consider the same as only some mathematical tools which increases there overheads only. The main reason of this perception is the lag of communication from the authors and teachers (with due respect to all of them) in terms of interconnections between the transformed domain realizations of signals and systems, interpretations and approach in solving the real world problems. Moreover, the students try to look at each of the transformations as a separate tool. The first thing I tried in my approach is interpreting the transformed domain representation used for analysis of signals and systems, physically. I have shown the physical interpretations of all the transformed coefficients so that they can just be derived by some very simple *geometric pattern matching*. I hope this will reduce the overhead of the readers and help them to understand the foundation and applicability of the different transformations, as well.

In this book, different transformations like Continuous and Discrete Time Fourier Series and Transforms (CTFS, CTFT, DTFS, DTFT), Discrete Fourier Transform (DFT), Fast Fourier Transform (FFT), Laplace Transform (LT), Z-transform, and Discrete Wavelet Transform (DWT) are presented in a collateral way like beads in a single necklace. As the research progressed, new transformations evolved from the existing one to meet the requirements, applicability and reduction of computational cost. In this book I have introduced each of the transformations in the same fashion, considering them as inherited or derived transform from the previous one or the fundamental one like Fourier series or transform. This approach denies the wrong idea of considering each of the transforms as lonely island. They all are closely interrelated and in this book their relationships are clearly projected by deriving one transform from another. The boundary conditions used in these inter-conversions will help the students to understand why one transformation is at all necessary to be introduced, what is its applicability and how does it help to solve a real time problems in an easier way. This will answer the very basic question, "Why should I study this?" in students' mind. Believe me the answer will not just be "It is in the syllables"; moreover it will be "This is an important and interesting knowledge for me which I can apply in my domain of interest." I hope this new approach will help the teachers also to deliver the knowledge in an elegant way.

In my student days I was very much interested to learn new things and validate my own interpretation from the real world. I think most of the students have this kind of tendency. Real researchers are hidden to be explored into them. When I started my career as a teacher in a degree engineering college, I tried to present the subject in my own way. The classes should not be a copy book one. After few years of teaching when I looked into my last few years, I found that I have delivered the same set subjects at least 20 times but each time I followed completely different approach. The teacher-student interaction is really a *value-zone*, according to me. Each time the teacher should enjoy discovering the new shade of the subject. Then only the students will also enjoy the *value-zone*. I have incorporated some derivations and physical interpretations of some fundamental theorems in non-conventional way. I have shown how a problem can be attacked from different end of knowledge pool. I am just trying to give one example of the approach using the statement, "Sampling results in periodic spectra". This statement can be mathematically derived from the frequency shifting property of Fourier transform in a conventional way. I have incorporated the physical interpretation of sampling in frequency domain using convolution also. The same idea will help the readers to understand the two fundamental transformations for discrete time (sampled) signals, namely DTFS and DTFT easily.

It is comparatively easier to understand the concept of frequency and phase in time domain representation of single dimensional signal than that of a spatial representation of a two dimensional image. It is really difficult to realize frequency and phase from a given image without understanding the concepts physically. I have tried to hit that area also. In the chapter of Discrete Fourier Transform (DFT) I have clearly described the concept of frequency in image, in Fourier Series chapter (Chap. 2) I have shown how from the phase congruency we can detect several features like edges and corners from an image, in the discussion of Discrete Wavelet Transform (DWT) in Chap. 10, I have shown how we can achieve good compression ratio ensuring quality preservation according to bit budget. The applications of signal conditioning including the understanding of different transformations are projected in the area of electrical signal processing, image enhancement and processing, speech analysis and synthesis. The readers' eyes of knowledge would definitely feel soothing by this wide spectrum of domain coverage.

The Chap. 11 comprises of one interesting application of the understanding of the entire subject. There I have introduce the concept of "perceptual redundancy". A new way of communication where the signal to noise ratio (SNR) is tried to be minimized so that the signal can be ensured to be unobserved by the intruders. Steganography is a technique of secret data hiding in multimedia. I have shown different real time application of steganography in all the multimedia covers (carriers) including text, image, audio, video and IP datagram.

Salient Features

1. The subject is introduced from the very basic understanding of signal and explored the entire gamut in an elegant way.
2. Several transformed domain representations and realizations like Fourier series and transform, Laplace and Z-transform, Wavelet transform are introduced and presented in a collateral way as the research progressed in time depending on requirements, applicability and reduction of computational cost.
3. All the transformations are considered as inherited or derived transform from the previous one or the fundamental one like Fourier series and Fourier transform.
4. The application areas included in this volume are rich and resemblance to the present trend of research.
5. A supplementary electronic material is included into this book in the form of CD and/ or online content (www.extras.springer.com) which includes a number of *codes* and *MATLAB*, with illuminating uncommon and common applications for better understanding of the subject.
6. Complete code for many state-of-art real life applications like wavelet based image compression (EZW) and data hiding (watermarking) in multimedia are included.

7. A GUI based Fourier Synthesis module (Version 1.1) is also supplied with the other supplementary electronic materials. Readers can play with the tool and observe the synthesized waveform by tuning the amplitudes of the *sine* and *cosine* Fourier coefficients.
8. Elegant worked out numerical problems are designed in such a way that, the readers can get the flavour of the subject and get attracted towards the future scopes of research and development in the domain of signal processing, communication, image processing and speech processing.
9. Unparallel tabular, flow chart based and pictorial methodology description is included for sustained impression of the proposed design/algorithms in mind.
10. Depending on the relationship and interdependencies between all the chapters, three distinct study flows are presented. It's now up to the readers to choose which study flow suits them according to their area/ domain of interest.

This volume is intended to fulfil the expectations of students as well as the teachers. I hope this book would not only be a favourite study material for the students but also can be a nice resource for teaching. In my career, I have been engaged in full time teaching, faculty upgradation training, served both government R & D, and hardcore public limited industry. Therefore I can feel the requirements of each and every sector closely. I expect detailed feedback from the readers of this book. My sincere efforts would be successful, if this volume meets the requirements and expectations of all the students, teachers, researchers, and professionals in the domain of communication, signal processing and image processing.

Chennai, 1st January 2012

Apurba Das

Acknowledgments

The first light of the sun—every morning it brings in the promise of new life. Each afternoon, it's a sunset to us. We cannot perceive the presence of the sun. But, originally that is the time for the sun to go to the other half of the earth with lots of new promises for all the dwellers there. I am just one of them who learn to live from that light. The light has taught me that, the morning of creativity must come however dark be the night. I acknowledge that first light of the sun for inspiring me for doing new things with enjoyment, again.

I like to express my heartiest thank to Dr. Christoph Baumann and Mrs. Carmen Wolf from Springer-Verlag, Germany for entrusting me to be the author for this new effort. I thank all the members of Springer-Verlag for such an wonderful production of this book.

I am thankful to my ex-colleagues Chandrani di and Washef from CDAC, Kolkata for keeping faith on me that I can complete the volume in time. Today, Chandrani di is no more with us. But, I can still see her happy face having this book volume in her hand.

I am grateful to my respected teacher Professor Kuntal Ghosh from Indian Statistical Institute, Kolkata. I am really obliged to him for allowing me to use the infrastructure of Centre for Soft Computing Research of Indian Statistical Institute, Kolkata so that I could undergo my research smoothly.

I like to thank my students Supriyo Srimani, Diptendu Kumar Kundu, Abhishek Das, Abhigayan Ganguly and Shubhojit Banerjee for their extreme support in content preparation, especially for the two chapters Amplitude Modulation and Z-transform. This book is written for the benefit of students, teachers, researchers and professionals. The persons from any discipline whoever reading this book, thank you all in advance for considering my book as your knowledge resource.

Just joining in a private industry, I thought it would be difficult for me to get accustomed in a completely new environment. My team-mates in HCL Technologies, Sankar, Bagavath, Thennarasan, Radhika, Karthikeyan, Kalps, Ganesh Babu always made a space for me where I can think freely, create freely, as before. Thank you all. I thank Dr. Bapu, Nandish and Vasudeva Rao from Imaging Tech Lab, Bangalore for their continuous encouragements.

In last one year I found two special persons in my life. I cannot say, they are only my neighbors, they are my relatives. Mr. Sandip Kanjilal and Mrs. Chandrani Kanjilal are now my real dada-boudi. I am honored to get their supports in my life.

Finally, I like to mention the name of my wife Mrs. Angana Das Ghosh. In my multi-dimensional activities like research, teaching, writing books I confess, I could not give sufficient time to my family. She never complained me, but probably she should. In spite of her cultural engagements, she had taken the pain of doing all the family works alone. This is her silent support. I could not do whatever I have done during last 4 years, without her support. I should not say thanks to her. Rather, I dedicate this book to her.

Chennai, India, 1st January 2012

The Flow of Study

Depending on the relationship and interdependencies between the eleven (11) chapters, three distinct study flows are presented. It's now up to the readers to choose which study flow suits them according to their domain of interest.

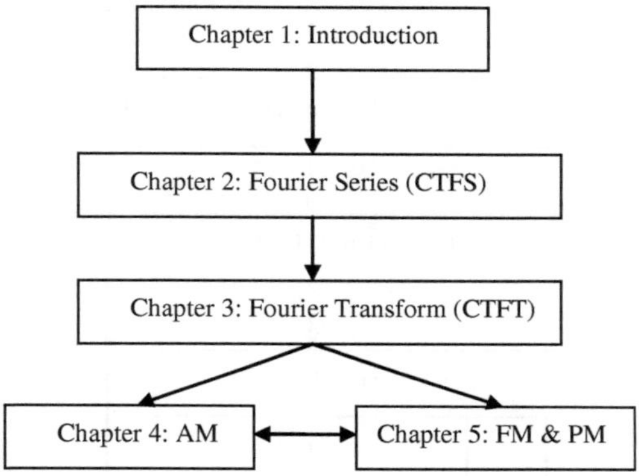

1st Study Flow

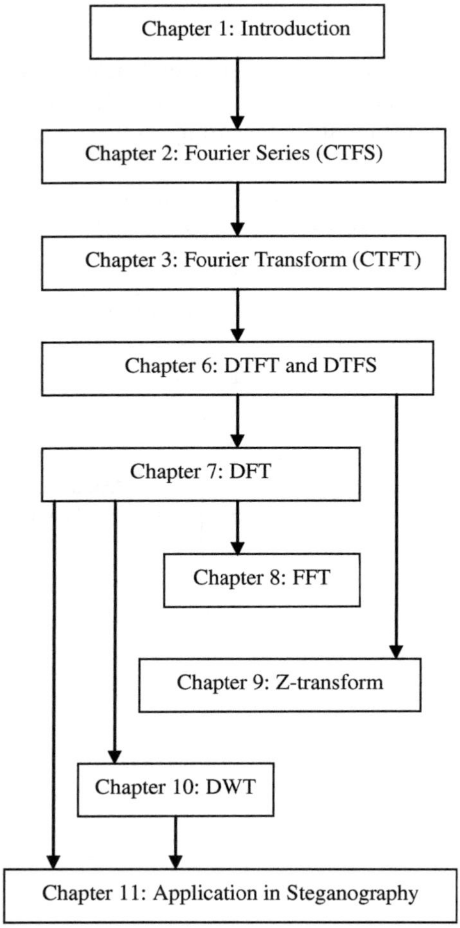

2nd Study Flow

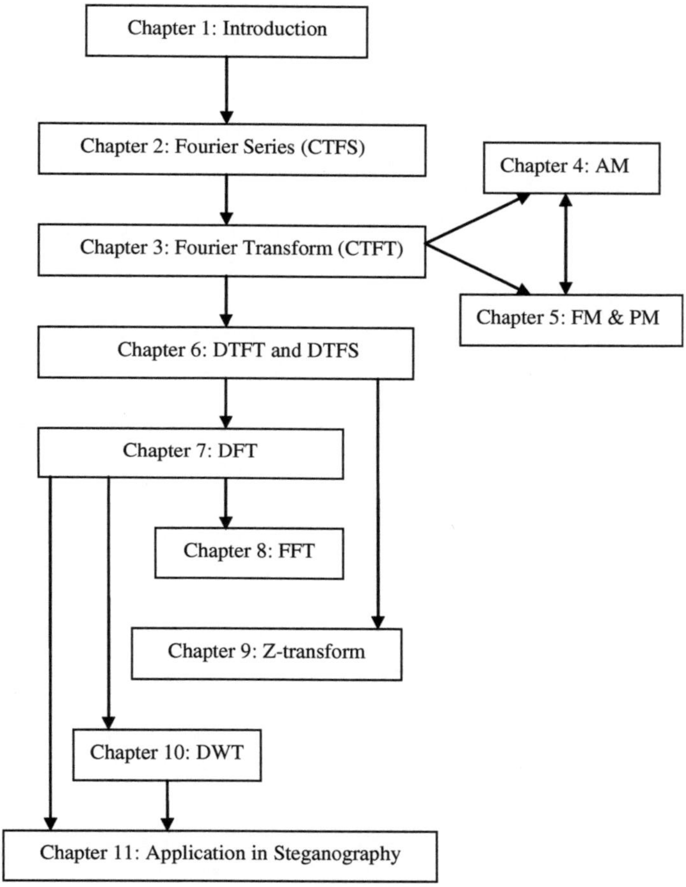

3rd Study Flow

Contents

Part II Discrete Signal Conditioning: 1D & 2D

Chapter 1
Preview and Introduction

The word "signal" carries a broad meaning in all the domains of knowledge gathering ranging from electronics and computer engineering to deaf-and-dumb communication. The word is uttered by every professionals ranging from journalists to linguists. In the present chapter we are trying to introduce signal and signal conditioning in a broader and elegant way. As the title of the book suggests, here we will take a close look not only on the aspects of analog to digital signal conversion as introduced in a pre-processor stage, but also we'll address different facets of continuous wave communication or analog communication system and general signal processing in time, spatial and frequency domain.

The present chapter would start with constructing a formal definition of signal in a logical way from its basic properties. The basic classifications of signal also have been discussed depending on time and amplitude continuity. The concept of signal conditioning is then formally addressed with brief discussion about the fundamental components of a signal conditioning system. Two very important operations titled as convolution and correlation, referred throughout the discussion of this book, are then discussed briefly.

1.1 Definition of Signal

Precisely, signal is the physical quantity by which information is carried between at least two entities. In other words by which physical quantity communication is done, is signal. Signal is not necessarily always represented as some electrical form of energy, it may be some gait, may be an image, and may be a video.

To construct a complete definition let us try to have a close look at the general properties of a signal.

Property 1: Representation of a physical quantity.

Property 2: Signal must carry some information. Meaning of a signal is not important in the property discussion. If a signal does not have meaning with respect to a particular communication, it may be in coded form or meaningful for another arena of

A. Das, *Signal Conditioning*, Signals and Communication Technology,
DOI: 10.1007/978-3-642-28818-0_1, © Springer-Verlag Berlin Heidelberg 2012

communication. It means the physical quantity is unwanted with respect to that specific application. Therefore, this is unwanted signal or noise.

Property 3: To measure or predict the response of a physical phenomena or system, mathematical modeling is important. Electrical (current) signal is generally characterized as

$$i(t) = I_0 \cos(\omega_0 t + \phi) \tag{1.1}$$

If we observe a small time window of that signal, the waveform obtained would be as shown in Fig. 1.1a.

Referring the Fig. 1.1, the time domain and frequency domain representations are shown. Obviously, the spectrum of the single tone signal is showing an impulse only, whereas the two tone DSB-SC type (to be discussed in Chap. 4) signal generates a pair of impulses. These representations of math models are satisfying the definition of signal but can only be considered as an example of a special class of signal, which is modeled as function of time or frequency as f(t) or $F(\omega)$. Let's consider now an image shown in the following figure. To represent the image as a discrete mathematical function, we have to first divide the entire image into a number of unit squares of homogeneous intensity value to each square. These smallest picture elements are called as pixels. Intensity or color of each smallest square block is now can be represented as function of ordinate and abscissa, i.e., function of space where the entire image plane is considered as a Cartesian co-ordinate system (Fig. 1.2).

Similarly, video images can be represented as a multivariate function as I(t,x,y), i.e., function of both time and space.

As per the discussion above, now we can characterize a signal mathematically as a function of independent variable/variables.

Now from the above three properties, we can define signal as, "**A physical quantity used for carrying information from one entity to another and mathematically modeled as a function of independent variable/variables**."

Our discussion would be mainly focused on the signals which can be modeled as a function of time. In the next section, we'll discuss the time-value definition and classification of signals. Hence, we'll clearly understand the time-value concept of analog and digital signals. Also we can find the step-by-step procedure of conversion of analog signal to digital signal.

1.2 Time-Value Definition of Signals: Analog and Digital

The word 'analog' comes from the word 'analogy' which means similarity. Here the "similarity" implies similarity with the nature. Directly we can say, any natural signal is analog signal. As nature, the starting time of any analog signal is also $t = -\infty$ and the ending time is $t = \infty$.

Let's consider the waveform as shown in the Fig. 1.3. We have taken one time window of the signal though the existence of the signal is beyond the time range

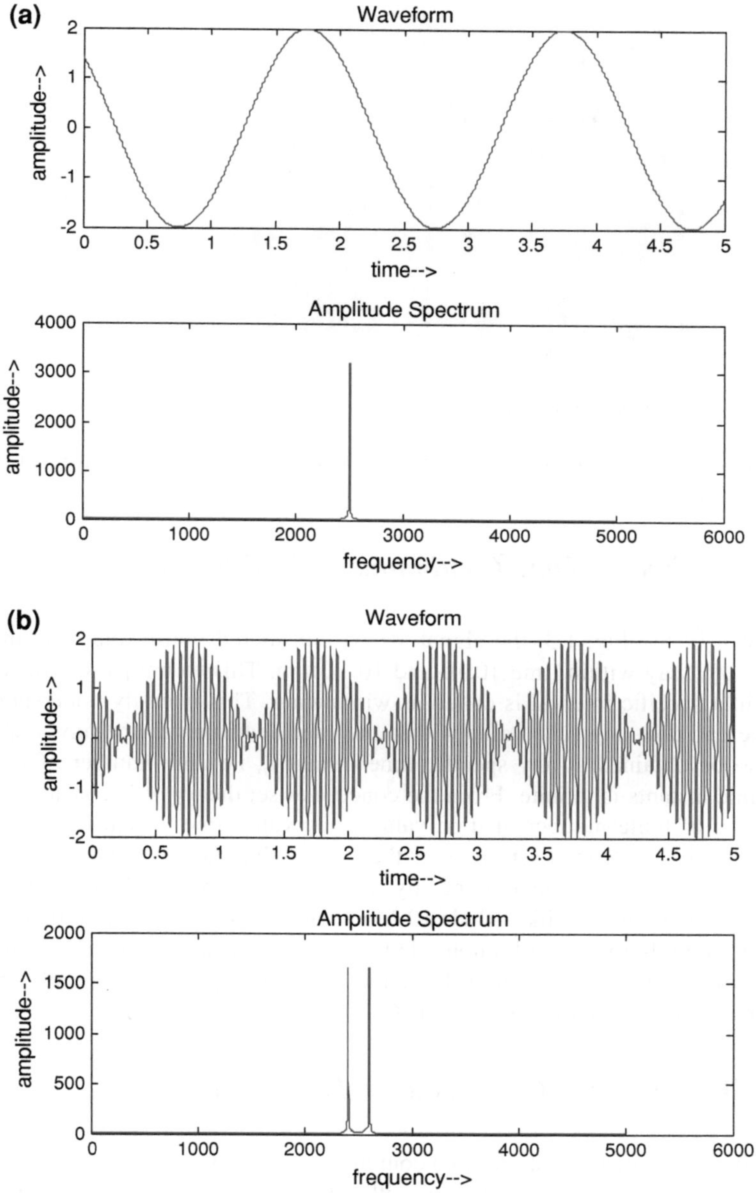

Fig. 1.1 **a** Time and frequency representation of a single tone signal. **b** Time and frequency domain representation of a two tone signal

$t = [t_1 \quad t_2]$. Within this time range, the signal $x(t)$ has maximum and minimum values X_{max} and X_{min} respectively. Now depending on the countability of the number of possible amplitude and possible time values, we can classify signal and this is in

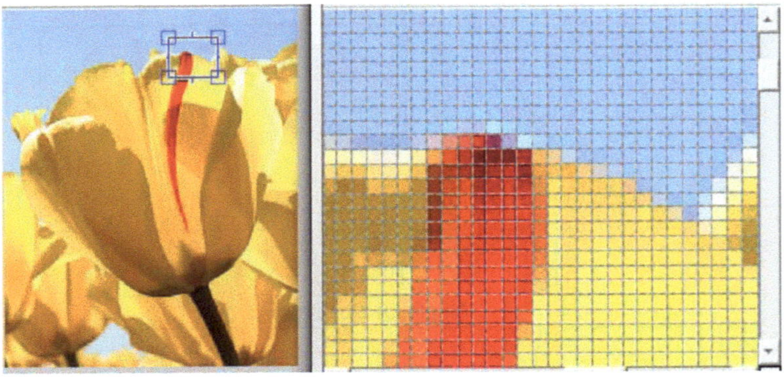

Fig. 1.2 Image representation as a function of 2D space

turn the step by step approach towards the procedure of ADC (analog to digital conversion).

1.2.1 Continuous Time Continuous Valued Signal

As shown in the Fig. 1.3, the signal $x(t)$ is considered within the time interval $t = [t_1 \quad t_2]$, say within time 10:10 and 10:11 a.m. This technique of watching a signal in a specific interval is called as windowing. This is easily understandable that, we can take any time value within the specified interval and any amplitude can also be obtained at any specific time instant t_i, i.e., the number of possible valid time instants is infinite. From the concept of set theory, it is considered that, from a set of finite number of time values, any valid time instant t_i must be the element of the predefined set, in order the declare that the set is discrete time set. In the present case we cannot define any set of finite number of elements. Because, we can have any instant like 10 Hr 10 Min 1 Sec 1.3 nS, so possible time values can be infinite. In the case of amplitude value also we can have infinite number of possible amplitude values within the range X_{max} and X_{min}. This type of signal is therefore called as continuous valued (CTCV) signal.

1.2.2 Discrete Time Continuous Valued Signal

Generally, the analog signals do hold continuous time and amplitude values. But no digital memory or processor can deal with infinite long words. Digital signals are represented using digits only. First, to express the signal, let's take the help of the equally spaced samples [1] of the signal. The equal time spacing between two adjacent samples is called as sampling interval, T_s. Then the second signal shown in the figure must have finite number of time values but, we can have infinite number of possible voltages values yet. This signal is called as discrete time continuous valued signal (DTCV).

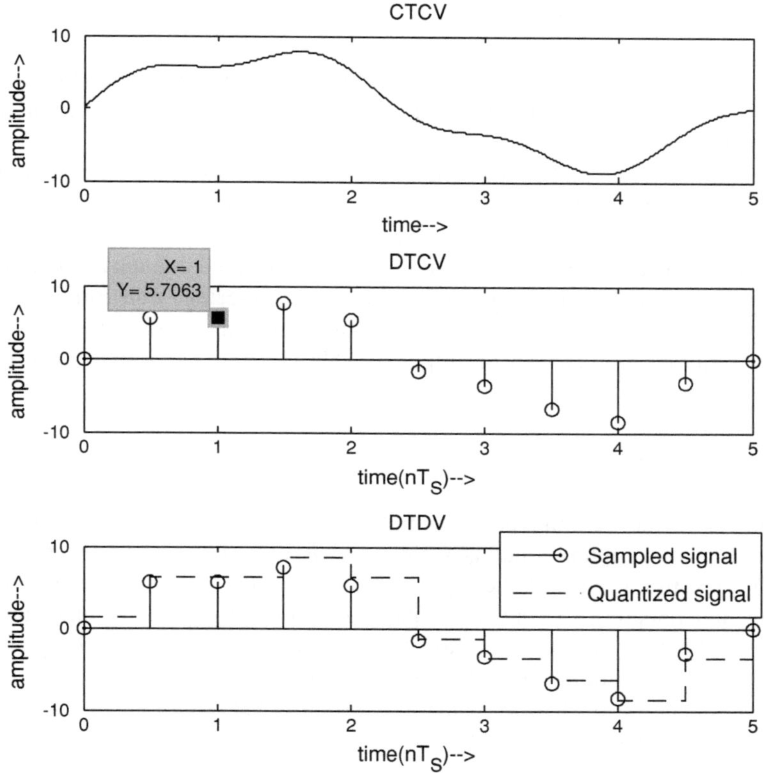

Fig. 1.3 Time valued definition of signal

1.2.3 Discrete Time Discrete Valued Signal

As shown in Fig. 1.3, at t = 1, the amplitude of the signal is 5.7063. It may be anything between X_{max} and X_{min}. Now, to get finite number of voltage values, a set of finite number of amplitude values with X_{max} and X_{min} as upper and lower limit, has to be defined. In the present case, within the amplitude range $10(X_{max})$ and $-10(X_{min})$, 8 values have been taken to form that closed set. The elements are, ±1.25, ± 3.75, ± 6.25, ± 8.75. Now, sampled ales can be approximated (i.e., quantized) to the closest defined element of the amplitude. As shown in Fig. 1.3, at time instant t = 1, the sampled value 5.7063 has been quantized to 6.25. The staircase form of the signal is having finite number of time and amplitude values. This signal is called as discrete time discrete valued (DTDV) signal. Now, that defined amplitude values can easily be **encoded** to get the digital bit stream.

1.3 Signal Conditioning

In the study of electronics, ***signal conditioning*** signifies manipulating an analog signal in such a way that it meets the requirements of the next stage for further processing.

In communication and control engineering applications, it is common to have a sensor, a signal conditioning stage (generally for amplification and calibration) and a processing stage which includes required analog-to-digital signal conversion, modulation, various operations like filtering in transformed domain and calibration.

Signal conditioning can include amplification, filtering, transformation, conversion, modulation, range matching, isolation and many more processes which may be required to make sensor output suitable for processing after conditioning.

1.3.1 Filtering

Filtering is the most common signal conditioning function, as not all the frequency components of a signal contains valid data, usually. Filtering is the process of ensuring the usable part of the signal to allow for the next stage of signal processing and rejecting the unwanted part. In electrical signal, to understand the process of filtering, it is essential to study the transformed domain representation of signal and their implications. In two dimensional signals like image a new term spatial frequency is dealt with, which is essentially the rate of change of intensity with respect to space.

1.3.2 Amplifying

Signal amplification performs two important functions: increases the resolution of the input signal, and increases its signal-to-noise ratio. For example, the output of an electronic temperature sensor, which is mostly in the millivolts range is too low for an Analog-to-digital converter (ADC) to process directly. In this case it is necessary to bring the voltage level up to that required by the ADC.

Commonly used amplifiers on signal conditioning include sample and hold amplifiers, Peak Detectors, Log amplifiers, Antilog amplifiers, Instrumentation amplifiers or programmable gain amplifiers [2].

1.3.3 Isolation

Signal isolation must be used in order to pass the signal from the source to the measurement device without a physical connection: it is often used to isolate possible sources of signal perturbations. It is important to isolate the potentially expensive equipment used to process the signal after conditioning from the sensor. For this purpose, generally magnetic, optical, or combination of resistive, inductive and capacitive isolations are used.

1.3.4 Modulation

To maintain realizable antenna height, to provide the facility of multiplexing and channel separation and to get higher level of noise immunity, a number of modulation schemes are employed both in the case of analog and digital signal conditioning like communication.

Modulation can be defined as a technique of modification of either or all the three basic properties of analog single tone[1] carrier signal with respect to the baseband signal. As the baseband signal (information) is solely responsible for the task of modulation, it is also known as the *modulating signal*. A single tone analog carrier signal can be expressed as:

$$s(t) = A\sin(\Omega t + \Phi) \tag{1.2}$$

where, A = Amplitude of the carrier signal
Ω = Frequency of the carrier signal
and, Φ = Initial Phase[2] of the carrier signal

The amplitude (A), frequency (Ω) and phase ($\theta(t) = (\Omega t + \Phi)$) are the three fundamental parameters of any signal. Therefore we can modify/modulate any one of these three parameters with respect to the analog modulating signal. Combining the frequency and initial phase part, the time dependent angle part $\theta(t) = (\Omega t + \Phi)$ is obtained. So, angle modulation is said to be the mother modulation scheme of frequency and phase modulation. Signal conditioning includes filters, instrument amplifiers, sample-and-hold amplifiers, isolation amplifiers, signal isolators, multiplexers, modulators, bridge conditioners, analog-to-digital converters, digital-to-analog converters, frequency converters or translators, voltage converters or inverters.

---

[1] Sinusoids and cosinusoids are only single tone (single frequency) signals.

[2] Initial phase angle: Phase at time $t = 0$.

1.4 Delayed and Advanced Signal

Time is an important part of any kind of systems used to sequence events, comparing processing time by each of the system module or event, analyzing signals and systems in various way, different kind of transformations. But we do not even know whether this is perfectly scalar or a vector. Generally for any system modeling or signal analysis, time $t = 0$ is taken as the reference time. For the ease of understanding let's consider this time $t = 0$ (say, 10.00 a.m. of Jan 1, 2012) as 'present'. Therefore anything before and after $t = 0$ would be considered as 'past' and 'future', respectively. Therefore, the positive maximum and negative maximum time range is $+\infty$ and $-\infty$ respectively.

According to the Fig. 1.4, the piecewise linear signal $f(t)$ has started at $t = 0$. So this reference time ("present") is the natural tendency of occurrence (start) of the signal $f(t)$. If the starting position of $f(t)$ is delayed by an amount τ, the starting position (from "present" to "future") of the signal would be shifted from $t = 0$ to $t = \tau$ i.e., $t - \tau = 0$. Therefore the delayed signal $f(t)$ would then be represented by $f(t - \tau)$. As the positive time axis is represented by "future", we can infer that the signal $f(t - \tau)$ occurred after its natural tendency with respect to $f(t)$.

Similarly, if the starting position of $f(t)$ is ***advanced*** by an amount of ξ, the starting position (from "present" to "past") of the signal would be shifted from $t = 0$ to $t = -\xi$ i.e., $t + \xi = 0$. Therefore the advanced signal $f(t)$ would then be represented by $f(t + \xi)$. As the negative time axis is represented by "past", we can infer that the signal $f(t + \xi)$ occurred before its natural tendency with respect to $f(t)$ (Fig. 1.4).

1.5 Even Signal and Odd Signal

From the discussion in previous section we now have a fair idea about the time axis. Time axis can be segmented as "present" $(t = 0)$, "past" $(t<0)$ and "future" $(t > 0)$. It is observed that the "present" is represented by a time *instant* whereas "past" and "future" are represented by two different *ranges* which is quite in agreement with the general concept of time.

A signal (or function) is said to be *even*, if the definition in the said two time ranges are identical. In other words, a signal is said to be even, which can be assumed to be synthesized by adding the positive time function and its mirror image (folded signal). If we consider the Y-axis as a mirror, realization of the even signal is obvious. As the virtual mirror is placed along Y-axis, the effect of imaging would be on time axis $(+t \rightarrow -t)$. Therefore, the condition of evenness of a signal is defined by

$$f(t) = f(-t) \tag{1.3}$$

Fig. 1.4 **a** Original signal, **b** delayed, and **c** advanced signal with respect to the original one

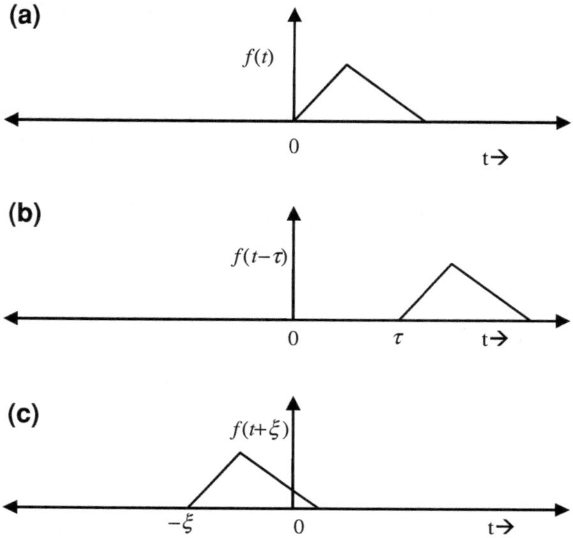

Examples of some even signals are as follows;

$$x(t) = 1;$$
$$y(t) = \cos \omega_0 t;$$
$$z(t) = rect\left(\frac{t}{\tau}\right)$$

A signal (or function) is said to be *odd*, if the definition in the said two time ranges are reciprocal. Unlike to the single mirror concept discussed in the even signal discussion, odd signal realization is a double-but-not-dual mirror concept. If the negative time axis portion of a signal is assumed to be realized by successive mirror effect along Y-axis and along X-axis respectively, then the signal is an odd signal. The first virtual mirror is placed along Y-axis, the effect of imaging would be on time axis $(+t \to -t)$. In the next mirror the effect of imaging would be on function axis $f \to -f$. Therefore, the condition of oddness of a signal is defined by

$$f(t) = -f(-t) \tag{1.4}$$

Examples of some even signals are as follows;

$$x(t) = t;$$
$$y(t) = \sin \omega_0 t;$$
$$z(t) = signum(t)$$

Some detail discussion on even and odd signal would be depicted in the chapters of Fourier series (Chap. 2) and Fourier transform (Chap. 3).

1.5.1 Even and Odd Components of a Signal

We have discussed about such types of signal which are having both positive and negative time axis presence. If a signal does not have the bothway time presence the signal can neither be odd nor be even. Even the both direction time presence cannot also ensure the signal to be either even or odd (Fig. 1.7). Therefore we can have a third kind of signal, which is neigher odd nor even. Let's assume the signal $x(n)$ is an example of such kind of signal expressed as a summation of two signals as

$$x(n) = x_1(n) + x_2(n) \tag{1.5}$$

Where,

$$x_1(n) = \frac{1}{2}\{x(n) + x(-n)\} \tag{1.6}$$

And,

$$x_2(n) = \frac{1}{2}\{x(n) - x(-n)\} \tag{1.7}$$

Let's observe now the folded version of the component signals $x_1(n)$ and $x_2(n)$.

$$x_1(-n) = \frac{1}{2}\{x(-n) + x(n)\} = x_1(n) \tag{1.8}$$

And

$$x_2(-n) = \frac{1}{2}\{x(-n) - x(n)\} = -\frac{1}{2}\{x(n) - x(-n)\} = -x_2(n) \tag{1.9}$$

From the definition of oddness and evenness of a signal, we can infer that, the component signals $x_1(n)$ and $x_2(n)$ are even and odd respectively.

Hence, from the discussion and expressions above we can state that, "*Any signal can be represented either as odd, or as even or as summation of odd and even signals.*"

Example 1.1 Derive and draw the even and odd component signal of the given function $f(t)$ as shown in the following Fig. 1.7.

As seen, the signal $f(t)$ satisfies neither the criteria of single mirror symmetry nor the criteria of double mirror symmetry as discussed in Figs. 1.5 and 1.6. Therefore the signal $f(t)$ falls under the third category of categorization which is neither odd nor even. Now, from the property of component decomposition as expressed in Eqs. 1.5, 1.6 and 1.7, we can represent the given signal as a summation of one odd signal and one even signal. The present problem thus can be solved by the use of simple geometric interpretation as follows.

As understood from the discussion of oddness and evenness we have folded the given signal (Fig. 1.7) with respect to a mirror aligned with Y-axis, as shown in

Fig. 1.5 **b** is the mirror image (folding operation) of the signal **a**; **c** an even signal which is symmetric along Y-axis

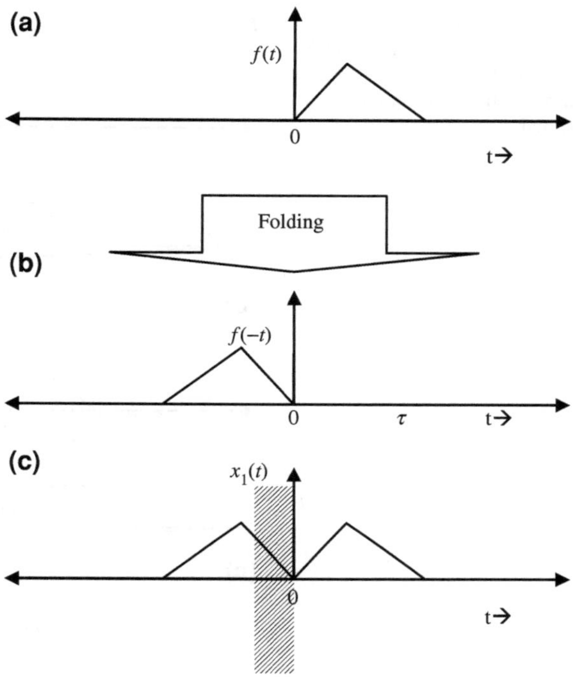

Fig. 1.8a. Next, the second mirror image is also obtained by considering a mirror aligned with X-axis as shown in Fig. 1.8b. By this way the signals $f(-t)$ and $-f(-t)$ have been generated from $f(t)$, respectively. Now, these two signals are added and subtracted with $f(t)$geometrically as shown in Fig. 1.8c and d and attenuated by a factor of 1/2. According to Eqs. 1.6 and 1.7 these are the even and odd components of the signal $f(t)$.

These representations can also be checked in two tier manner. In the first layer of checking, we can observe that the two signals $f_E(t)$ and$f_o(t)$ (as shown in Fig. 1.8c, d respectively) follows the single mirror and double mirror concepts respectively as discussed in Figs. 1.5 and 1.6. Therefore they are obviously even and odd signals.

In the second layer of testing we can observe that, just be adding $f_E(t)$ and $f_o(t)$ we can retrieve back the original signal $f(t)$.

1.6 Convolution

Generally, we define a system as a function or mapping which maps the input sequence (sampled signal) to output sequence. The mapping rule may be defined clearly or not. Say, the output or response $y(n)$ of a system τ with input or excitation $x(n)$ is clearly defined as $y(n) = 0.5x^2(n)$. If that is the case, then we can

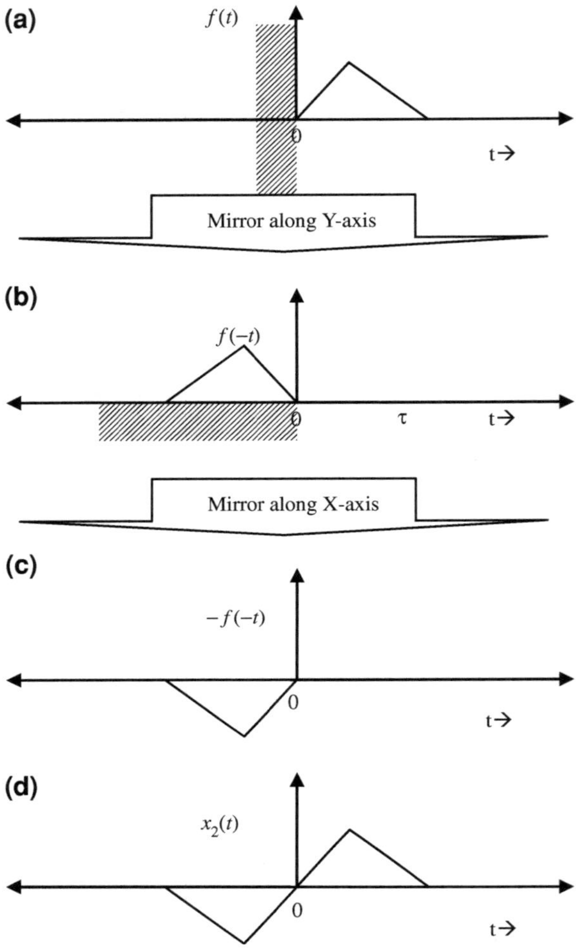

Fig. 1.6 **b** and **c** are the mirror images along Y and X mirror of signal **a** respectively, **d** an odd signal

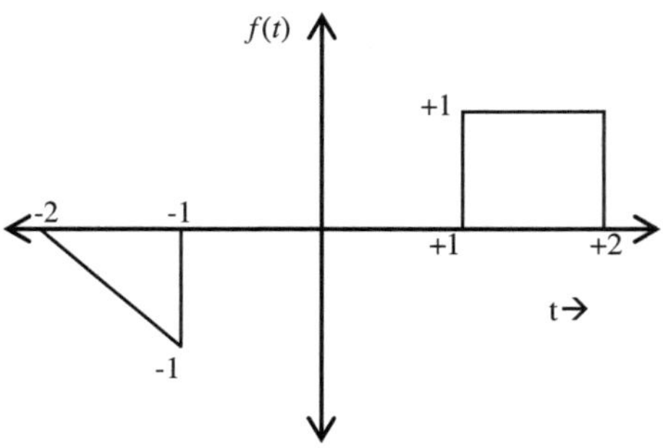

Fig. 1.7 A signal which is neither odd nor even

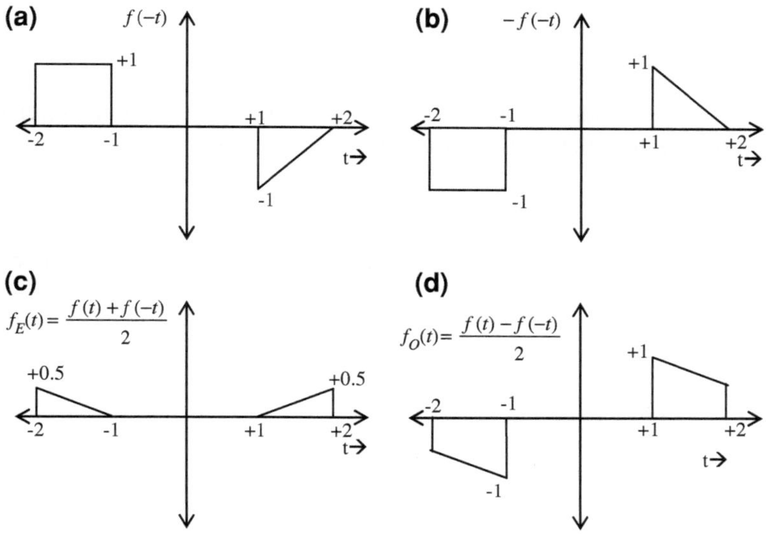

Fig. 1.8 Representing a signal $f(t)$ (shown in Fig. 1.7) as a summation of even **c** and odd **d** components. **a** is folded $f(t)$, i.e., $f(-t)$; **b** is the X-mirror image (second mirror concept) of the signal $f(-t)$, i.e., $-f(-t)$

Fig. 1.9 A system with unknown transfer function

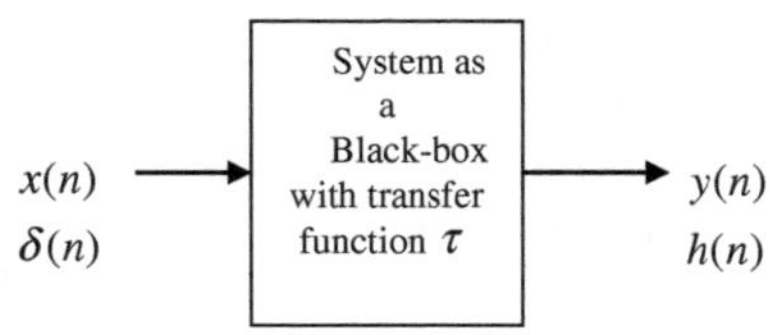

predict the output for any query input. But in practice, the situation is not so easy. The mapping τ is unknown in most of the cases and we have to consider the system as a black box as shown in the figure above (Fig. 1.9).

Now, if one asks for the predicted output for any arbitrary input, there must also be some procedure taking the help of the black-box. With this objective let us try to define any given sequence $x(n)$ as Eq. 1.10

$$
\begin{aligned}
x(n) &= x(-\infty)\delta(n+\infty) + \cdots\cdots\cdots\cdots\cdots + x(-1)\delta(n+1) + x(0)\delta(n) \\
&\quad + x(1)\delta(n-1) + \ldots x(\infty)\delta(n-\infty) \\
&\Rightarrow x(n) = \sum_{k=-\infty}^{\infty} x(k)\delta(n-k)
\end{aligned}
\tag{1.10}
$$

This is true for any sequence of infinite length, for finite length sequence the limits of summation of Eq. 1.10 will be finite. We know that $\delta(n)$ is the unit impulse defined as

$$\delta(n) = \begin{cases} 0; n \neq 0 \\ 1; n = 0 \end{cases} \tag{1.11}$$

And, the impulse response of a particular system is defined as the output (response) of the system with input (excitation) unit impulse, given by

$$h(n) = \tau[\delta(n)] \tag{1.12}$$

Now, if this sequence is allowed to excite the system 'τ', the output must be given by

$$\begin{aligned} y(n) &= \tau[x(n)] \\ &= \tau[\sum_{k=-\infty}^{\infty} x(k)\delta(n-k)] \\ &= \sum_{k=-\infty}^{\infty} x(k)\tau[\delta(n-k)] \quad \text{[If the system is Linear]} \\ &= \sum_{k=-\infty}^{\infty} x(k)h(n-k) \qquad \text{[If the system is Time Invariant]} \\ &\Rightarrow x(n) * h(n) \end{aligned} \tag{1.13}$$

The operation given by Eq. 1.13 is the convolution. This is not similar to simple multiplication. Here four operations are involved namely *fold, multiply, add, shift.* The operations are understandable from the Eq. 1.13

First the impulse response is obtained from the system (black-box) with the unit impulse excitation. Next, the impulse response is folded to get $h(-k)$ with n = 0 and taking k as the index. Then all the samples are multiplied with $x(k)$ in sample by sample basis and added up to get $y(0)$. To get other output samples $y(n \neq 0)$, $h(-k)$ is to be shifted by amount 'n' to both of the sides of time axis. The operations are described using an example as shown in Fig. 1.10 Say the impulse response $h(n) = \{1, -1\}$ and the input sequence is $x(n) = \{-1, -1, -1, 1, 1, 1\}$. In this example, $y(n)$ is 0 except $n = 2$. The physical interpretation of this is very important for system analysis and synthesis. Here, the impulse response $h(n)$ itself is behaving as the transfer function of the system. We can notice that the impulse response is actually combination of a $+ 1$ followed by a $- 1$. The combination results in a derivative, as in digital domain differential equation become difference equation. If any input having that differential property is feed into that particular system, the system $(h(n))$ will make the property prominent at the output. Therefore, the system characteristic suggests the system $h(n)$ to be a *differentiator.* Similarly, the impulse response can be of filter type, of modulator type and so on. The convolution operation is applicable for linear time invariant (LTI) systems only.

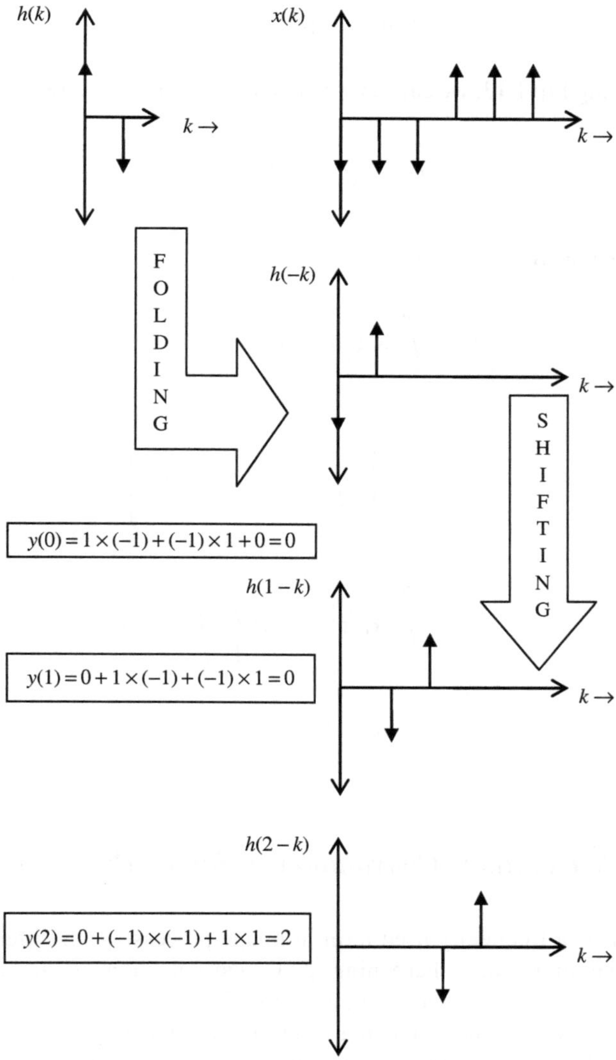

Fig. 1.10 Convolution operation between two sequences

1.6.1 Transformed Domain Simplicity

Convolution in time domain is actually dual to multiplication in frequency domain. Say, given two signals are $x_1(t)$ and $_2(t)$. The convolution sum between these two sequences is $x_3(t) = x_1(t) * x_2(t)$.

The duality theorem states that, if the Fourier Transforms (Chap. 3) of $x_1(t)$, $x_2(t)$ and $x_3(t)$ are $X_1(\omega)$, $X_2(\omega)$ and $X_3(\omega)$, and $x_3(t) = x_1(t) * x_2(t)$, then

$$X_3(\omega) = X_1(\omega) \times X_2(\omega)$$

Proof Referring Eq. 1.13, we can also express convolution in continuous domain as

$$x_3(t) = x_1(t) * x_2(t) = \int_{-\infty}^{\infty} x_1(\tau)x_2(t-\tau)d\tau$$

Now, Fourier transform of $x_3(t)$ is

$$\begin{aligned} X_3(\omega) &= \int_{-\infty}^{\infty} x_3(t)e^{-j\omega t}dt \\ &= \int_{-\infty}^{\infty}\left\{\int_{-\infty}^{\infty} x_1(\tau)x_2(t-\tau)d\tau\right\}e^{-j\omega t}dt \\ &= \int_{-\infty}^{\infty} x_1(\tau)e^{-j\omega\tau}d\tau\left\{\int_{-\infty}^{\infty} x_2(t-\tau)e^{-j\omega(t-\tau)}d(t-\tau)\right\} \\ &= X_1(\omega) \times X_2(\omega) \end{aligned} \tag{1.14}$$

1.6.2 2D Convolution: Convolution in Image Processing

'Convolution' is a frequently used term in image processing, in different types of operations like smoothing, sharpening, and edge detection of images. A digital image can be well expressed as 2 dimensional (2D) discrete signal represented in spatial domain. It is to be noted that, 2D convolution is just extension of 1D convolution which is discussed in the previous sub-section. This is done by convolving the mask (or system) both in horizontal and vertical directions in 2 dimensional spatial domain.

The impulse (delta) function is also represented in 2D space as shown in Fig. 1.11. This impulse function $\delta[m, n]$ has 1 where m and n is zero and zeros at $m \neq 0; n \neq 0$. The impulse response (h[m,n]) in 2D is usually called "*mask*" or "*kernel*" or "*filter*" in image processing.

The kernel (impulse response) in 2D is center originated in most cases, which means the center point of a kernel is $h[0, 0]$. For example, if the kernel size is 5, then the array index of 5 elements will be $-2, -1, 0, 1$, and 2. The origin is located at the middle of kernel as shown in Fig. 1.12.

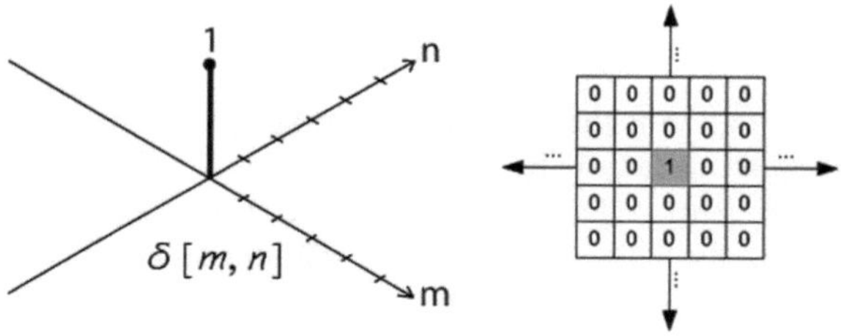

Fig. 1.11 Impulse function in two dimension (2D)

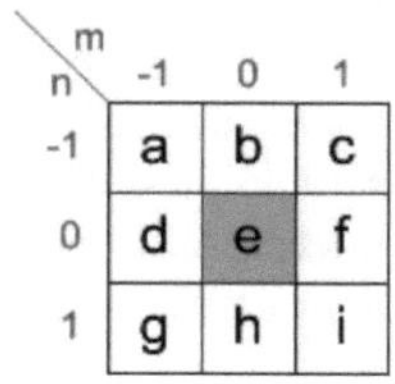

Fig. 1.12 A 2D kernel to be used for different kind of image processing

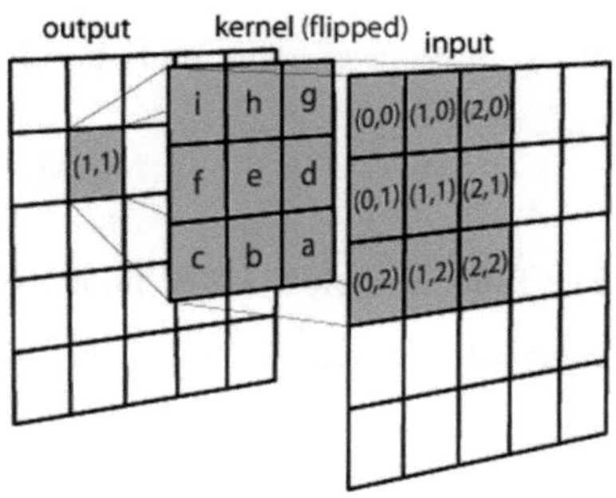

Fig. 1.13 Illustration of 2D convolution (applicable to image processing)

Referring to the basic expression of discrete convolution in 1D (Eq. 1.13), we can also realize the expression of two dimensional discrete convolution as

$$y(m,n) = x(m,n) * h(m,n) = \sum_{i=-\infty}^{\infty} \sum_{j=-\infty}^{\infty} x(i,j)h(m-i,n-j) \tag{1.15}$$

Just by considering the mask shown in Fig. 1.12, the method of doing convolution is shown in Fig. 1.13. For instance, the output at (1, 1) will be

$$\begin{aligned} y(1,1) &= \sum_{i=0}^{2}\sum_{j=0}^{2} x(i,j)h(1-i,1-j) \\ &= x(0,0)h(1,1) + x(1,0)h(0,1) + x(2,0)h(-1,1) \\ &\quad + x(0,1)h(1,0) + x(1,1)h(0,0) + x(2,1)h(-1,0) \\ &\quad + x(0,2)h(1,-1) + x(1,2)h(0,-1) + x(2,2)h(-1,-1) \end{aligned} \tag{1.16}$$

1.7 Correlation

Correlation is the measure of similarity. It determines how much alike two signals are. Correlation by itself generates a single number, called a *correlation coefficient*. A large, positive correlation coefficient indicates a close relation or similarity between the signals, while zero correlation coefficient signifies that the two signals are not at all similar. A negative number indicates a negative correlation, meaning that one signal tends to decrease as the other increases.

Correlation between two different sequences is the cross-correlation and correlation between two identical sequences is the auto-correlation. Correlation can be viewed as convolution, by reversing one signal. Instead of considering signal y[k] being indexed from 0 to some positive limit K, we will *fold* it around the zeroth sample, and consider that it starts at −K. For example, if y[n] = [5, 3, 1, 6], we interpret this as y[0] = 5; y[1] = 3; y[2] = 1, and y[3] = 6. If we fold this signal around, we would have y[−3] = 6, y[−2] = 1, y[−1] = 3, and y[0] = 5. When we convolve y with x, we would sum $x[n] * y[n-k]$. Using the folded version of y, the only thing to change is the index, e.g., $x[n] * y[n-(-k)]$ or simply $x[n] * y[n+k]$.

As the number of data samples affects the cross-correlation measure, so the cross-correlation value must be divided by the number of data samples.

$$S_{xy}[k] = \frac{1}{N}\sum_{n=0}^{N-1} x[n]y[n+k] \tag{1.17}$$

The function below [3, 4] finds the cross-correlation between two signals.

correlate_apu.m

```
% Measure of similarity between two signals.
%
% Usage: [rho] = correlate_apu(k, x, y);
% inputs: k = shift (integer between 0 and length(y)-1)
% x, y = the signals to correlate
% output: rho = the correlation coefficient -1..0..1
%
% Note: This function assumes length(x) = length(y).
% Apurba Das, HCL Technologies, India

function [rho] = correlate_apu(k, x, orig_y)
% Shift y by k units
y = [orig_y(k+1:length(orig_y)), orig_y(1:k)];
N = length(x);
sxx = x*x.' - sum(x)*sum(x)/N;
syy = y*y.' - sum(y)*sum(y)/N;
sxy = x*y.' - sum(x)*sum(y)/N;
rho = sxy / sqrt(sxx*syy);
```

References

1. Das, A.: Line Communication System- Telecommunication Switching Approach. New Age International Publishers, New Delhi (2010)
2. Data Acquisition Techniques Using PCs, pp. 44–47. Academic Press
3. Das, A.: Digital Communication—Principles and System Modelling. Springer, New York (2010)
4. Weeks, M.: Digital Signal Processing Using MATLAB and Wavelets. Infinity Science Press LLC, Hingham (2007)

Part I
Continuous Wave Communication and Analog Signal Conditioning

Chapter 2
Fourier Series

2.1 Introduction

The formula

$$\frac{x}{2} = \sin x - \frac{1}{2}\sin 2x + \frac{1}{3}\sin 3x\ldots \tag{2.1}$$

was published by Leonhard Euler (1707–1783) before Fourier's work began, so you might like to ponder the question why Euler did not receive the credit of Fourier series [1]. The Eq. 2.1 is really interesting in terms of types of the functions placed in left and right hand side. The function of x in left hand side is just a definite number, whereas functions used in the right hand side are all infinite series. Sum of all the sinusoids of x in *radian* will be equal to exactly $x/2$ (*if* $x = 1$, $x/2 = 0.5$). As we allow the number of terms for addition in right hand side, it will closer to the value of $x/2$. It can be easily checked by putting any finite value for x.

Basic interpretation of Fourier series also suggests to express the periodic signals as summation of sine and cosines of integral multiple of frequencies. The concept is therefore to some extent analogous to Eq. 2.1. The detail understanding will be obtained from discussion in the subsequent sections of the present chapter.

We, as the readers of communication engineering, shall look at Fourier series as an efficient tool for signal conditioning. From Chap. 1 it is understood that in the world of signals, only sinusoids (and obviously co sinusoids) are mono-tone (single frequency) signals. If we can successfully express any composite periodic signal into sinusoids, we can directly analyze the components of different signal, i.e., we can obtain the mono tone signal components with definite amount of amplitude (either voltage or current or energy) and phase. This representation is called as *spectral representation* like amplitude spectrum, phase spectrum, energy spectrum respectively. The representation can solve a lot of problems in communication engineering like choice of efficient bandwidth of a composite signal transmission. During transmission, the signal components with high energy

A. Das, *Signal Conditioning*, Signals and Communication Technology,
DOI: 10.1007/978-3-642-28818-0_2, © Springer-Verlag Berlin Heidelberg 2012

are chosen depending upon the *power spectral density* (*PSD*) representation, which gives the average power distribution of signal with respect to unit bandwidth.

In this chapter, we will discuss the foundation of Fourier series with elegant support of mathematics and our own understanding. In the light of Fourier series, the concept of signal translation, scaling and many more geometric operations are discussed. The applicability of Fourier series in terms of Dirichlet's condition is also discussed. Signal energy and power are also measured in frequency domain and in time domain, as well using Perseval's theorem. The reasons behind the physical representation of Fourier spectrum of periodic signals or power signal to be (i.e., Fourier series representation) a discrete line representation is also discussed in the following sections.

Some interesting worked out problems are presented in this chapter for better understanding of the subject. The visual attacks on the problems are from the knowledge of basic Euclidian geometry also presented with supporting logics. Finally, an application on feature extraction from image using phase congruency concept of Fourier series is discussed.

2.2 Statement and Interpretation

Fourier series states that,

> Any periodic function (or signal) can be expressed as a summation of orthogonal pair of matrices with one fundamental frequency and infinite number of harmonics.

From the statement, the first thing which needs to be clarified is the term "orthogonality". In one sentence, the word orthogonality means the measure of similarity or the correlation coefficient equal to zero, i.e., two vectors or *phasors* (vector representation of signal) are said to be *orthogonal*, if the cross-correlation coefficient between those two vectors is zero. The concept is discussed incorporating boundary conditions as follows.

From Fig. 2.1, we can see two vectors $\vec{x}$ and $\vec{g}$ of same magnitude make an angle θ between them. If we apply our knowledge of engineering drawing and graphics, we can see the vector $\vec{g}$ from the top to make a "plan" or top-view. If the angle θ is very small, we can say, $\vec{x}$ and $\vec{g}$ are more or less similar. As the angle θ increases, the amount of similarity decreases. We can also measure the amount of similarity by the factor c, when the *projection* (or plan) of vector $\vec{g}$ on $\vec{x}$ is defined by $c\vec{x}$. It is also observed that, when $\vec{x}$ and $\vec{g}$ are perfectly aligned, the value of $c = 1$. As the angle increases, c goes far from 1. It physically signifies that the length of the line given by projection of $\vec{g}$ on $\vec{x}$, decreases. At an instant, when $\theta = 90°$, projection of $\vec{g}$ on $\vec{x}$ would be a point instead of a line. As $\vec{x}$ is a vector of finite magnitude, whereas the approximated/projected vector becomes a point, i.e. a vector of length zero, therefore the measure of similarity or correlation coefficient becomes $c = 0$. Hence, we can say, $\vec{g}$ and $\vec{x}$ are *orthogonal* to each other. If we now keep on increasing the angle θ,

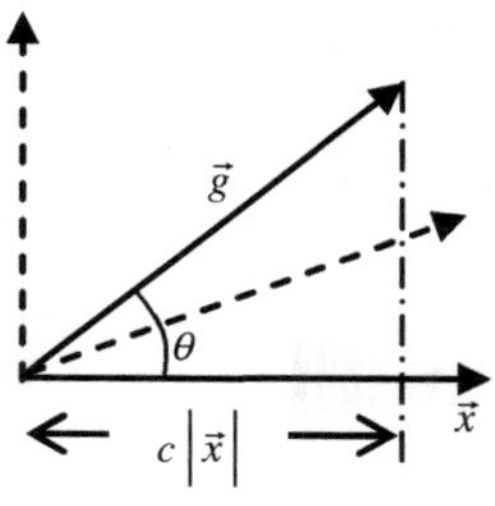

Fig. 2.1 Measure of similarity between two vectors

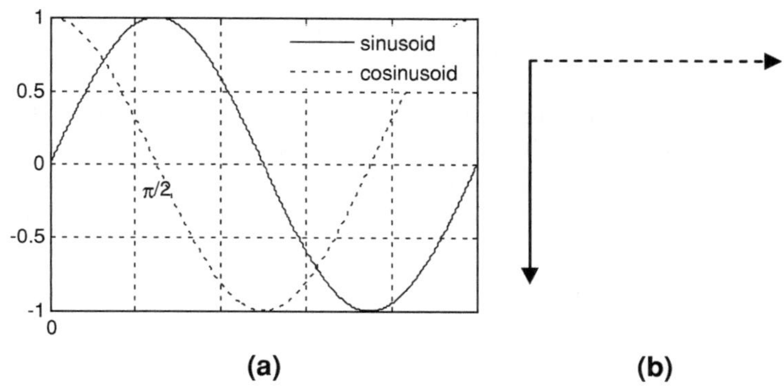

Fig. 2.2 Trigonometric orthogonality: **a** sine and cosine waveforms, **b** sine and cosine phasors

the amount of similarity will again increase but in opposite direction. Therefore, range of correlation coefficient is always $-1 < c < 1$.

Let's consider now a pair of signals/functions which are orthogonal. As shown in Fig. 2.2, one sinusoid signal is just $\pi/2$ behind the cosinusoid signal of same frequency. The phasor (vector representation of signal) also shows the same effect. Therefore, from the previous discussion, we can choose Sine and Cosine of same angle, i.e., frequency as a pair of *orthogonal set*. When a number of harmonics are used to form the set of frequencies and thereafter angles, the orthogonal set becomes *orthogonal set of matrices*. Table 2.1 completes our understanding of Fourier series applying trigonometric functions.

It means any periodic function/signal $f(t)$ can be expressed using trigonometric orthogonal signals sine and cosine from the Table 2.1, as

$$\begin{aligned} f(t) =& a_0 \cos 2\pi(0)t + b_0 \sin 2\pi(0)t \\ &+ a_1 \cos 2\pi(f_0)t + b_1 \sin 2\pi(f_0)t \\ &+ a_2 \cos 2\pi(2f_0)t + b_2 \sin 2\pi(2f_0)t \\ &+ a_3 \cos 2\pi(3f_0)t + b_3 \sin 2\pi(3f_0)t \\ & \cdots \end{aligned} \tag{2.2}$$

Table 2.1 Expression of Fourier series

Frequency	Coefficient of in-phase (cos) component	Coefficient of quadrature-phase (sin) component	In-phase (cos) component	Quadrature-phase (sin) component
0	a_0	b_0	$a_0\cos 2\pi(0)t$	$b_0\sin 2\pi(0)t$
f_0	a_1	b_1	$a_1\cos 2\pi(f_0)t$	$b_1\sin 2\pi(f_0)t$
$2f_0$	a_2	b_2	$a_2\cos 2\pi(2f_0)t$	$b_2\sin 2\pi(2f_0)t$
$3f_0$	a_3	b_3	$a_3\cos 2\pi(3f_0)t$	$b_3\sin 2\pi(3f_0)t$
–	–	–	–	–
∞	a_∞	b_∞	$a_\infty\cos 2\pi(\infty f_0)t$	$b_\infty\sin 2\pi(\infty f_0)t$

i.e.,

$$f(t) = a_0 + \sum_{n=1}^{\infty} (a_n\cos n\omega_0 t + b_n\sin n\omega_0 t) \tag{2.3}$$

It is clear from Fig. 2.3 that the synthesized signal by adding a number of sine and/or cosine waves becomes different depending upon the amplitudes/coefficients of the sinusoids and co sinusoids. In other words, the amplitudes a_0, a_n and b_n should be function of the given periodic signal $f(t)$, in the process of signal analysis.

2.3 Fourier Coefficients

Signals are not *like* vectors, instead signals are vectors. There is a strong analogy between signals and vectors [2]. As vectors, signals also have measurable magnitude and direction. As vector, we can also express signal as a sum of components. The vector representation of signal is called as *phasor*. The representation of signals as summation its components in various ways, eventually supports the statement of Fourier series.

2.3.1 Component of a Vector

A vector is defined using its magnitude and direction. According to the Fig. 2.4, a vector $\vec{x}$ is presented. The direction is aligned with the X-axis. Now, if we virtually rotate the vector by an angle θ, it forms another vector $\vec{g}$ which is having equal magnitude with that of vector $\vec{x}$, whereas the directions are not same. From the preliminary concept of engineering drawing/graphics/drafting, if we see from top to take a *plan* (top view), we have a reduced magnitude vector. Physically, taking top view of the vector is nothing but taking the component of vector $\vec{g}$ along X-axis or along the direction of $\vec{x}$. It means, approximating the magnitude of the vector, if it would be of direction along X-axis. From the Fig. 2.4, it is clear that the vector $\vec{g}$ is approximated by vector $c\vec{x}$, where $c < 1$. The approximation is not

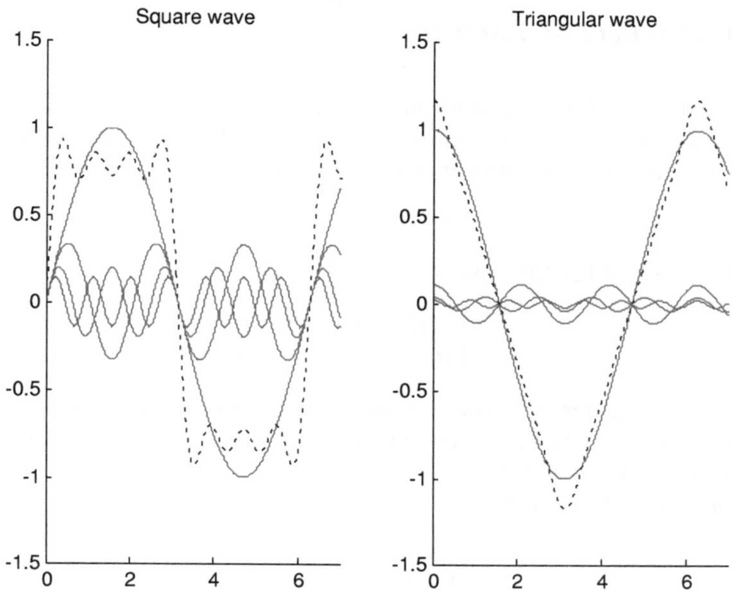

Fig. 2.3 Fourier synthesized signals by varying the amplitudes of the component signals

Fig. 2.4 Approximation of a vector in terms of another vector

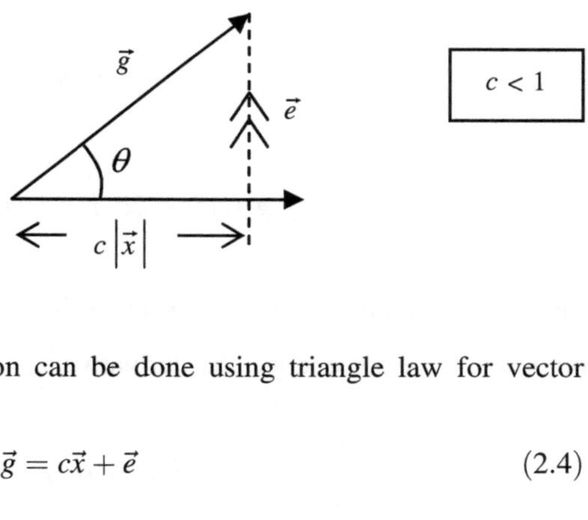

cent percent correct. The correction can be done using triangle law for vector summation. The figure says,

$$\vec{g} = c\vec{x} + \vec{e} \tag{2.4}$$

In other words, we can interpret the observation. If we approximate the vector $\vec{g}$ by vector $c\vec{x}$,

$$\text{i.e.,} \, \vec{g} \approx c\vec{x} \tag{2.5}$$

the error in the approximation is the vector $\vec{e} = \vec{g} - c\vec{x}$. Here, one may ask, why to find out the component of a vector, the looking direction must be exactly 90°. The answer is, we need to minimize the *error*; and from a fixed point the minimum distance to a line is always the perpendicular distance.

2.3.2 Component of a Signal

The concepts of vector component and orthogonality may be extended to signals. Making analogy with the statement in the previous sub-section, let's consider the problem of approximating a real signal g(t) in terms of x(t) over an interval [t_1, t_2]:

$$g(t) \approx cx(t) \qquad t_1 < t < t_2 \tag{2.6}$$

The error *e(t)* in this approximation is

$$e(t) = \begin{cases} g(t) - cx(t) & t_1 < t < t_2 \\ 0; & otherwise \end{cases} \tag{2.7}$$

For best approximation, we need to minimize the error. As the error is expressed here essentially as a signal, for minimum error, the error energy, E_e in the interval (t_1, t_2) needs to be minimized.
The error energy is given by:

$$E_e = \int_{t_1}^{t_2} |e(t)|^2 dt$$

or

$$\begin{aligned} E_e &= \int_{t_1}^{t_2} \{g(t) - cx(t)\}^2 dt \\ &= \int_{t_1}^{t_2} g^2(t)dt - 2c\int_{t_1}^{t_2} g(t)x(t)dt + c^2\int_{t_1}^{t_2} x^2(t)dt \end{aligned} \tag{2.8}$$

The error is minimized with respect to the correlation coefficient *c*. therefore, to minimize the error energy, the derivative $\frac{dE_g}{dc}$ must be equal to zero.

$$\frac{dE_g}{dc} = 0 - 2\int_{t_1}^{t_2} g(t)x(t)dt + 2c\int_{t_1}^{t_2} x^2(t)dt = 0 \tag{2.9}$$

$$\therefore c = \frac{\int_{t_1}^{t_2} g(t)x(t)dt}{\int_{t_1}^{t_2} x^2(t)dt} \tag{2.10}$$

The Eq. 2.10 suggests the final formation of correlation coefficient i.e., similarity measure between these two signals *g(t)* and *x(t)*. This expression for deriving the similarity coefficient c will be useful for derivation of Fourier coefficients in trigonometric (a_0, a_n, b_n), compact (C_0, C_n, θ_n) and complex (D_n) Fourier series. In the next section, the Trigonometric Fourier coefficients are derived using Eq. 2.10.

2.3.3 Coefficients of Trigonometric Fourier Series

Due to the property of orthogonality, just like unit vectors, we can express any periodic signal by summation of sine and cosine, i.e., choosing the sinusoids and cosinusoids as the *basis*. For orthogonality, we find the following properties between sine and cosine given by Eq. 2.11a–2.11d [3].

$$\int_T \cos m\theta \cos n\theta d\theta = \begin{cases} 0; & m \neq n \\ T/2; & m = n \neq 0 \\ T; & m = n = 0 \end{cases} \tag{2.11a}$$

$$\int_T \sin m\theta \sin n\theta d\theta = \begin{cases} 0; & m \neq n \\ T/2; & m = n \neq 0 \\ 0; & m = n = 0 \end{cases} \tag{2.11b}$$

$$\int_T \cos m\theta \sin n\theta d\theta = 0; \tag{2.11c}$$

$$\int_T e^{j(n-m)\theta} d\theta = \begin{cases} 0; & m \neq n \\ T; & m = n \end{cases} \tag{2.11d}$$

where, $\int_T$ signifies definite integral with T distance between upper and lower bound. One time period signal can be taken out from any time window, but the width of the window must be T, i.e., the range can be $[0, T]$ or $[-T/2, T/2]$ or $[-T/4, 3T/4]$ or any other window with width T. Depending on the pattern and formation of the signal, the window is chosen. The choice of time window is discussed using a worked out problem in Example 2.1.

From Eq. 2.3, let's take out the terms in the right hand side one by for minimum error of approximation as shown below to find the expressions of correlation coefficients. The terms can be taken out one by one as sine and cosine holds the criterion of orthogonality as shown in Eq. 2.11a–2.11d.

$$f(t) \approx a_0 \times 1 \tag{2.12}$$

Comparing Eq. 2.6 and 2.12, from Eq. 2.10, we get

$$a_0 = \frac{\int_T f(t) \times 1 dt}{\int_T 1 dt} = \frac{\int_T f(t) dt}{T} = \frac{1}{T} \int_T f(t) dt \tag{2.13}$$

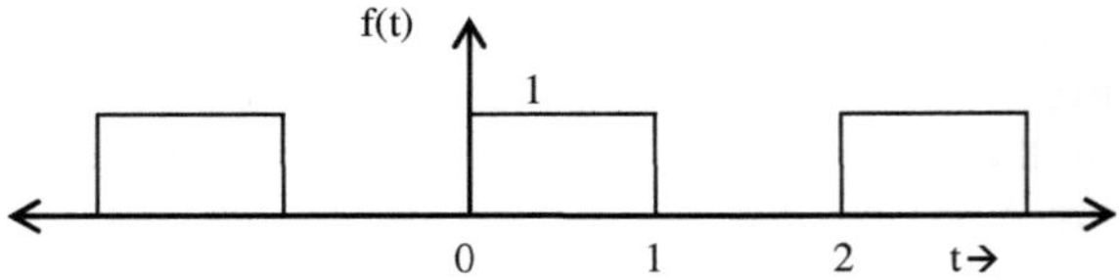

Fig. 2.5 The signal f(t) in time domain

If we approximate

$$f(t) \approx a_n \times \cos n\omega_0 t \tag{2.14}$$

Comparing Eq. 2.6 and 2.14, from Eq. 2.10, we get

$$a_n = \frac{\int_T f(t) \times \cos n\omega_0 t dt}{\int_T \cos^2 n\omega_0 t dt} = \frac{\int_T f(t) \cos n\omega_0 t dt}{T/2}$$

or,

$$a_n = \frac{2}{T} \int_T f(t) \cos n\omega_0 t dt \tag{2.15}$$

Similarly,

$$b_n = \frac{\int_T f(t) \times \sin n\omega_0 t dt}{\int_T \sin^2 n\omega_0 t dt} = \frac{\int_T f(t) \sin n\omega_0 t dt}{T/2}$$

or,

$$b_n = \frac{2}{T} \int_T f(t) \sin n\omega_0 t dt \tag{2.16}$$

Example 2.1 Express the signal f(t) in Fourier series and draw the amplitude spectrum.

From the Fig. 2.5, we can define the function f(t) as

$$f(t) = \begin{cases} 1, & 0 < t < 1 \\ 0, & 1 < t < 2 \end{cases}$$

and,

$$f(t+2) = f(t)$$

Therefore, the periodicity of the function f(t) is T = 2. To chose one time period (T) window, we should find the time region, where minimum number of pieces are to be handles during ‘T’. If we take the region [0 T], it is having two regions. From 0 to T/2, f(t) is 1 and from T/2 to T, the value is 0.
Then, we obtain the coefficients a_0, a_n and b_n

$$a_0 = \frac{1}{T}\int_0^T f(t)dt = \frac{1}{2}\int_0^2 f(t)dt = 0.5\int_0^1 1dt + 0.5\int_1^2 0dt = 0.5 - 0 = 0.5$$

Or, since $\int_a^b f(t)dt$ total area under the graph $y = f(t)$ over the interval $[a, b]$. Hence,

$$a_0 = \frac{1}{T}\int_0^T f(t)dt = \frac{1}{T} \times \begin{pmatrix} \text{Area under graph} \\ \text{over } [0, T] \end{pmatrix} = \frac{1}{2} \times (1 \times 1) = 0.5$$

$$\begin{aligned} a_n &= \frac{2}{T}\int_0^2 f(t)\omega_0 t dt \\ &= \int_0^1 1\cos n\pi t dt + \int_1^2 0dt = \left[\frac{\sin n\pi t}{n\pi}\right]_0^1 = \frac{\sin n\pi}{n\pi} \end{aligned}$$

It is to be noted that, n is an integer which leads $\sin n\pi = 0$ since $\sin \pi = \sin 2\pi = \sin 3\pi = \ldots = 0$
Therefore,

$$a_n = 0$$

$$\begin{aligned} b_n &= \frac{2}{T}\int_0^2 f(t)\omega_0 t dt \\ &= \int_0^1 1\sin n\pi t dt + \int_1^2 0dt = \left[-\frac{\cos n\pi t}{n\pi}\right]_0^1 = \frac{1 - \cos n\pi}{n\pi} \end{aligned}$$

It can be viewed that,

$$\cos \pi = \cos 3\pi = \cos 5\pi = \ldots = -1$$
$$\cos 2\pi = \cos 4\pi = \cos 6\pi = \ldots = 1$$

Or,

$$\cos n\pi = (-1)^n$$

Therefore,

$$b_n = \frac{1 - (-1)^n}{n\pi} = \begin{cases} 2/n\pi, & n \text{ odd} \\ 0, & n \text{ even} \end{cases}$$

Finally,

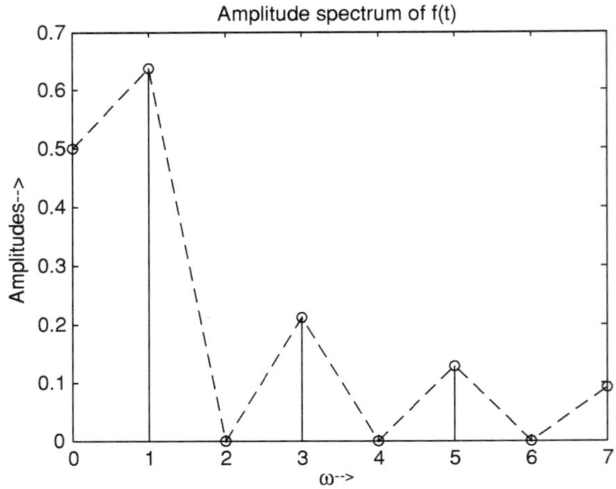

Fig. 2.6 Amplitude spectrum of f(t)

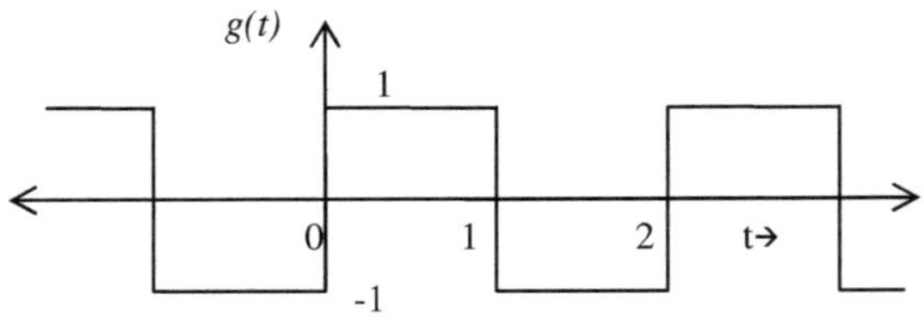

Fig. 2.7 The signal g(t) in time domain

$$
\begin{aligned}
f(t) &= a_0 + \sum_{n=1}^{\infty} (a_n \cos n\omega t + b_n \sin n\omega t) \\
&= \frac{1}{2} + \sum_{n=1}^{\infty} \left[\frac{1-(-1)^n}{n\pi} \right] \sin n\omega_0 t \\
&= \frac{1}{2} + \frac{2}{\pi} \sin \omega_0 t + \frac{2}{3\pi} \sin 3\omega_0 t + \frac{2}{5\pi} \sin 5\omega_0 t + \ldots \qquad (2.17)
\end{aligned}
$$

Example 2.2 Express the signal g(t) in Fourier series and draw the amplitude spectrum.

We can approach to the problem similarly as approached to the previous problem. Otherwise, we can also solve the problem utilizing the result of the previous problem (Fig. 2.6).
From the Fig. 2.7, we can define the function $g(t)$ as

$$
g(t) = \begin{cases} 1, & 0 < t < 1 \\ -1, & 1 < t < 2 \end{cases}
$$

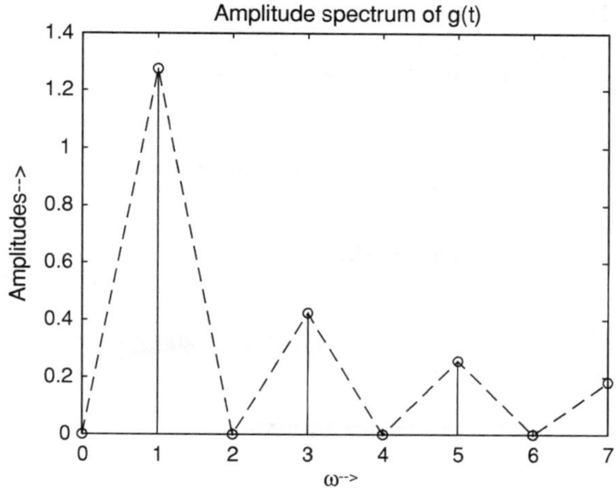

Fig. 2.8 Amplitude spectrum of $g(t)$

If we just take a look over the two function $f(t)$ and $g(t)$ (Figs. 2.6 and 2.7), we can observe some similarities and dissimilarities as follows.
Similarities:

(1) Time periods, i.e., fundamental frequencies for both the signals are same.
(2) Geometric patterns of both the signals are similar, i.e., square wave.

Dissimilarities:

(1) Peak-to-peak amplitude of $g(t)$ is just double with respect to that of $f(t)$.
(2) The average value over one time period is 0.5 for $f(t)$ whereas, average value over time period is zero for $g(t)$

Now, to relating the dissimilarities we can express $g(t)$ in terms of $f(t)$ as, they are having some important signal parameter similarities.

We have just removed the DC component 0.5 and then multiplied the entire signal by 2 (Fig. 2.8).

$$g(t) = 2\{f(t) - 0.5\}$$

From Eq. 2.17

$$g(t) = 2\left\{\left(\frac{1}{2} + \frac{2}{\pi}\sin\omega_0 t + \frac{2}{3\pi}\sin 3\omega_0 t + \frac{2}{5\pi}\sin 5\omega_0 t + \ldots\right) - \frac{1}{2}\right\}$$

$$\therefore g(t) = \left(\frac{4}{\pi}\sin\omega_0 t + \frac{4}{3\pi}\sin 3\omega_0 t + \frac{4}{5\pi}\sin 5\omega_0 t + \ldots\right) \quad (2.18)$$

Table 2.2 Derivation of a_0 as extraction of DC offset

Signal	a_0
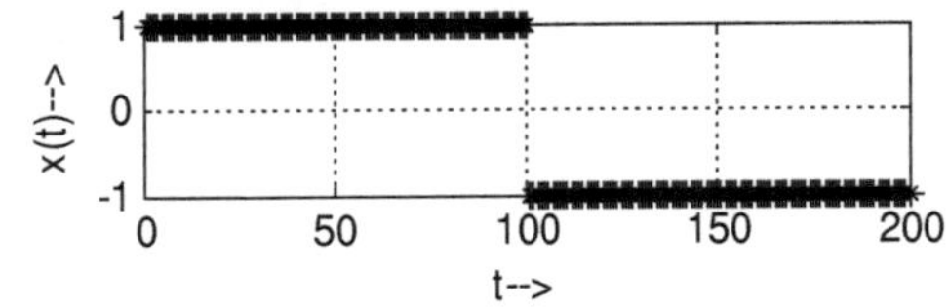	$a_0 = 0$ as, areas over and under x-axis over one time period (T = 200) are exactly equal.
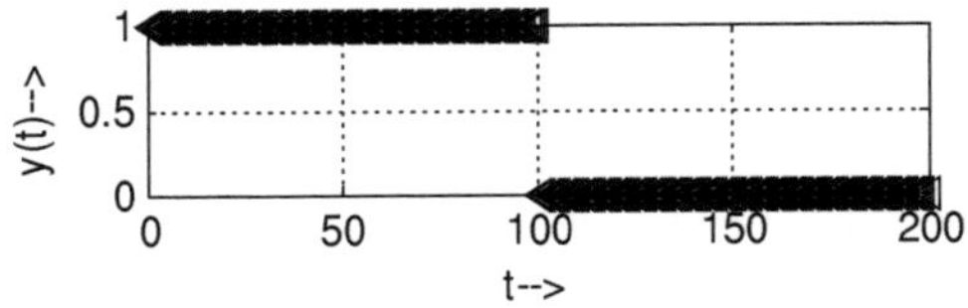	$a_0 = 0.5$ as, if we shift the entire signal down by subtracting a DC of amount 0.5 (or shifting the time axis up to 0.5), we can get equal areas over and under the curve.
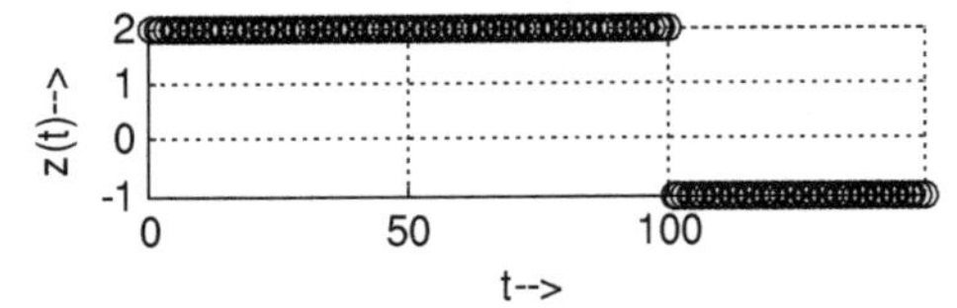	$a_0 = \dfrac{100 \times 2 + (-50)}{150} = 1$
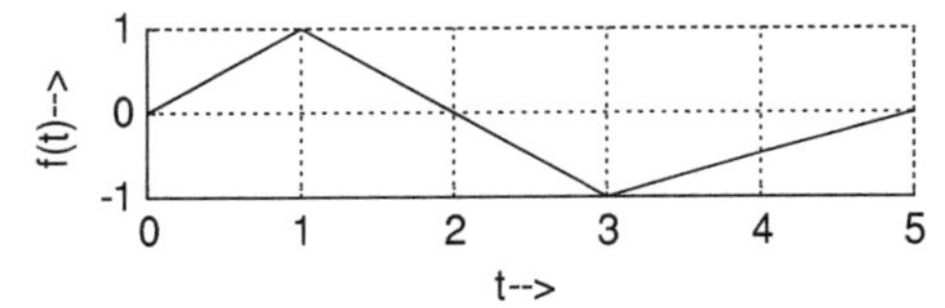	$a_0 = \dfrac{0.5 \times 1 \times 2 + (-0.5 \times 3 \times 1)}{5} = \dfrac{1 - 1.5}{5} = -\dfrac{0.5}{5} = -\dfrac{1}{10}$

2.3.4 Physical Existences of the Coefficients

The existence and measurement of the coefficients of Fourier series is dependent on the shape or pattern of the signal. If we can recognize the pattern in a signal, we can infer to some extent on the existence of coefficients a_0, a_n and b_n. In this subsection we will discus the physical meaning of the coefficients and then we will infer the possible existence of those coefficients. As seen in the Example 2.1 and 2.2, in both the cases, only sine components are there, i.e., a_n for all n is 0. It is to be noted that in the spectral representation (Figs. 2.6 and 2.8) we have used the index 'n' of frequency. In Example 2.1, $a_0 = 0.5$, whereas $a_0 = 0$ in Example 2.2. The reasons behind will be thoroughly understood after going through this subsection.

2.3.4.1 Physical interpretation of a_0

If we look at the Eq. 2.13, it clearly says, this coefficient is actually the average magnitude of the signal over one time period. In other words, this is the DC component of the signal. From Table 2.1 and Eq. 2.3 it is observed that, no *cosine* or *sine* component of a signal is attached with the coefficient a_0. Moreover, it is also seen from Eq. 2.13 that the derivation of a_0 indicates the average area formed by the signal in both the polarities over one complete time period. Therefore, when a signal comes for Fourier analysis, the first step employed is to describe the signal mathematically over one time period. If the areas formed by the signal over and under X-axis are same, the value of a_0 is obviously zero. If the areas are unequal, then we should try to predict the amount of up or down shift of the signal to make the areas over and under x-axis equal. The amount of shift of the time axis is actually the value of a_0. In some cases, the geometric prediction or visualization is not so easy. Then we need to calculate the areas over and below the x-axis over one time period. The total area is then averaged by the time period to get a_0. Some examples of getting a_0 without much calculations are given in Table 2.2.In the four chosen different examples, it has been shown how a_0 is derived by extracting the DC offset of the signal.

2.3.4.2 Physical interpretation of a_n and b_n

From understanding of signal, we can classify signal into three types like (1) odd, (2) even and (3) odd + even. We can also understand that, the odd part of the signal contains sinusoids; even part of the signal contains co-sinusoids. Therefore, if a signal can be clearly identified as odd or even using the two-mirror process as discussed in Chap. 1, we may need to find either the *cos* or the *sin* components for even and odd signals respectively. To measure the fitness of sine or cos, we need to make the signal DC free. Then we need to try to fit cos or sine wave into the signal for closest approximation. As for an example, we can refer Table 2.2. The signal f(t) is of some DC value. If we remove the DC and look at the signal, drawing at least three periods in right and left side of the Y-axis, we can easily feel that *sine* can be fitted into the signal suitably. As obtained from the Eqs. 2.15 and 2.16, a_n and b_n are the cosine and sine components of a signal. Therefore, b_n will have some definite values, whereas $a_n = 0$ as no cosine components should be there in trigonometric Fourier series expression, as a_n be the coefficient of cosine component of the signal. It also inferred that, for odd signals, $a_n = 0$ and in even signal $b_n = 0$, only for the signal which is neither odd nor even, i.e., which can be expressed as summation of odd and even signal, have non-zero a_n and b_n values. The idea would be more clear when we discuss odd and even symmetry in the next section of study.

2.4 Even and Odd Symmetry

Any function *f*(*t*) is even if its plot is symmetrical about the vertical axis, i.e. $f(t) = f(-t)$. We can guess a mirror placed in the y-axis to get the image on the negative x-plane, as shown in the examples below.

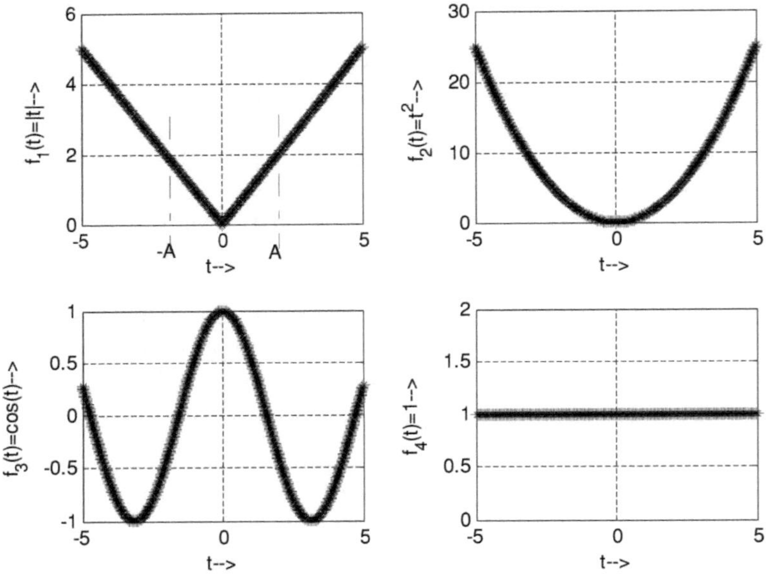

Fig. 2.9 Examples of even functions

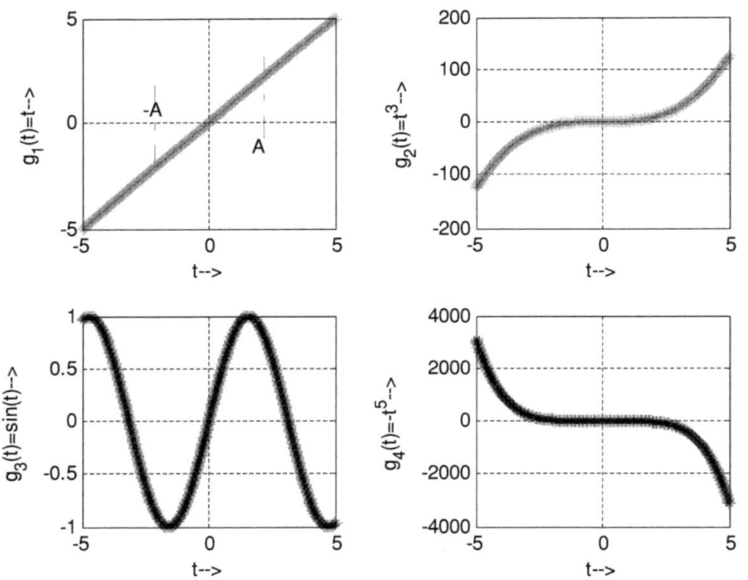

Fig. 2.10 Examples of odd functions

The integral of an **even** function from $-A$ to $+A$ is twice the integral from 0 to $+A$ as shown in $f_1(t)$.

$$\int_{-A}^{+A} f_{\text{even}}(t)dt = 2\int_{0}^{+A} f_{\text{even}}(t)dt \tag{2.19}$$

On the other hand, any function $f(t)$ is **odd** if its plot is anti-symmetrical about the vertical axis, i.e. $f(t) = -f(-t)$. We can understand the formation of the function as double reflection by one mirror placed on the y-axis followed by another mirror placed on the x-axis, as shown in some examples in Figs. 2.9 and 2.10.

The integral of an **odd** function from $-A$ to $+A$ is zero as shown in $g_1(t)$.

$$\int_{-A}^{+A} g_{\text{odd}}(t)dt = 0 \tag{2.20}$$

Therefore, the calculations for derivation of Fourier coefficients become easy (and in some cases unnecessary), after understanding the physical interpretations. The ***even–odd properties*** of functions are as follows. It differs from the general concept of even–odd properties of numbers.

$$\begin{aligned} Even \times Even &= Even \\ Odd \times Odd &= Even \\ Odd \times Even &= Odd \\ Even \times Odd &= Odd \end{aligned} \tag{2.21}$$

From the even–odd properties of even symmetry,

$$a_n = \underbrace{\frac{2}{T}\int_{-T/2}^{T/2} f(t)\cos n\omega t dt}_{Even \times Even = Even} = \frac{4}{T}\int_{0}^{T/2} f(t)\cos n\omega t dt \tag{2.22}$$

$$b_n = \underbrace{\frac{2}{T}\int_{-T/2}^{T/2} f(t)\sin n\omega t dt}_{Even \times Odd = Odd} = 0 \tag{2.23}$$

Similarly, from the even–odd properties of odd symmetry,

$$a_0 = \underbrace{\frac{2}{T}\int_{-T/2}^{T/2} f(t)dt}_{Odd} = 0 \tag{2.24}$$

$$a_n = \frac{2}{T}\underbrace{\int_{-T/2}^{T/2} f(t)\cos n\omega t dt}_{Odd\times Even=Odd} = 0 \tag{2.25}$$

$$b_n = \frac{2}{T}\underbrace{\int_{-T/2}^{T/2} f(t)\sin n\omega t dt}_{Odd\times Odd=Even} = \frac{4}{T}\int_{0}^{T/2} f(t)\sin n\omega t dt \tag{2.26}$$

2.5 Compact Fourier Series

In basic formation of trigonometric Fourier series, as seen from the previous section, even function gives the valid a_n magnitudes whereas, odd function (signal) gives valid b_n magnitudes only. So, while drawing the magnitude and phase spectrum, we can just take a_n or b_n with respect to the odd or even property of the signal, and then the set (either a_n or b_n) are plotted with respect to n (As Figs. 2.6 and 2.8). But, when we look at those signals which are neither odd nor even, both the magnitudes are non-zero, i.e., valid. Then we need to think about another coefficient which is the sole representative of both a_n and b_n. For this purpose, another now coefficient is proposed as $C_n = \sqrt{a_n^2 + b_n^2}$. The new Fourier series expression incorporating the new coefficient is called as compact Fourier series.

From Eq. 2.3,

$$\begin{aligned} f(t) &= a_0 + \sum_{n=1}^{\infty}(a_n\cos n\omega_0 t + b_n\sin n\omega_0 t) \\ &= a_0 + \sum_{n=1}^{\infty}\sqrt{a_n^2 + b_n^2}\left(\frac{a_n}{\sqrt{a_n^2 + b_n^2}}\cos n\omega_0 t + \frac{b_n}{\sqrt{a_n^2 + b_n^2}}\sin n\omega_0 t\right) \end{aligned} \tag{2.27}$$

Now, taking

$$C_0 = a_0 \tag{2.28}$$

$$C_n = \sqrt{a_n^2 + b_n^2} \tag{2.29}$$

$$\cos\theta_n = \frac{a_n}{\sqrt{a_n^2 + b_n^2}}$$

$$\therefore \sin\theta_n = \sqrt{1 - \cos^2\theta_n} = \sqrt{1 - \frac{a_n^2}{a_n^2 + b_n^2}} = \frac{b_n}{\sqrt{a_n^2 + b_n^2}}$$

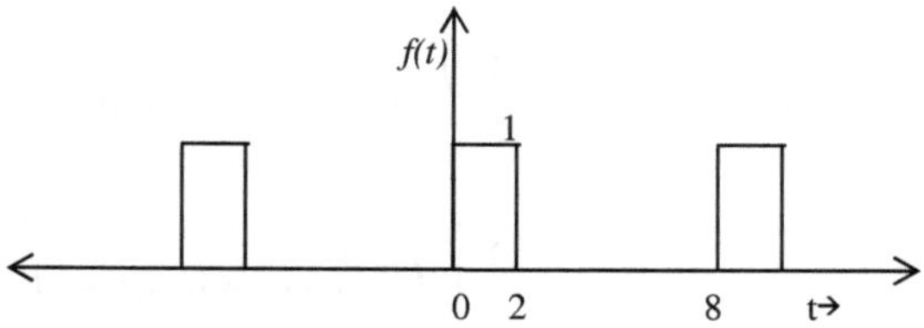

Fig. 2.11 A signal which is neither odd nor even

Therefore, Eq. 2.17 becomes

$$f(t) = C_0 + \sum_{n=1}^{\infty} C_n\left(\cos\theta_n \cos n\omega_0 t + \sin\theta_n \sin n\omega_0 t\right)$$
$$= C_0 + \sum_{n=1}^{\infty} C_n \cos\left(n\omega_0 t - \theta_n\right) \tag{2.30}$$

Now,

$$\tan\theta_n = \frac{\sin\theta_n}{\cos\theta_n} = \frac{b_n}{a_n}$$
$$\therefore \theta_n = \tan^{-1}\frac{b_n}{a_n} \tag{2.31}$$

Therefore, Eq. 2.30 can be re-written as

$$f(t) = C_0 + \sum_{n=1}^{\infty} C_n \cos\left(n\omega_0 t - \tan^{-1}\frac{b_n}{a_n}\right) \tag{2.32}$$

This is the generalized or compact form of the trigonometric Fourier series.

Example 2.3 Express the function f(t) is Fourier series and draw the magnitude and phase spectrum (Fig. 2.11)

One period of the signal f(t) is taken out. It is obvious from the figure that, the time period is $T = 8$.

Therefore, fundamental frequency $\omega_0 = 2\pi/8 = 0.7854$ rad/s. From the signal pattern, we have chosen the limit of integration for finding out the coefficients of Fourier series, to be [0,8].

One time period window of the signal now can be defined as

$$f(t) = \begin{cases} 1; & (0 \geq t > 2) \\ 0; & (2 \geq t > 8) \end{cases} \tag{2.33}$$

From the definition of Fourier coefficients,

$$a_0 = \frac{1 \times 2 + 0 \times 6}{8} = \frac{1}{4} \tag{2.34}$$

$$\begin{aligned}
a_n &= \frac{2}{T}\int_0^8 f(x) \times \cos n\omega_0 t dt \\
&= \frac{2}{T}\left[\int_0^2 1 \times \cos n\omega_0 t dt + \int_2^8 0 \times \cos n\omega_0 t dt\right] \\
&= \frac{2}{T} \times \frac{1}{n\omega_0}\left\{\sin n\omega_0 t|_0^2 + 0\right\} \\
&= \frac{2}{T} \times \frac{T}{2\pi n}\sin\frac{4\pi n}{8} \\
&= \frac{1}{n\pi}\sin\frac{n\pi}{2} = \frac{1}{2}sinc\frac{n\pi}{2}
\end{aligned} \tag{2.35}$$

Thus the a_n values will be zero for even values of n and non-zero with alternative sign for odd values of n.

Therefore, alternatively a_n can be expressed as

$$a_n = \begin{cases} 0; & for \quad n \rightarrow Even \\ \dfrac{1}{n\pi}; & for \quad n = 1, 5, 9, 13, \ldots \\ -\dfrac{1}{n\pi}; & for \quad n = 3, 7, 11, 15, \ldots \end{cases} \tag{2.36}$$

$$\begin{aligned}
b_n &= \frac{2}{T}\int_0^8 f(x) \times \sin n\omega_0 t dt \\
&= \frac{2}{T}\left[\int_0^2 1 \times \sin n\omega_0 t dt + \int_2^8 0 \times \sin n\omega_0 t dt\right] \\
&= \frac{2}{T} \times \frac{-1}{n\omega_0}\left\{\cos n\omega_0 t|_0^2 + 0\right\} \\
&= \frac{2}{T} \times \frac{-T}{2\pi n}\left\{\cos\frac{4\pi n}{8} - 1\right\} \\
&= \frac{1}{n\pi}\left(1 - \cos\frac{n\pi}{2}\right)
\end{aligned} \tag{2.37}$$

The b_n coefficients will be zero whenever $\cos(n\pi/2) = 1$, which occurs for $n = 4, 8, 12, 16, \ldots..$

The signal given in this example is following neither the single mirror concept nor the double mirror concept to satisfy the evenness or oddness of a signal respectively. Therefore from the discussion in Chap. 1, we can conclude that the given signal f(t) must be expressed as a summation of an odd and an even signal. a_n and b_n are the representatives of the even and odd component of the given signal, respectively. Now, to draw the amplitude spectrum, we need to consider

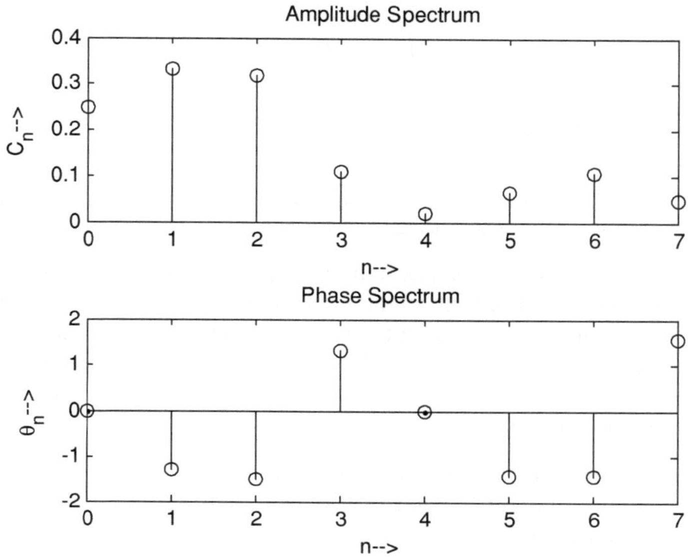

Fig. 2.12 Amplitude and phase spectrum of a general signal (neither odd nor even) using compact Fourier series

both a_n and b_n. As they are orthogonal pair, we can express resultant amplitude using Pythagoras theorem for right angle triangle and express the compact amplitude as hypotenuse (Eq. 2.29).

From Eqs. 2.28, 2.29 and 2.31 we can compute the compact Fourier series coefficients as

$$C_0 = a_0$$
$$C_n = \sqrt{a_n^2 + b_n^2}$$
$$\theta_n = \tan^{-1}\frac{b_n}{a_n}$$

Next, C_n and θ_n are plotted with respect to n to represent amplitude spectrum and phase spectrum respectively as shown in Fig. 2.12.

2.6 Dirichlet Conditions

For the existence of Fourier series, there are two basic conditions as follows.

1. **Necessary condition:** The series exists if and only if the coefficients $a_0, a_n, b_n$are finite (Eqs. 2.13, 2.15, 2.16). From Eq. 2.13 it is understandable that, the existence of the coefficients are guaranteed if $g(t)$ is absolutely integrable over one time period, i.e.,

$$\int_T |g(t)|dt < \infty \tag{2.38}$$

This is known as the *weak Dirichlet condition.*

If a function satisfies weak Dirichlet condition, the existence of Fourier series is guaranteed, but the series may not converge at every point.

As for an example, if a function has infinite number of maxima and minima in one time period, then the function contains an appreciable amount of components of frequencies approaching to infinity. Thus, the higher coefficients in the series do not decay rapidly. Therefore the series will not converge uniformly. For convergence of the Fourier series, we need another condition over weak Dirichlet condition as follows.

2. **Sufficient condition:** The function have only finite number of maxima and minima in one time period, and it may have only finite number of discontinuities in one time period.
   Combining these two conditions we can form complete condition for applicability and convergence of Fourier series named as *strong Dirichlet condition.* Signal obeying the strong Dirichlet condition can be expressed in Fourier series and the series will be convergent. This is the sufficient condition for convergence of Fourier series.

2.7 Exponential Fourier Series

Exponential Fourier series is the modified form of trigonometric Fourier series. This expression can also be called as *most generalized* or *double sided Fourier series.* As we know sine and cosine can be expressed using summation of exponential functions. From Eq. 2.11d, it is also observed that, exponential function also obeys the rule of orthogonality. From trigonometric Fourier series,

$$\begin{aligned} f(t) &= a_0 + \sum_{n=1}^{\infty} (a_n \cos \mathrm{n}\omega_0 t + b_n \sin \mathrm{n}\omega_0 t) \\ &= a_0 + \sum_{n=1}^{\infty} \left(a_n \frac{\mathrm{e}^{\mathrm{jn}\omega_0 t} + \mathrm{e}^{-\mathrm{jn}\omega_0 t}}{2} + b_n \frac{\mathrm{e}^{\mathrm{jn}\omega_0 t} - \mathrm{e}^{-\mathrm{jn}\omega_0 t}}{2\mathrm{j}} \right) \\ &= a_0 + \sum_{n=1}^{\infty} \frac{(a_n - jb_n)}{2} \mathrm{e}^{\mathrm{jn}\omega_0 t} + \frac{(a_n + jb_n)}{2} \mathrm{e}^{-\mathrm{jn}\omega_0 t} \end{aligned} \tag{2.39}$$

Now, if we assume that, the new coefficients for this new type of Fourier series representation are given by

$$D_0 = a_0 \tag{2.40}$$

$$D_n = \frac{(a_n - jb_n)}{2} \tag{2.41}$$

$$D_{-n} = D_n^* = \frac{(a_n + jb_n)}{2} \tag{2.42}$$

Therefore, Eq. 2.3 can be re-written as

$$f(t) = D_0 + \sum_{n=1}^{\infty} \left(D_n \mathrm{e}^{\mathrm{j}n\omega_0 t} + D_{-n} \mathrm{e}^{-\mathrm{j}n\omega_0 t}\right) \tag{2.43}$$

The first term within the summation signifies positive going of ordered variable *n*. Similarly, the second term within summation signifies positive going of –*n*, i.e., negative going of *n*. We can therefore express Eq. 2.43 as

$$f(t) = D_0 + \sum_{\substack{n=-\infty \\ n\neq 0}}^{\infty} D_n \mathrm{e}^{\mathrm{j}n\omega_0 t} \tag{2.44}$$

We can also include the zeroth index easily. The first term D_0 can be written as $D_0 e^{j0\omega_0 t}$. Therefore, the final equation is

$$f(t) = \sum_{n=-\infty}^{\infty} D_n \mathrm{e}^{\mathrm{j}n\omega_0 t} \tag{2.45}$$

Eq. 2.45 is the expression of *complex Fourier series*. Similar to the other Fourier series expression, in complex Fourier series also we can determine the Fourier coefficient. From the relationship between the coefficients of classical trigonometric Fourier series and that of the complex Fourier series illustrated in Eq. 2.35, and the derived expression from Eq. 2.13, 2.15 and 2.16, we have

$$\begin{aligned}
D_n &= \frac{1}{2}(a_n - jb_n) \\
&= \frac{1}{2}\left\{\frac{2}{T}\int_T f(t)\cos n\omega_0 t dt - j\frac{2}{T}\int_T f(t)\sin n\omega_0 t dt\right\} \\
&= \frac{1}{\mathrm{T}}\left\{\int_T f(t)\cos n\omega_0 t dt - \int_T f(t) j \sin n\omega_0 t dt\right\} \\
&= \frac{1}{\mathrm{T}}\int_T f(t)\{\cos n\omega_0 t - j\sin n\omega_0 t\} dt \\
&= \frac{1}{\mathrm{T}}\int_T f(t) e^{-jn\omega_0 t} dt
\end{aligned} \tag{2.46}$$

This is really interesting to see that (Eq. 2.39, 2.40), unlike to the other forms of Fourier series; here the sequence index n can run from $-\infty$ to ∞. So, this form of Fourier series representation is defined as *double sided Fourier series* or universal Fourier series.

2.8 Parseval's Theorem for Power

Parserval's theorem states that, the average power in a periodic signal is equal to the sum of the average power in its DC component and the average powers in its harmonics.

Let's consider a periodic signal $x(t)$.

As we know that

$$|x(t)|^2 = x(t)x^*(t)$$

($x^*(t)$ is the complex conjugate of $x(t)$)
The power of a signal $x(t)$ over a cycle is expressed as

$$P = \frac{1}{T}\int_{-\frac{T}{2}}^{+\frac{T}{2}} |x(t)|^2 \tag{2.47}$$

$$= \frac{1}{T}\int_{-\frac{T}{2}}^{+\frac{T}{2}} x(t)x^*(t)dt$$

$$= \frac{1}{T}\int_{-\frac{T}{2}}^{+\frac{T}{2}} x^*(t)\left[\sum_{n=-\infty}^{\infty} D_n e^{jn\omega_0 t}\right] dt$$

Here, $\omega_0 = 2\pi/T$
Interchanging the order of integration and summation, we get

$$P = \frac{1}{T}\sum_{n=-\infty}^{\infty} D_n . \int_{-\frac{T}{2}}^{+\frac{T}{2}} x^*(t)e^{jn\omega_0 t}dt \tag{2.48}$$

As,

$$D_n = \frac{1}{T}\int_{-\frac{T}{2}}^{+\frac{T}{2}} x(t)e^{-jn\omega_0 t}dt \tag{2.49}$$

$$D_n^* = \frac{1}{T}\int_{-\frac{T}{2}}^{+\frac{T}{2}} x(t)e^{jn\omega_0 t}dt \tag{2.50}$$

So, from above three equations, power can be expressed as

$$\begin{aligned} P &= \frac{1}{T}\sum_{n=-\infty}^{\infty} D_n.\left(TD_n^*\right) \\ &= \sum_{n=-\infty}^{\infty} D_n.\left(D_n^*\right) \\ &= \sum_{n=-\infty}^{\infty} |D_n|^2 \end{aligned} \tag{2.51}$$

From the expressions given in Eqs. 2.47 and 2.51, Perseval's Theorem of power is validated.

2.9 Phase Congruency: Application of Fourier Series in 1D and 2D Signal Processing (Image Processing)

Till the previous section we have discussed a lot on frequency domain representation of periodic signals. As understood from the statement of Fourier series, periodic signals always hold sinusoidal components of different frequencies which are integral multiple of a fundamental frequency. In the present section we'll concentrate on the phase criteria or rather the relative phase property of the component signals.

If we just take a look on the Example 2.1, it has derived infinite number of sinusoid components from a given square wave. To look closely, we see that the sinusoid components are all odd harmonics of the fundamental. The amplitudes of the even harmonics are zero, in this case. In the present section we will investigate the "why" of the result and from the understanding we will discuss a real life application using this concept.

In Fig. 2.13 we have shown the two sets of sinusoidal components and their respective synthesized signal. In Fig. 2.13a, a circular region is shown. For both the signals (with frequency f_0 and $2f_0$), the circular region is the region of zero crossing. The interesting thing to observe in Fig. 2.13a is that, the lower frequency signal goes from positive to negative polarity (i.e., negative going) at zero-crossing where as the higher frequency signal is positive going at the region of zero crossing. These opposite going of the component signals form a cross-like (X pattern) appearance at zero crossing. Therefore these two signals cannot contribute in synthesizing a new signal like square wave which is further negative going at the region of zero crossing. In other words all the component signals must support the fundamental frequency sinusoid to increase the sloe so that it approaches towards a perfect vertical line.

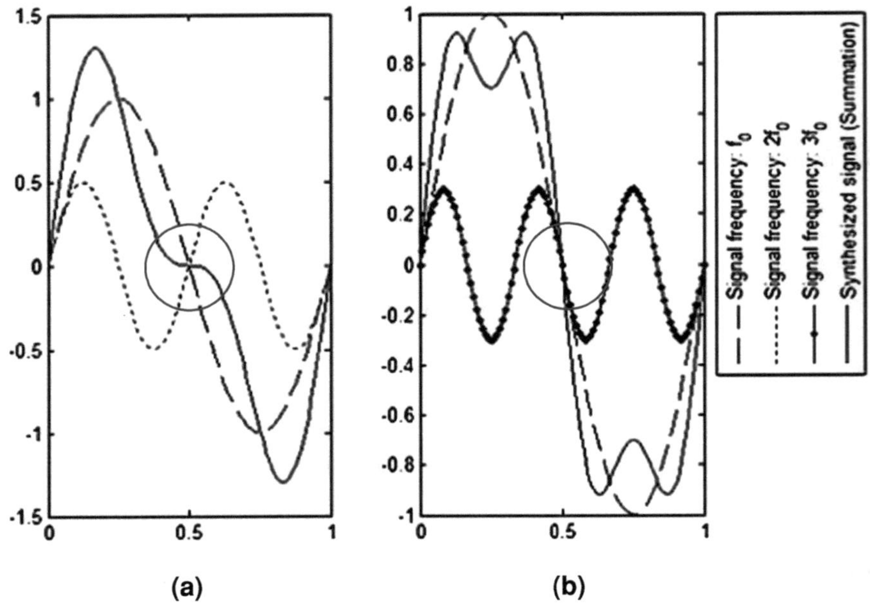

Fig. 2.13 Illustration of phase congruency in terms of Fourier components

On the contrary, as seen in Fig. 2.13b, we have used the third harmonic $(3f_0)$ along with the fundamental signal for Fourier synthesis. It has successfully generated a square like signal with higher slope with respect to the fundamental signal. Actually at the zero crossing, both the component signal are negative going. Therefore in synthesis the higher frequency component strengthens the square wave synthesis. That's why we did not get any even harmonics in Fourier synthesis of a square wave. This is the concept of *phase congruency*. In the present example, the region of amplitude transition of the square wave from +1 to −1 is detected in terms of congruent phase, i.e., the region where are the component signals would meet in same phase. The idea can also be realized from Fig. 2.15. The readers can also design their own experiments for the realization of phase congruency by using the GUI based Fourier synthesizer (Version 1.1), enclosed herewith as supplementary electronic material.

The step-like transition in one dimensional signal is equivalent to sharp intensity transition with respect to space in image like two dimensional signals. This sharp intensity-transition can also be characterized as an "edge" in image. Therefore, from the concept of phase congruency we can also detect edge like features from image.

In the classical approach of edge detection researchers always seared for high intensity gradients in an image. Rather than think of features in differential terms an alternative approach is to think of features in the *frequency domain*. Image profiles can be thought of as being formed by a Fourier series as shown in

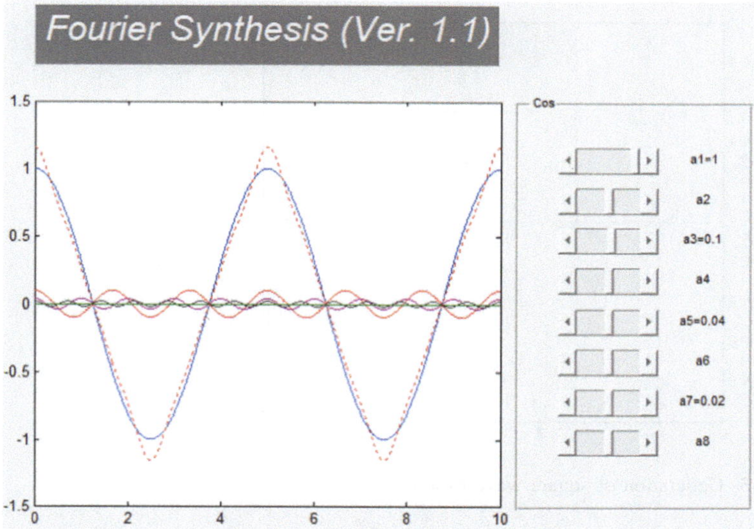

Fig. 2.14 Generation of triangular wave by Fourier synthesis

Figs. 2.14 and 2.15. The triangular and square spatial waveforms can be expressed in terms of Fourier components as I_1 and I_2 respectively as follows. Here only one direction (either row wise or column wise) is considered for the ease of understanding.

$$
\begin{aligned}
I_1(x) &= \sum_{n=0}^{\infty} \frac{1}{(2n+1)^2} \cos\{(2n+1)x\} \\
I_1(x) &= \sum_{n=0}^{\infty} \frac{1}{(2n+1)^2} \sin\left\{(2n+1)x + \frac{\pi}{2}\right\}
\end{aligned}
\tag{2.52}
$$

and

$$
I_2(x) = \sum_{n=0}^{\infty} \frac{1}{(2n+1)} \sin\{(2n+1)x\} \tag{2.53}
$$

Here also it is to be noted that the Fourier components are all *in phase* at the point of the step in the square wave (Fig. 2.15), and at the peaks and troughs of the triangular wave (Fig. 2.14). Congruency of phase at any angle produces a clearly perceived feature. We can generalize our Fourier series expression to generate a wide range of waveforms with the equation

$$
I(x) = \sum_{n=0}^{\infty} \frac{1}{(2n+1)^p} \sin\{(2n+1)x + \phi\} \tag{2.54}
$$

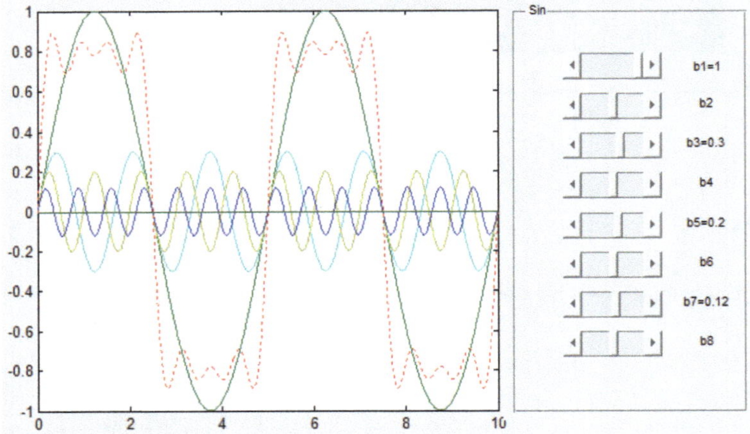

Fig. 2.15 Generation of square wave by Fourier synthesis

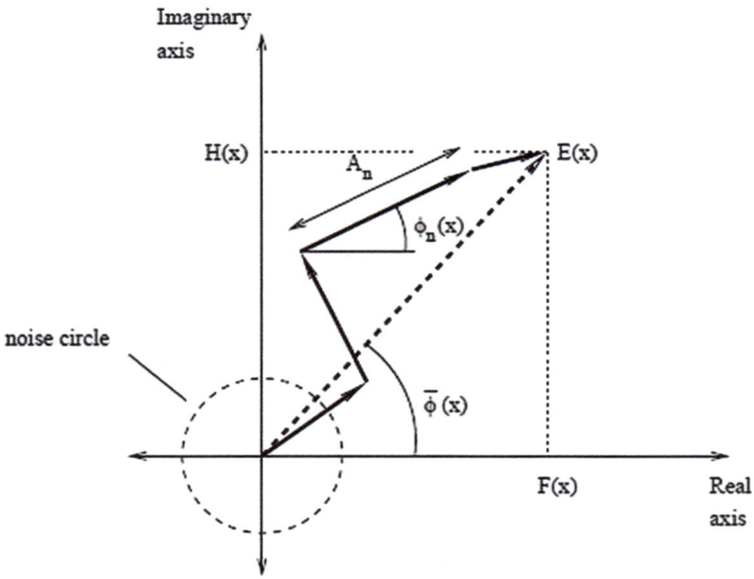

Fig. 2.16 Polar diagram showing the Fourier components at a location in the signal plotted head to tail. The weighted mean phase angle is given by $A(x)$. The noise circle represents the level of $E(x)$ one can expect just from the noise in the signal

where ϕ is the phase offset defining the angle at which phase congruency occurs at features, and p is the exponent that describes the rate of amplitude decay with frequency in the Fourier series.

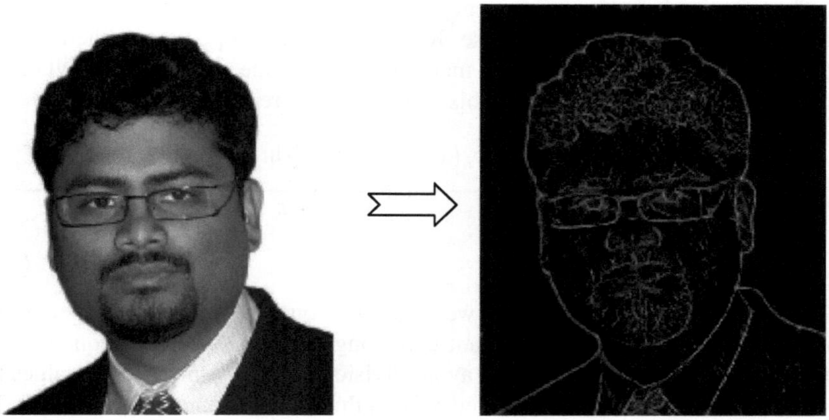

Fig. 2.17 Edge-like feature extraction from an image using Kovesi's measure of phase congruency

The measurement of phase congruency at a point in a signal can be seen geometrically in Fig. 2.16. The local Fourier components at a location X in the signal will each have amplitude $A_n(x)$ and a phase angle $\phi_n(x)$. Figure 2.16 plots these local Fourier components as complex vectors adding head to tail. The sum of these components projected onto the real axis represents $F(x)$, the original signal. The magnitude of the vector from the origin to the end point is the *Local Energy*, $|E(x)|$.

The measure of phase congruency developed by Morrone et al. [4] is

$$PC_{Morrone} = \frac{|E(x)|}{\sum_n A_n(x)} \tag{2.55}$$

Under this definition phase congruency is the ratio of $|E(x)|$ to the overall path length taken by the local Fourier components in reaching the end point. If all the Fourier components are in phase all the complex vectors would be aligned and the ratio $\frac{|E(x)|}{\sum_n A_n(x)}$ would be one. If there is no coherence of phase, the ratio falls to a minimum of zero. Phase congruency provides a measure that is independent of the overall magnitude of the signal making it invariant to variations in image illumination and/or contrast. Fixed threshold values of feature significance can then be used over wide classes of images. It can be shown that this measure of phase congruency is a function of the cosine of the deviation of each phase component from the mean

$$PC_{derived} = \frac{\sum_n A_n\left(\cos\left(\phi_n(x) - \bar{\phi}(x)\right)\right)}{\sum_n A_n(x)} \tag{2.56}$$

This measure of phase congruency (Eq. 2.55) does not provide good localization and it is also sensitive to noise. Kovesi [5–8] developed a modified measure consisting of the cosine minus the magnitude of the sine of the phase deviation as given in the following equation; this produces a more localized response.

$$PC_{Kovesi} = \frac{\sum_n W(x)\lfloor A_n\left(\cos\left(\phi_n(x) - \bar{\phi}(x)\right) - \left|\sin\left(\phi_n(x) - \bar{\phi}(x)\right)\right|\right) - T\rfloor}{\sum_n A_n(x) + \varepsilon} \tag{2.57}$$

The term $W(x)$ is a factor that weights for frequency spread (congruency over many frequencies is more significant than congruency over a few frequencies). A small constant ε is incorporated to avoid division by zero. Only energy values that exceed the threshold T, the estimated noise influence, are counted in the result. The symbols $\lfloor\ \rfloor$ denote that the enclosed quantity is equal to itself when its value is positive, and zero otherwise. Edge detection using Kovesi's measure of phase congruency [5–8] is presented in Fig. 2.17.

References

1. Bracewell, R.N.: The Fourier Transform and Its Applications, 2nd edn. McGraw-Hill Book Company, NY (1987)
2. Lathi, B.P.: Modern Digital and Analog Communication Systems, 3rd edn. Oxford University Press, NY (2005)
3. Hardy, G.H., Rogosinski, W.W.: Fourier Series. Dover Publications, INC., NY (1999)
4. Morrone, M.C., Ross, J.R., Burr, D.C., Owens, R.A.: Mach bands are phase dependent. Nature **324**, 250–253 (1986)
5. Kovesi, P.: Symmetry and asymmetry from local phase. In: AI'97, 10th Australian Joint Conference on Artificial Intelligence. Proceedings - Poster Papers, pp. 185–190, 2–4 Dec 1997
6. Kovesi, P.: Image features from phase congruency. Videre: J. Comput. Vision Res. **1**(3), 1–26 (1999) (MIT Press)
7. Kovesi, P.: Edges are not just steps. In: Proceedings of ACCV2002 the 5th Asian Conference on Computer Vision, Melbourne, pp. 822–827, 22–25 Jan 2002
8. Kovesi, P.: Phase congruency detects corners and edges. In: The Australian Pattern Recognition Society Conference: DICTA 2003, December 2003, pp. 309–318. Sydney (2003)

Chapter 3
Fourier Transform

3.1 Introduction

To study a signal (general signals) in frequency domain or to study a system in terms of its frequency response characteristics is an important and unavoidable task of electrical and electronics engineers. Even to the layman, the designations of their favorite radio channels are understood by their respective carrier frequencies. In the previous chapter of Fourier series, we have already seen how any periodic signal can be represented in terms of its single tone components. The understanding of unique frequency present in any periodic signal has been changed there to fundamental frequency of principle frequency component.

In the present chapter we'll extend our idea of spectral representation for aperiodic signals also to get a feel of frequency response characteristics of natural systems as well as frequency domain distribution of natural signals. In the next section the extension of idea of Fourier series to Fourier transform is discussed. The mathematical interpretation helps us to understand the physical significance of Fourier transform. Signals which have finite phase response can mostly be defined as complex function of time. These types of complex signals with evenness and oddness are addressed and studied in transformed domain. Convergence of Fourier transform into cosine or sine transform has also been studied. From different properties of Fourier transform we have also discussed the method of modulation, verified the realizability of ideal filters and searched the reason to be the phase response of a filter to be negative linear.

This chapter along with the chapter of Fourier series can be considered as the fundamental study before approaching towards Discrete Time Fourier Series (DTFS), Discrete Time Fourier Transform (DTFT), Discrete Fourier Transform (DFT), Fast Fourier Transform (FFT) and Discrete Wavelet Transform (DWT).

A. Das, *Signal Conditioning*, Signals and Communication Technology,
DOI: 10.1007/978-3-642-28818-0_3, © Springer-Verlag Berlin Heidelberg 2012

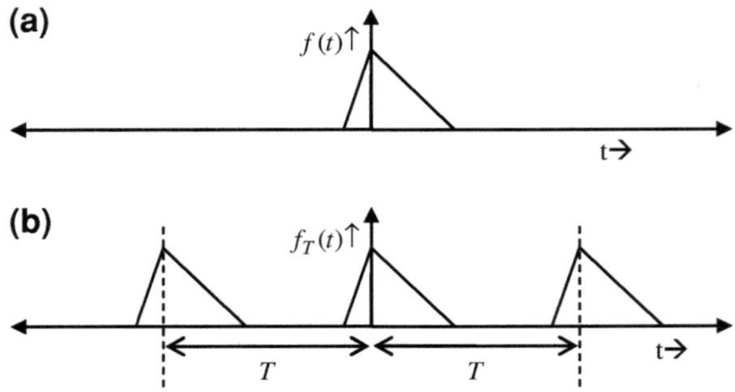

Fig. 3.1 Construction of periodic signal by aperiodic extension

3.2 Mathematical Interpretation

As seen from the previous chapter, Fourier series and all its calculations of coefficients are valid for periodic signals. Therefore for aperiodic signals we must put on some limit or condition for periodic signals by which it can be inferred as aperiodic signals. Let's consider the signal $f(t)$ shown in the Fig. 3.1a which is essentially an aperiodic signal. Another signal $f_T(t)$ is considered which is a periodic one with periodicity T and with unit function exactly matching with $f(t)$.

It is obvious that if T approaches infinity than the function is effectively aperiodic. Therefore, we can establish a relationship between $f(t)$ and $f_T(t)$ as

$$\underset{T\to\infty}{\ell t}\, f_T(t) = f(t) \tag{3.1}$$

Thus the Fourier series representing $f_T(t)$ will also represent $f(t)$ in the limit $T \to \infty$. From the concept of complex Fourier series (Eq. 2.45),

$$f_T(t) = \sum_{n=-\infty}^{\infty} D_n e^{jn\omega_0 t} = \sum_{n=-\infty}^{\infty} D_n e^{j\frac{2\pi n t}{T}} \tag{3.2}$$

If the product (n/T) in the exponent is replaced by a variable k_n then the equation becomes

$$f_T(t) = \sum_{n=-\infty}^{\infty} D_n e^{j2\pi k_n t} \tag{3.3}$$

It is clear that for large T $(T \to \infty)$ the summation contains a large number of waves each with wavelength difference.

$$\Delta k = k_{n+1} - k_n = \frac{1}{T} \tag{3.4}$$

The discrete summation tends to a continuous integration (as 1/T becomes 0) with$_n$ replaced by a variable k and D_n becoming a function of k represented as D_k. That is the equation evolves as shown:

$$f(t) = \int_{-\infty}^{\infty} D_k e^{j2\pi kt} dk \tag{3.5}$$

Now, if we consider, $D_k = F(k)$, Eqs. 2.45 and 2.46 can be re-written as

$$f(t) = \int_{-\infty}^{\infty} F(k) e^{j2\pi kt} dk \tag{3.6}$$

$$F(k) = \int_{-\infty}^{\infty} f(t) e^{-j2\pi kt} dt \tag{3.7}$$

The above two expressions are collectively called as Fourier transform pair. Equations 3.7 and 3.6 represents the forward and inverse Fourier transform, respectively for the continuous time signals. As understood, Fourier transform is applicable for aperiodic signals only; we can be more accurate saying that, the equations are valid for *continuous time aperiodic signals*.

In these equations F(k) is the Fourier transform of f(t). The variables, for this example of t and k are called conjugate variables. When conjugate variables are multiplied the product should be unit less. If the variables are t = time (say seconds) and k = frequency say cycles per second then the equations shown are appropriate.

The variable k in Eqs. 3.7 and 3.6 can be converted to our familiar angular frequency term ω by very simple conversion as given in Eqs. 3.8 and 3.9, respectively.

$$F(\omega) = \int_{-\infty}^{\infty} f(t) e^{-j\omega t} dt \tag{3.8}$$

$$f(t) = \frac{1}{2\pi} \int_{-\infty}^{\infty} F(\omega) e^{j\omega t} d\omega \tag{3.9}$$

Equations 3.8 and 3.9 are the expressions of forward and inverse Fourier transform respectively.

Example 3.1 Compute Fourier transform and draw the amplitude and phase spectrum of the given function

$$f(t) = e^{-at} u(t)$$

$$F(\omega) = \int_{-\infty}^{\infty} f(t) e^{-j\omega t} dt = \int_{0}^{\infty} e^{-at} e^{-j\omega t} dt = \int_{0}^{\infty} e^{-(a+j\omega)t} dt$$

$$\Rightarrow F(\omega) = -\frac{1}{a + j\omega}$$

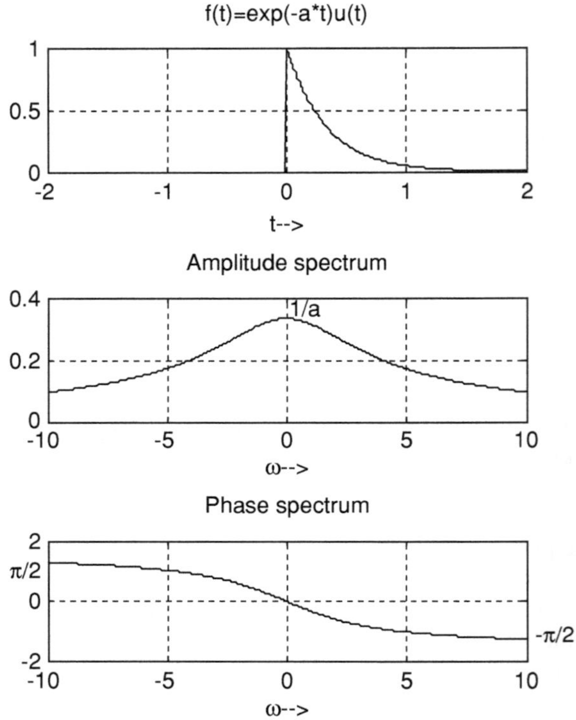

Fig. 3.2 $e^{-at}u(t)$ and its spectra

To get the amplitude spectrum,

$$|F(\omega)| = \frac{1}{\sqrt{a^2 + \omega^2}} \tag{3.10}$$

Similarly, to get the phase spectrum,

$$\angle F(\omega) = -\tan^{-1}\frac{\omega}{a} \tag{3.11}$$

As seen from Fig. 3.2, the amplitude spectrum is an even function of frequency whereas the phase spectrum is an odd function of frequency. It is quite in agreement with our understanding of transformed domain representation of signals and systems.

Example 3.2 Compute Fourier transform of a gate pulse.

The gate function can be defined as

$$g(t) = \begin{cases} 1; & -\frac{\tau}{2} \leq t \leq \frac{\tau}{2} \\ 0; & elsewhere \end{cases}$$

Therefore, $g(t)$ can be characterized in terms of the well known *rect* function as $g(t) = rect\left(\frac{t}{\tau}\right)$. The function is shown in the Fig. 3.3.

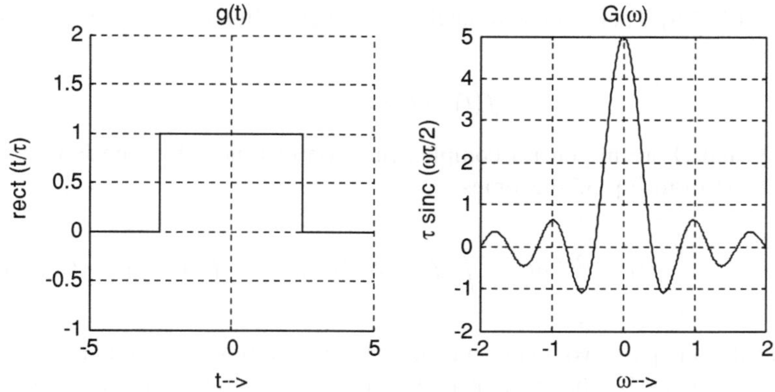

Fig. 3.3 Fourier transform of a gate function

We know,

$$G(\omega) = \int_{-\infty}^{\infty} g(t)e^{-j\omega t}dt$$

$$\begin{aligned} G(\omega) &= \int_{-\frac{\tau}{2}}^{\frac{\tau}{2}} e^{-j\omega t}dt \\ &= -\frac{1}{j\omega}\left(e^{-j\omega\tau/2} + e^{j\omega\tau/2}\right) \\ &= \frac{2}{\omega}\frac{e^{-j\omega\tau/2} + e^{j\omega\tau/2}}{2j} \\ &= \frac{2}{\omega}\frac{\omega\tau}{2}\frac{\sin\frac{\omega\tau}{2}}{\frac{\omega\tau}{2}} \\ &= \tau \sin c\frac{\omega\tau}{2} \end{aligned}$$

Therefore,

$$rect\left(\frac{t}{\tau}\right) \Leftrightarrow \tau \sin c\frac{\omega\tau}{2} \tag{3.12}$$

3.3 Significance of Oddness and Evenness in Complex Plane

From the understanding of oddness and evenness of signals as studied in the introductory chapter and the odd–even multiplication rule in Fourier series, we can now be sure that, any function (here, referred as a signal) can be expressed as either an odd function or an even function or summation of an odd (O) and an even (E) function

[Eq. 1.5, 1.8, 1.9]. Therefore, without the loss of generality, we can express any signal $f(t)$ as

$$f(t) = E(t) + O(t) \tag{3.13}$$

where E and O are in general complex function of time. Then the Fourier transform of f(t) from Eq. 3.7 becomes

$$F(k) = \int_{-\infty}^{\infty} f(t)e^{-j2\pi kt}dt = \int_{-\infty}^{\infty} f(t)\cos 2\pi ktdt - j\int_{-\infty}^{\infty} f(t)\sin 2\pi ktdt \tag{3.14}$$

Now, as the odd parts would be vanished after integration and even parts would be doubly calculated by the integral rule of odd–even signal multiplication as discussed in the previous chapter, the above equation reduces to [1]

$$F(k) = 2\int_{0}^{\infty} E(t)\cos 2\pi ktdt - 2j\int_{0}^{\infty} O(t)\sin 2\pi ktdt \tag{3.15}$$

The above equation suggests that if a signal is even, its transfer function (Fourier transform) is also even and if a signal is odd, the transfer function is again odd. The property can be summarized from the relational equation as follows:

$$f(t) = o(t) + e(t) = \mathrm{Re}\{o(t)\} + j\,\mathrm{Im}\{o(t)\} + \mathrm{Re}\{e(t)\} + j\,\mathrm{Im}\{e(t)\}$$

$$F(k) = O(k) + E(k) = \mathrm{Re}\{O(k)\} + j\,\mathrm{Im}\{O(k)\} + \mathrm{Re}\{E(k)\} + j\,\mathrm{Im}\{E(k)\} \tag{3.16}$$

In the Fig. 3.4, the *'relative sense of oddness'* is examined. When f(t) is real and odd with a *positive* moment, the odd part of F(k) has j[1] times a *negative* moment; and when f(t) is real but not necessarily odd, we also find the opposite sense of oddness. In the present study the real even functions/signals and imaginary even signals remain unchanged in transformed domain (after doing Fourier transform) as shown in Eq. 3.15 In the Fig. 3.4, the left column shows the time domain and right column represents frequency domain representation of same signal.[2] The transformations of real odd, imaginary odd and imaginary even signals are also important in this regard. A set of special kind of symmetry is proposed by the signals like f(t) whose real part is

---

[1] $j = \sqrt{-1}$.

[2] The patterns of the signals are not accurate; the oddness is represented by *sinusoid* and evenness by *co-sinusoid* for the ease of understanding.

even and imaginary part is odd. Such a function is described as hermitian (Fig. 3.5). It is often defined by the property

$$f(t) = f^*(-t) \tag{3.17}$$

As mentioned above, the Fourier transform is real. As an example of algebraic procedure for handling matters of this kind, we consider

$$f(t) = E + O + j\hat{E} + j\hat{O} \tag{3.18}$$

Then,

$$f(-t) = E - O + j\hat{E} - j\hat{O} \tag{3.19}$$

And

$$f^*(-t) = E + O - j\hat{E} - j\hat{O} \tag{3.20}$$

If we now suggest $f(t) = f^*(-t)$, then from the above three equations, we must have $O = 0$ and $\hat{E} = 0$. Hence,

$$f(t) = E + j\hat{O} \tag{3.21}$$

The relationship between the Fourier transforms pairs of complex signals referring to the Fig. 3.4 are listed in the Table 3.1.

3.4 Cosine and Sine Transform

The cosine transform of a signal $f(t)$ can be defined inspiring from the Eq. 3.15, as

$$F_c(k) = 2\int_0^{\infty} f(t)\cos 2\pi kt dt \tag{3.22}$$

The transformation shown in the above expression is the same as Fourier transform if $f(t)$ is an even function. In general, we can say the even part of the Fourier transform of $f(t)$ is the cosine transform of the even part of $f(t)$. The inverse of cosine transform by which $f(t)$ can be obtained from $F_c(k)$ would be identical to the above equation as

$$f(t) = 2\int_0^{\infty} F_c(k)\cos 2\pi kt dk \tag{3.23}$$

The sine transform and inverse of sine transform can also be defined in a similar fashion as

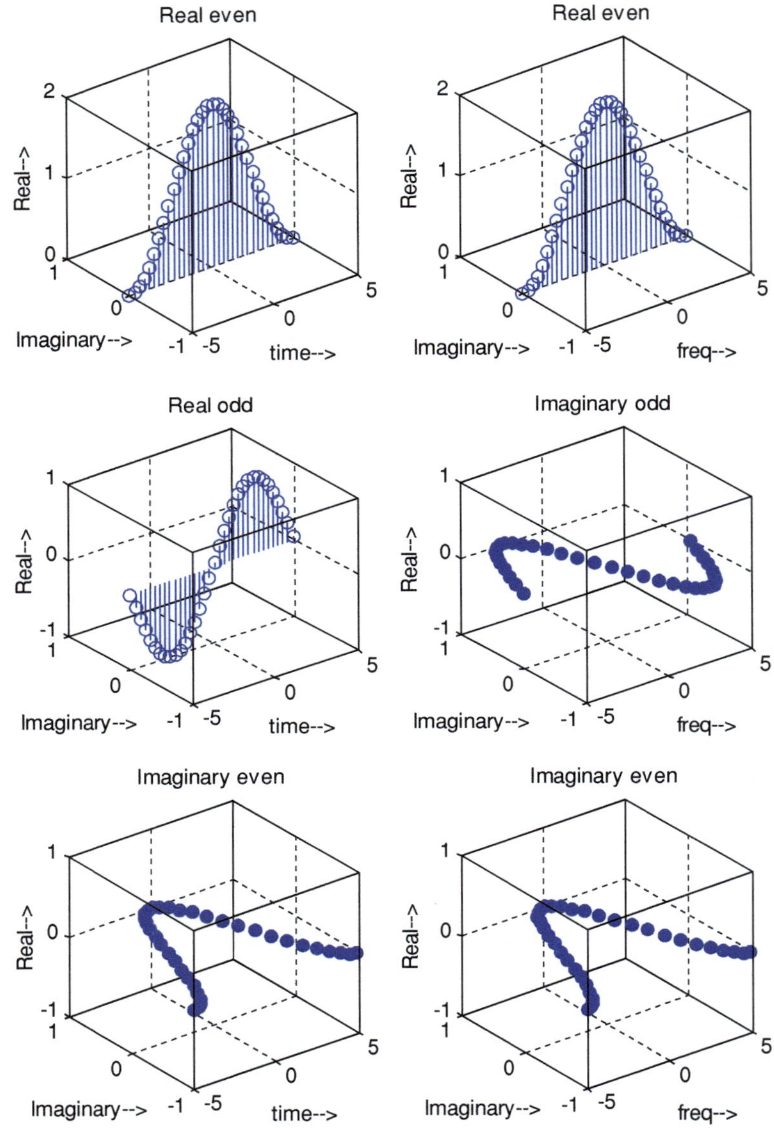

Fig. 3.4 Symmetry properties of complex signals (*first column*) and their Fourier transform (*second column*)

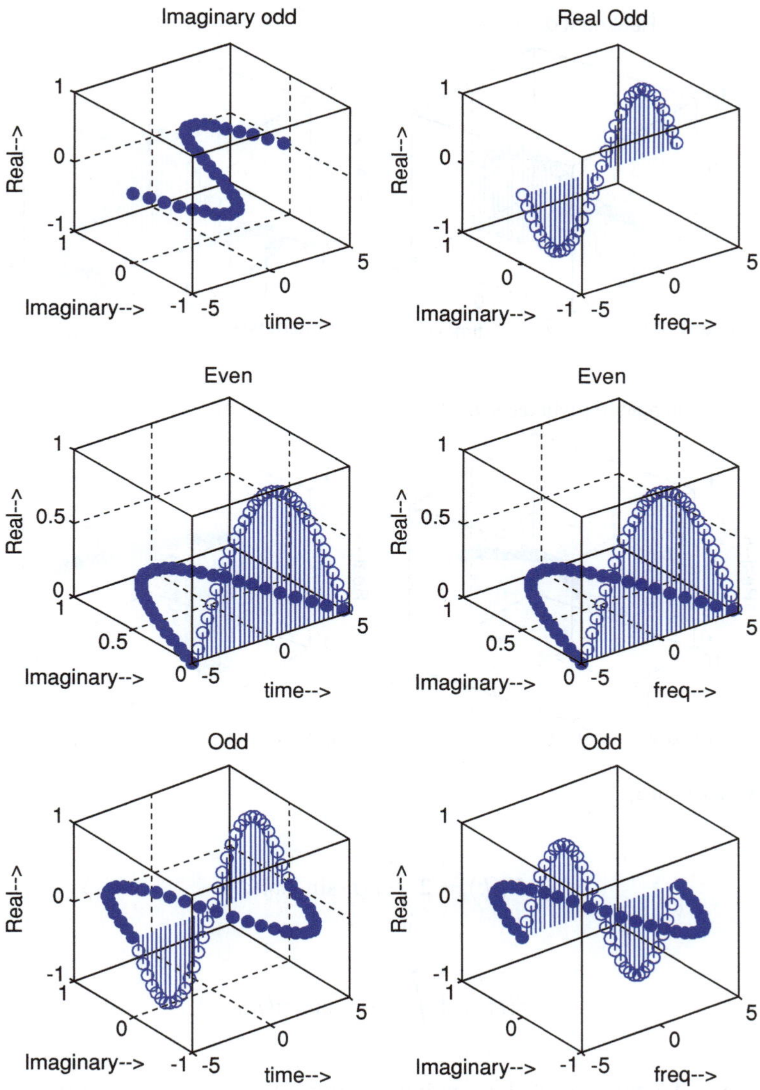

Fig. 3.4 (continued)

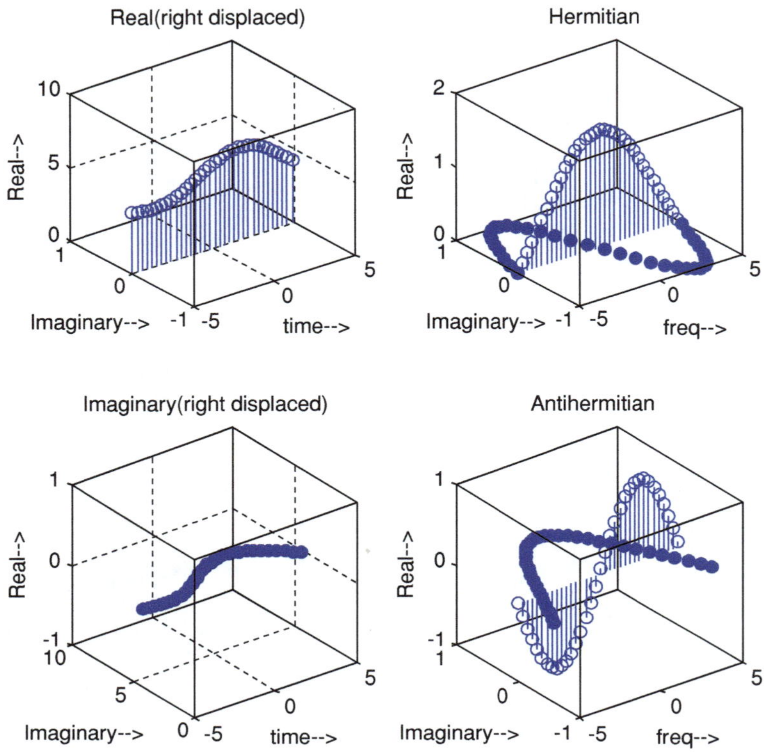

Fig. 3.4 (continued)

$$F_s(k) = 2\int_0^\infty f(t)\sin 2\pi ktdt \tag{3.24}$$

$$f(t) = 2\int_0^\infty F_s(k)\sin 2\pi ktdk \tag{3.25}$$

From the understanding of the previous section of symmetry and anti-symmetry, we can say that, j times the odd part of the Fourier transform of f(t) is sine transform of the odd part of f(t). Combining the sine and cosine transform of even and odd parts leads to the complete Fourier transform as below.

$$F(f(t)) = F_C(e(t)) - jF_S(o(t)) \tag{3.26}$$

If f(t) is zero for t < 0, then the above equation converges to

$$F(k) = \frac{1}{2}F_C(k) - j\frac{1}{2}F_S(k) \tag{3.27}$$

Table 3.1 Relationships between the time and frequency domain representation of complex signals

Time domain signal f(t)		⟶	Frequency domain (F(k))	
Real	Imaginary		Real	Imaginary
Even	Not present		Even	Not present
Odd	Not present		Not present	Odd
Not present	Even		Not present	Even
Even	Even		Even	Even
Odd	Odd		Odd	Odd
Asymmetric	Not present		Even	Odd
Not present	Asymmetric		Odd	Even

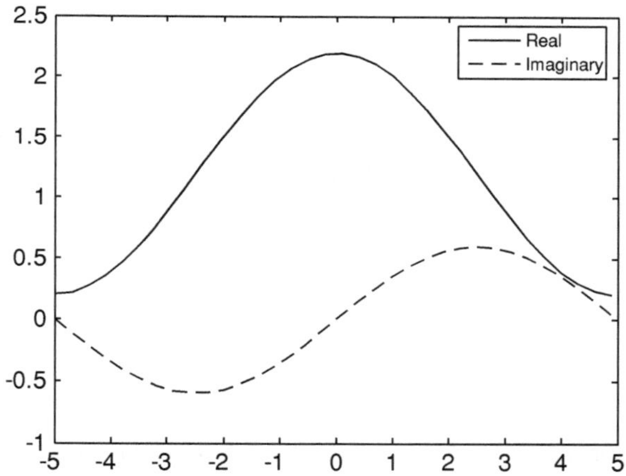

Fig. 3.5 Hermitian function having real even and imaginary odd

or, to be precise,

$$\frac{1}{2}F_C(k) - j\frac{1}{2}F_S(k) = F(f(t) \times u(t)) \tag{3.28}$$

where u(t) is the unit step signal.

3.4.1 Interpretation of the Formula

By the basic statement of Fourier series, as described in the previous chapter, we are now conscious of graphical interpretations of Fourier integral. Since the integral contains a complex factor, the simpler cosine and sine signals are more often considered [3]. Thus given f(t) we illustrated $f(t)\cos 2\pi kt$ as an oscillation laying within the envelope $f(t)$ and $-f(t)$. Twice the area under the $f(t)\cos 2\pi kt$ is then $F_c(k)$ (Eq. 3.22). In Fig. 3.6b the area is virtually zero, but high value of frequency (k) is implied. In Fig. 3.6c low value of k is implied and the area obtained is finite positive.

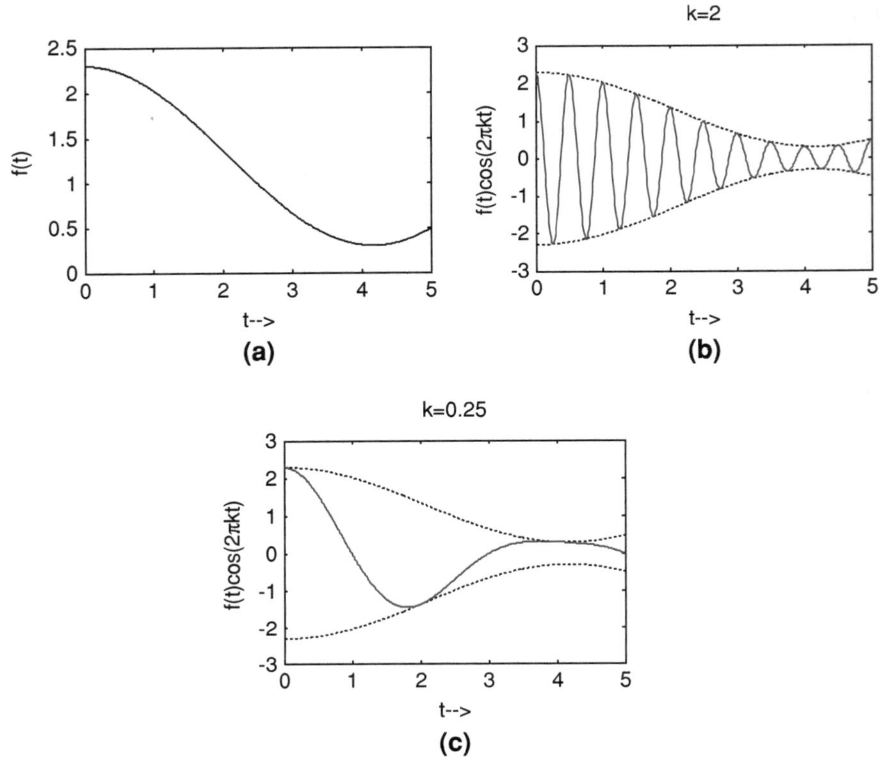

Fig. 3.6 The product of $f(t)$ with $\cos 2\pi kt$ as a function of t

Thus the Fourier integral can be visualized for discrete values of k (Fourier transform converges to Fourier series in a limiting condition). Therefore the frequency factor k can be interpreted as the frequency of co-sinusoids is equal to the number of cycles per unit of t. The dependency of k on the value of $F_c(k)$ is as follows.

$$F_C(k)\begin{cases} \to 0 & k \to \infty \\ = 2\int\limits_0^\infty f(t)dt & k = 0 \end{cases} \tag{3.29}$$

The sine transform may also be illustrated in the same way. Finally the complex transform would be illustrated as combination of odd and even parts.

3.5 Properties of Fourier Transform

Some important properties which are related to our concerned area of study are depicted here. Their implications and applications will be understood either in the present chapter or in subsequent studies.

3.5.1 Time–Frequency Duality

This property states that if $g(t)$ and $G(\omega)$ are Fourier Transform pair, the shape or pattern of the functions are interchangeable in folded manner [2]. That is, if

$$g(t) \Leftrightarrow G(\omega)$$

then

$$G(t) \Leftrightarrow 2\pi g(-\omega) \tag{3.30}$$

Proof From Eq. 3.9,

$$g(t) = \frac{1}{2\pi} \int_{-\infty}^{\infty} G(x) e^{jxt} dx$$

Hence,

$$g(-t) = \frac{1}{2\pi} \int_{-\infty}^{\infty} G(x) e^{-jxt} dx$$

Changing t to ω yields,

$$g(-\omega) = \frac{1}{2\pi} \int_{-\infty}^{\infty} G(x) e^{-j\omega x} dx = \frac{1}{2\pi} F[G(t)]$$

or,

$$G(t) \Leftrightarrow 2\pi g(-\omega)$$

Example 3.3 Show time–frequency duality for rect–sinc Fourier pair.

By applying time–frequency duality property, Fourier transform of a sinc function is obtained from rect–sinc Fourier pair as seen in Eq. 3.12 (Fig. 3.7).

3.5.2 Scaling Property

This property states that if $g(t)$ and $G(\omega)$ are Fourier Transform pair, the contraction in $g(t)$ results in expansion in $G(\omega)$ and vice versa. That is, if

$$g(t) \Leftrightarrow G(\omega)$$

then for an arbitrary scalar a

$$g(at) \Leftrightarrow \frac{1}{|a|} G\left(\frac{\omega}{a}\right) \tag{3.31}$$

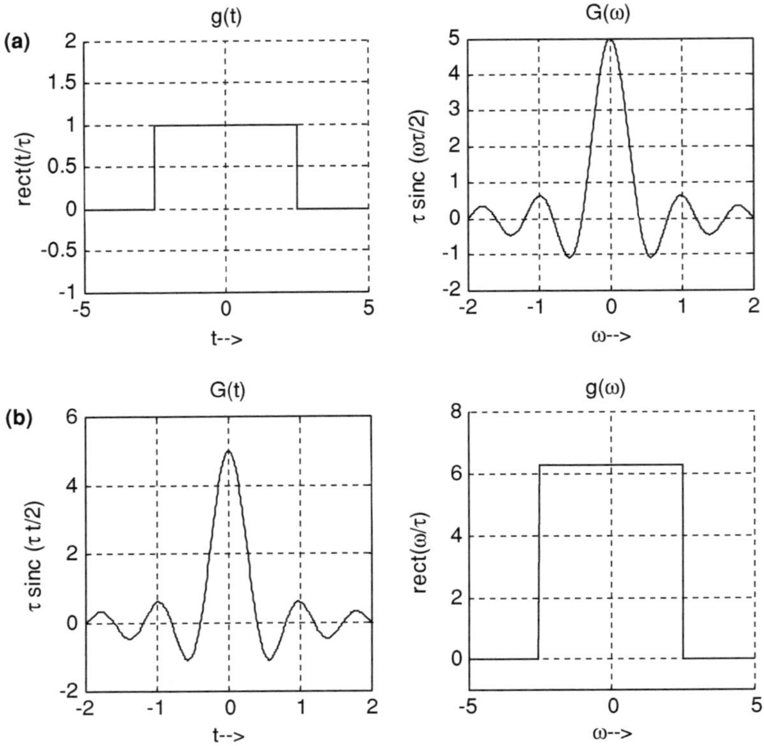

Fig. 3.7 Duality property of Fourier transform

Proof For a positive real constant a,

$$F[g(at)] = \int_{-\infty}^{\infty} g(at)e^{-j\omega t}dt$$

$$= \frac{1}{a}\int_{-\infty}^{\infty} g(x)e^{(-j\omega/a)x}dx = \frac{1}{a}G\left(\frac{\omega}{a}\right)$$

Similarly, for $a < 0$,

$$F[g(at)] = -\frac{1}{a}G\left(\frac{\omega}{a}\right)$$

Hence,

$$g(at) \Leftrightarrow \frac{1}{|a|}G\left(\frac{\omega}{a}\right)$$

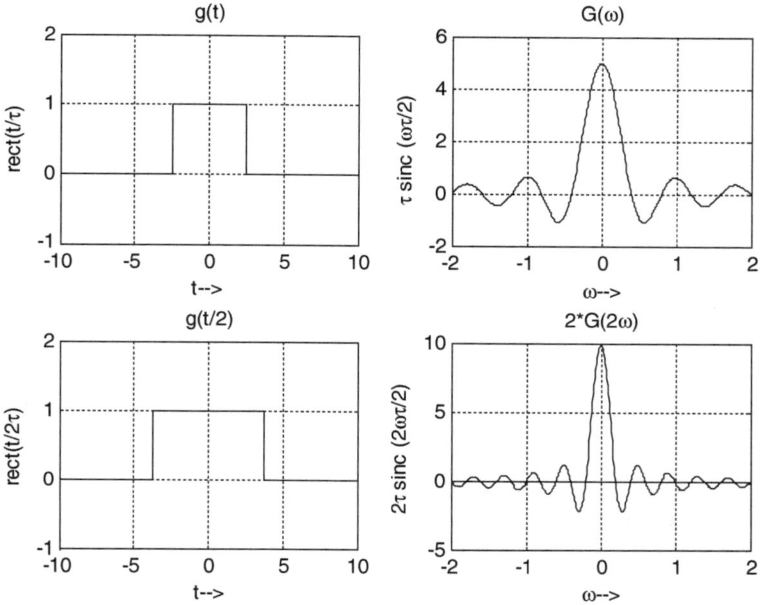

Fig. 3.8 Scaling property of Fourier transform

The amount of positive scaling (compression) in time or space domain is directly proportional to the amount of negative scaling (expansion) in spectra and vice versa. In Fig. 3.8, we have illustrated the concept using the popular rect-sinc Fourier pair. Increase in width of the rect function in time domain results contraction in width of the respective sinc in frequency domain. The discussion is supplemented by another discussion in the coming chapter of DFT (Discrete Fourier Transform).

3.5.3 Time Shifting Property

This property states that if $g(t)$ and $G(\omega)$ are Fourier Transform pair, delaying in the signal $g(t)$ by an amount of τ seconds does not change its amplitude spectrum. The phase spectrum however is changed to $-\omega\tau$. That is, if

$$g(t) \Leftrightarrow G(\omega)$$

then for any time delay τ

$$g(t-\tau) \Leftrightarrow G(\omega)e^{-j\omega\tau} \tag{3.32}$$

Proof From the definition of Fourier transform,

$$F[g(t-\tau)] = \int_{-\infty}^{\infty} g(t-\tau)e^{-j\omega t}dt$$

Considering $(t-\tau)=x$, we have

$$F[g(t-\tau)] = \int_{-\infty}^{\infty} g(x)e^{-j\omega(x+\tau)}dx$$

or,

$$F[g(t-\tau)] = e^{-j\omega\tau}\int_{-\infty}^{\infty} g(x)e^{-j\omega x}dx = G(\omega)e^{-j\omega\tau}$$

3.5.3.1 Linear Phase

Even for ideal systems (or filters) the phase response plays an important role as suggested by the time shifting criterion of Fourier transform. For natural delay due to time constant (τ) of a particular system, the obvious relationship between the input and output would be $y(t) = x(t - t_d)$. It does not effect on the amplitude response; on the contrary effect on the phase response of the filter is effectively influenced by the time shifting property of Fourier transform. In the present subsection, we would try to understand the obviousness of linear and negative phase response not from the mathematics as given in Eq. 3.32. The said equation is the evident of being the phase $\theta = -\omega t_d$, if t_d is considered to be the delay introduced by the system itself.

Here we'll use the *GUI based Fourier synthesis* tool[3] supplemented with this text. From the basic understanding of Fourier series, we are convinced that a square wave $(f(t))$ can be expressed as a summation of sinusoids which includes the fundamental frequency sinusoid and its odd harmonics as

$$f(t) = \frac{2}{\pi}\left\{\sin\omega_0 t + \frac{1}{3}\sin 3\omega_0 t + \frac{1}{5}\sin 5\omega_0 t + \frac{1}{7}\sin 7\omega_0 t + \cdots\right\}$$

By some weight factor of $\frac{\pi}{2}$, same square wave can be synthesized by adding sinusoids of amplitudes 1, 0.33, 0.2, 0.14,... respectively as shown in Fig. 3.9. The dotted line is the synthesized signal which is merely a square wave with some ripples due to approximation up to seventh harmonics. This phenomenon can be inferred in reverse way. If the square wave is passed through a low pass filter with cut-off frequency ω_0, then the fundamental frequency sinusoid would be retrieved.

[3] Fourier Synthesis (Ver 1.1) is downloadable from www.extras.springer.com.

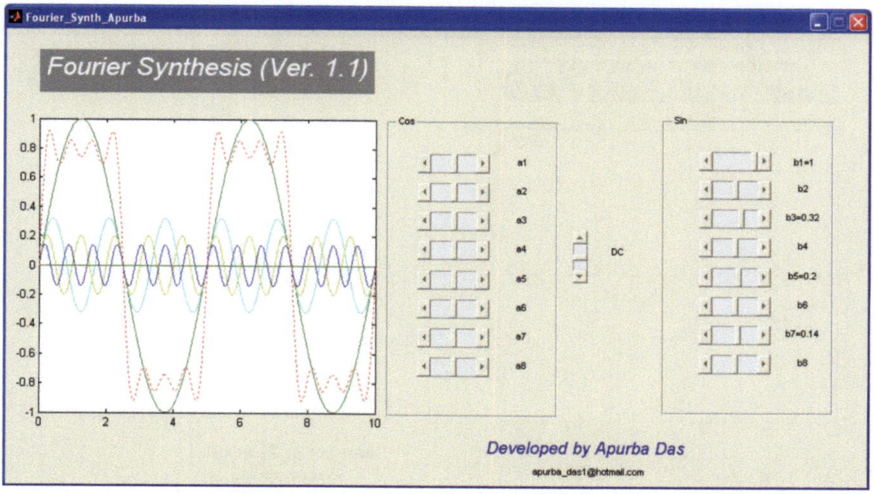

Fig. 3.9 A square wave is synthesized by adding sinusoids of odd harmonics

Now, let's assume the phase response of the filter is constant (or uniform). If the assumption is correct, the same reverse procedure of retrieving the fundamental frequency sinusoid should work. We have introduced a constant amount of delay by $\phi = \frac{\pi}{2}$ to each of the sinusoids. As seen in the Fig. 3.10, the synthesized signal (shown using dotted line) is not even a square wave. Therefore the reverse process of retrieving the fundamental component of the sinusoid is impossible with the filter which such type of (constant) phase response (Fig. 3.10). In the next experiment, we have varied the amount of delay introduced into the harmonics in some intended manner. The sinusoid with fundamental frequency (ω_0) is delayed by $\phi = \frac{\pi}{2}$. The third harmonic (sinusoid with frequency $3\omega_0$) is delayed by an amount $\phi = \frac{3\pi}{2}$. The fifth and seventh harmonics have been taken care of by the same way by introducing delays of amount $\phi = \frac{5\pi}{2}$ and $\frac{7\pi}{2}$ respectively. Therefore the patterns of the sinusoids have been converted following the expressions as given

$$\sin\left(\omega_0 t - \frac{\pi}{2}\right) = -\cos\omega_0 t \tag{3.33a}$$

$$\sin\left(3\omega_0 t - \frac{3\pi}{2}\right) = \cos 3\omega_0 t \tag{3.33b}$$

$$\sin\left(5\omega_0 t - \frac{5\pi}{2}\right) = -\cos 5\omega_0 t \tag{3.33c}$$

$$\sin\left(7\omega_0 t - \frac{7\pi}{2}\right) = \cos 7\omega_0 t \tag{3.33d}$$

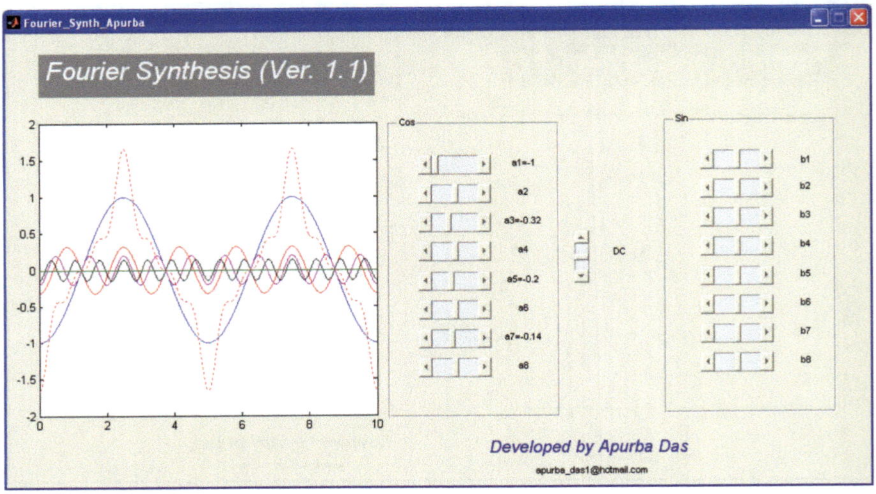

Fig. 3.10 Failure in square wave synthesis for uniform phase delay of the component sinusoids

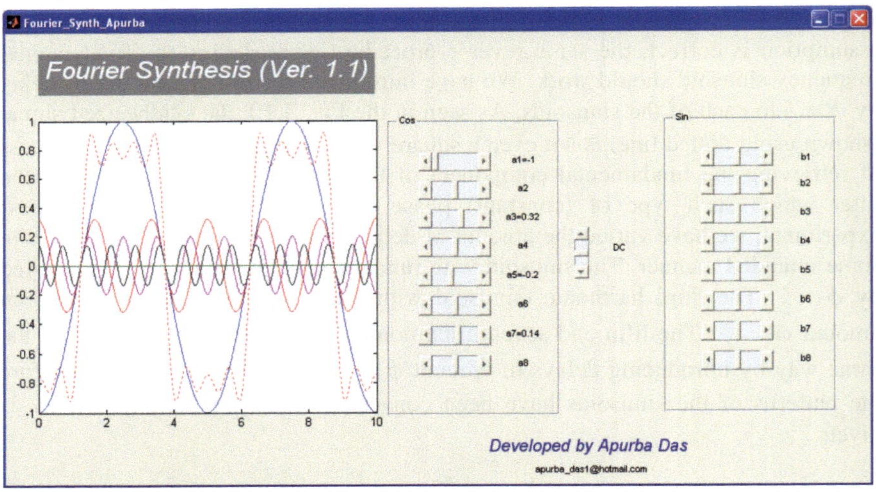

Fig. 3.11 Success in square wave synthesis for linear phase delay of the component sinusoids

The polarities of the cosinusoids are chosen according to the Eq. 3.33a and fit into the GUI based Fourier synthesis software. As a consequence the resultant waveform obtained (dotted line) as seen in Fig. 3.11 becomes exactly square as obtained in Fig. 3.9 with a delay of amount $\phi = \frac{\pi}{2}$. Hence the reverse process of retrieving the fundamental frequency sinusoid will be successful with filters having this kind of phase response characteristics. The phase response of the studied filter can now be plotted as in Fig. 3.12. It is very clear from the graph that the phase response has linear and negative slope with respect to the frequencies.

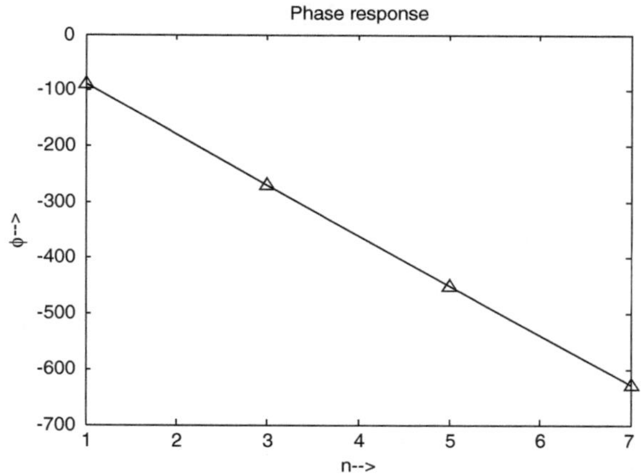

Fig. 3.12 Linear phase response of a filter shown for nth harmonics

3.5.4 Frequency Shifting Property

This property states that if $g(t)$ and $G(\omega)$ are Fourier Transform pair, multiplication of the signal $g(t)$ by $e^{j\Omega t}$ shifts its spectrum $G(\omega)$ by $\omega = \Omega$. It is interesting to note that, still the *duality* is observed in time shifting and frequency shifting. That is, if

$$g(t) \Leftrightarrow G(\omega)$$

then

$$g(t)e^{j\Omega t} \Leftrightarrow G(\omega - \Omega) \tag{3.34}$$

Proof From the definition of Fourier transform,

$$F\left[g(t)e^{j\Omega t}\right] = \int_{-\infty}^{\infty} g(t)e^{j\Omega t}e^{-j\omega t}dt = \int_{-\infty}^{\infty} g(t)e^{-j(\omega-\Omega)t}dt = G(\omega - \Omega)$$

Replacing Ω by $-\Omega$, the above equation can be re-written as

$$g(t)e^{-j\Omega t} \Leftrightarrow G(\omega + \Omega) \tag{3.35}$$

This is now an additional understanding through the journey staring from phasor representation to time domain representation of single tone signal as shown in the following equation.

$$g(t)\cos\Omega t = \frac{1}{2}\left\{g(t)e^{j\Omega t} + g(t)e^{-j\Omega t}\right\} \Leftrightarrow \frac{1}{2}\{G(\omega - \Omega) + G(\omega + \Omega)\} \tag{3.36}$$

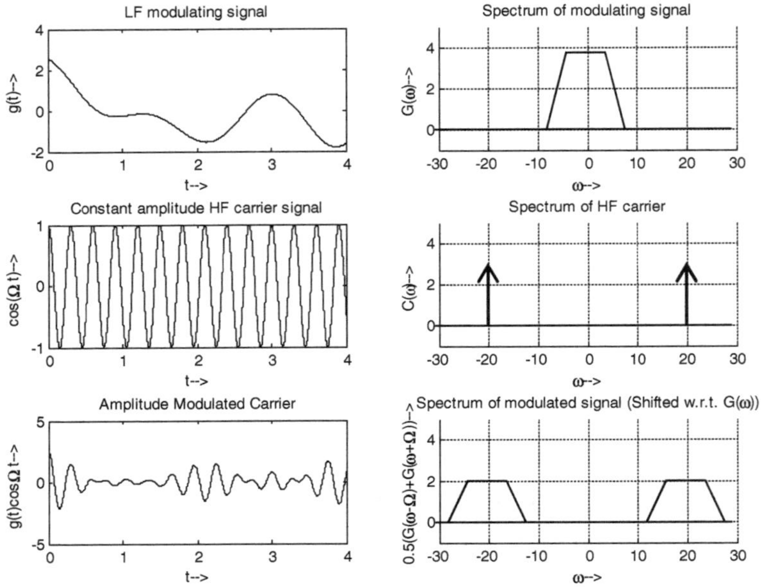

Fig. 3.13 Interpretation of AM using frequency shifting property of Fourier transform

3.5.4.1 Shifting of Spectra: Amplitude Modulation

Let's pose a problem of sending low frequency signal $g(t)$ though high frequency single tone carrier $\cos \Omega t$. Just by multiplying the carrier $\cos \Omega t$ by the modulating signal $g(t)$, the amplitude of the sinusoid would be modified or modulated influenced by the shape of the low frequency signal $g(t)$. This process is known as amplitude modulation (*AM*). According to the frequency shifting property of continuous time Fourier transform (CTFT), the spectrum of $g(t)$ would be shifted to $\omega = \pm\Omega$, and obviously the bandwidth would be doubled as shown in Fig. 3.13.

3.5.5 Transformed Convolution Property

This property states that if $g_1(t)$ & $G_1(\omega)$ and $g_2(t)$ & $G_2(\omega)$ are two sets of Fourier Transform pairs, then convolution between the signals $g_1(t)$ and $g_2(t)$ would form Fourier transform pair with $G_1(\omega) \times G_1(\omega)$. Convolution of two functions $g_1(t)$ and $g_2(t)$, denoted by $g_1(t) * g_2(t)$, is defined by the integral (as discussed in the introductory chapter)

$$g_1(t) * g_2(t) = \int_{-\infty}^{\infty} g_1(\tau) g_2(t - \tau) d\tau$$

The time–frequency convolution duality states that if

$$g_1(t) \Leftrightarrow G_1(\omega) \text{ and } g_2(t) \Leftrightarrow G_2(\omega)$$

then *(time convolution)*

$$g_1(t) * g_2(t) \Leftrightarrow G_1(\omega)G_2(\omega) \tag{3.37}$$

and *(frequency convolution)*

$$G_1(\omega) * G_2(\omega) \Leftrightarrow 2\pi \times g_1(t)g_2(t) \tag{3.38}$$

Proof By definition,

$$\begin{aligned} F[g_1(t) * g_2(t)] &= \int_{-\infty}^{\infty} \left\{ \int_{-\infty}^{\infty} g_1(\tau)g_2(t-\tau)d\tau \right\} e^{-j\omega t}dt \\ &= \int_{-\infty}^{\infty} g_1(\tau)e^{-j\omega\tau}d\tau \int_{-\infty}^{\infty} g_2(t-\tau)e^{-j\omega(t-\tau)}d(t-\tau) \\ &= G_1(\omega)G_2(\omega) \end{aligned}$$

Similarly,

$$\begin{aligned} F^{-1}[G_1(\omega) * G_2(\omega)] &= \frac{1}{2\pi} \int_{-\infty}^{\infty} \left\{ \int_{-\infty}^{\infty} G_1(\Omega) * G_2(\omega-\Omega)d\Omega \right\} e^{j\omega t}d\omega \\ &= \frac{1}{2\pi} \int_{-\infty}^{\infty} G_1(\Omega)e^{j\Omega t}d\Omega \int_{-\infty}^{\infty} G_2(\omega-\Omega)e^{j(\omega-\Omega)t}d(\omega-\Omega) \\ &= g_1(t) \int_{-\infty}^{\infty} G_2(\omega-\Omega)e^{j(\omega-\Omega)t}d(\omega-\Omega) \\ &= 2\pi g_1(t) \frac{1}{2\pi} \int_{-\infty}^{\infty} G_2(\omega-\Omega)e^{j(\omega-\Omega)t}d(\omega-\Omega) \\ &= 2\pi g_1(t)g_2(t) \end{aligned}$$

3.5.5.1 Bandwidth of Product of Two Signals

As understood from the discussion of convolution in the introductory chapter, in continuous convolution (convolution between two continuous signals) results in a new signal whose length or size is exactly the summation of the lengths of the two inputs. If $g_1(t)$ and $g_2(t)$ are two signals with bandwidths B_1 and B_2 Hz respectively, then by multiplying $g_1(t)$ and $g_2(t)$ we are ultimately convolving their spectra. So, the

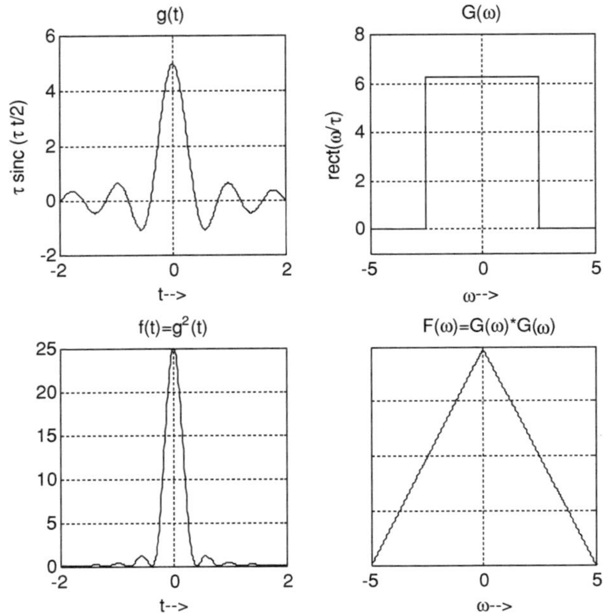

Fig. 3.14 Squaring of a signal gets the bandwidth doubled

width of the convolved spectra, i.e., the bandwidth of $g_1(t) * g_2(t)$ would be $B_I + B_2$ Hz. For the same reason, if the bandwidth of a signal *f*(*t*) is *B* Hz, then the bandwidth of $f^n(t)$ would be *nB* Hz. In Fig. 3.14 two sinc signals (g(t)) are multiplied in time domain. The Fourier transform of sinc, i.e., rect spectrum are convolved to get spectrum of f(t) as a triangular spectrum.

3.5.5.2 AM Spectra: Interpretation by the Convolution Property

As seen in the previous subsection, the process amplitude modulation (AM) generates shifted spectra with respect to the principle spectrum of the modulating signal. By the frequency convolution property also the process can be interpreted. As we have done multiplication in time domain, that must generate its Fourier pair which will be convolution in frequency domain. The spectrum of single tone carrier signal is two impulses at frequencies $\omega = \pm\Omega$. Therefore, the convolution between the trapezoidal spectrum of modulating signal and the two impulses (considering the negative frequency concept) would form the same trapezoidal spectra at $\omega = \pm\Omega$ with half the height of the principle spectrum of the modulating signal as shown in Fig. 3.13.

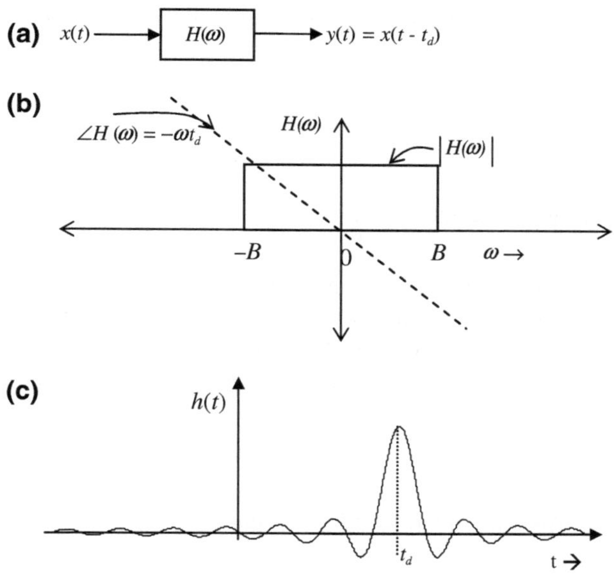

Fig. 3.15 Ideal low pass filter: **a** System representation. **b** Transfer function: amplitude and phase response. **c** Unit impulse response

3.6 System Realization: Ideal and Practical Filter

3.6.1 System Causality

Causality of a system is defined as a property that ensures the relationship between the input (or excitation) and output (or response) to be like *cause-effect*. To elaborate, a system is said to be causal if a cause-effect relationship holds between the input–output pairs in terms of time. If the time of existences/start of input and output of a system be t_1 and t_2 respectively, to be the system causal, t_1 must be less than or equal to t_2, i.e., $t_1 \leq t_2$. It signifies, effect (output) cannot come before cause (input). That's why the causal systems are also called as physically *realizable* system.

As we understand from the process of convolution studied in Chap. 1, the input$[x(t)]$–output$[y(t)]$ relationship holds in terms of the impulse response of the system$[h(t)]$ as

$$y(t) = x(t) * h(t)$$

The rule of convolution suggests that the starting time of the output signal $[y(t)]$ must be the summation of the staring time of input $[x(t)]$ signal and the impulse response $[h(t)]$. Even if the starting time of $y(t)$ and $x(t)$ to be identical, the starting time of $h(t)$ must be zero. Therefore, the condition of causality hence the condition of realizability of a system is

$$h(t) = 0;\ \text{for}\, t < 0 \tag{3.39}$$

3.6.2 Causality of Ideal Filter

Let's consider an ideal filter with transfer function $H(\omega)$ as shown in Fig. 3.15 Delay introduced into the system is the reason of getting negative linear phase response of the filter as expressed in the following equations. The mapping/relation between the input ($x(t)$) and output ($y(t)$) of the system within the frequency range $|\omega| \leq B$ ($\because$ LPF) is

$$y(t) = x(t - t_d) \tag{3.40}$$

Therefore, the magnitude and phase response (transfer function) of the system can be observed from the relation shown in the Eq. 3.32 is

$$|H(\omega)| = \begin{cases} 1; & if \quad \omega \leq B \\ 0; & elsewhere \end{cases} \tag{3.41}$$

$$\angle H(\omega) = \begin{cases} -\omega t_d; & if \quad \omega \leq B \\ 0; & elsewhere \end{cases} \tag{3.42}$$

By considering the negative frequency concept, the evenness of the amplitude response and oddness of the phase spectrum are understood and depicted in Fig. 3.15(b).

By combining Eqs. 3.41 and 3.42, we get

$$\begin{aligned} H(\omega) &= |H(\omega)| e^{j\angle H(\omega)} \\ H(\omega) &= rect\left(\frac{\omega}{2B}\right) e^{-j\omega t_d} \end{aligned} \tag{3.43}$$

The unit impulse response of the filter can be found by inverse Fourier transform using the frequency shifting property.

$$h(t) = F^{-1}[H(\omega)] = F^{-1}\left[rect\left(\frac{\omega}{2B}\right) e^{-j\omega t_d}\right]$$

or,

$$h(t) = \frac{B}{\pi} \sin c\{B(t - t_d)\} \tag{3.44}$$

As seen from Fig. 3.15(c), the impulse response of the ideal filter is having non-zero values even for negative values of time. From our understanding of system causality, i.e., realizability (previous subsection) the criteria posed in Eq. 3.39 does not hold. Therefore *ideal filters are not realizable*.

3.7 Parseval's Theorem for Energy

The signal energy E_g for the signal $g(t)$ is defined as the area under the curve $|g(t)|^2$ with respect to time. The signal energy E_g can also be determined from the frequency domain representation (Fourier transform) $G(\omega)$ using the Parseval's theorem for energy.

$$E_g = \int_{-\infty}^{\infty} |g(t)|^2 dt = \int_{-\infty}^{\infty} g(t)g^*(t)dt = \int_{-\infty}^{\infty} g(t)\left[\frac{1}{2\pi}\int_{-\infty}^{\infty} G^*(\omega)e^{-j\omega t}d\omega\right]dt \tag{3.45}$$

Here the expression g^* represents the complex conjugate of g. Interchanging the order of integration of Eq. 3.45, we get

$$E_g = \frac{1}{2\pi}\int_{-\infty}^{\infty} G^*(\omega)\left[\int_{-\infty}^{\infty} g(t)e^{-j\omega t}dt\right]d\omega = \frac{1}{2\pi}\int_{-\infty}^{\infty} G^*(\omega)G(\omega)d\omega$$

or,

$$E_g = \frac{1}{2\pi}\int_{-\infty}^{\infty} |G(\omega)|^2 d\omega \tag{3.46}$$

Example 3.4 Verify Parseval's theorem for the signal considered in Example 3.1, as shown in Fig. 3.2; $a > 0$.

Firstly, the energy has been determined from the time domain representation (waveform) of the signal $g(t)$

$$E_g = \int_{-\infty}^{\infty} |g(t)|^2 dt = \int_{0}^{\infty} e^{-2at} dt = \frac{1}{2a} \tag{3.47}$$

Now, the Fourier transform of $g(t)$ is obtained from Eq. 3.10

$$G(\omega) = \frac{1}{a + j\omega}$$

$$\therefore |G(\omega)|^2 = \frac{1}{a^2 + \omega^2}$$

From Eq. 3.46,

$$E_g = \frac{1}{2\pi} \int_{-\infty}^{\infty} |G(\omega)|^2 d\omega = \frac{1}{2\pi} \int_{-\infty}^{\infty} \frac{1}{a^2 + \omega^2} d\omega$$
$$= \frac{1}{2\pi a} \tan^{-1} \frac{\omega}{a} \Big|_{-\infty}^{\infty} = \frac{1}{2a} \quad (3.48)$$

Equations 3.47 and 3.48 result in the same energy. Therefore the Parseval's theorem is verified.

References

1. Bracewell, R.N.: The Fourier Transform and its Applications, 2nd edn. McGraw-Hill Book Company, New York (1987)
2. Lathi, B.P.: Modern Digital and Analog Communication Systems, 3rd edn. Oxford Universal Press, Oxford (2005)
3. Hardy, G.H., Rogosinski, W.W.: Fourier Series. Dover Publications Inc, New York (1999)

Chapter 4
Amplitude Modulation

4.1 Introduction

Telecommunication suffers from a plenty of problems both in the transmitter and in the receiver end. To maintain realizable antenna height, to provide the facility of multiplexing and channel separation and to get higher level of noise immunity, a number of modulation schemes are employed both in the case of analog and digital communication. Analog modulation signifies modification of three basic properties of analog single tone[1] carrier signal according to the analog baseband signal. As the baseband signal (information) is solely responsible for the task of modulation, it is also known as the *modulating signal*. A single tone analog carrier signal can be expressed as:

$$s(t) = A\sin(\Omega t + \Phi) \tag{4.1}$$

where, A = Amplitude of the carrier signal
Ω = Frequency of the carrier signal
and, Φ = Initial phase[2] of the carrier signal

The amplitude (A), frequency (Ω) and phase $(\theta(t) = (\Omega t + \Phi))$ are the three fundamental parameters of any signal. Therefore we can modify/modulate any one of these three parameters with respect to the modulating signal. Combining the frequency and initial phase part, the time dependent angle part $\theta(t) = (\Omega t + \Phi)$ is obtained. So, angle modulation is said to be the mother modulation scheme of frequency and phase modulation. The classification is as shown in Fig. 4.1.

It is interesting to note that, while modeling the method of modulation mathematically, we are going to express all the aforesaid modulations as a single function, say $f_1(\circ)$ of the modulating signal. The function would be applicable to amplitude, frequency and phase modulation. The modification on "frequency"

[1] Sinusoids and co sinusoids are only single tone signals.
[2] Initial phase angle: Phase at time $t = 0$.

A. Das, *Signal Conditioning*, Signals and Communication Technology,
DOI: 10.1007/978-3-642-28818-0_4, © Springer-Verlag Berlin Heidelberg 2012

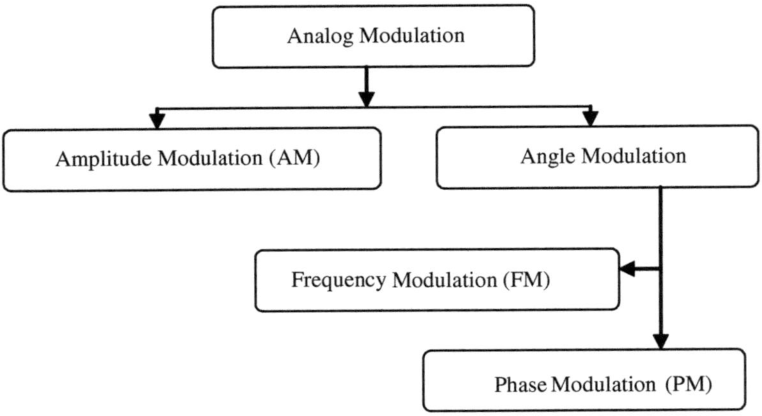

Fig. 4.1 Classification of analog modulation schemes

ultimately would be reflected to change of the entire angle $\theta(t)$. But, this change of $\theta(t)$ would be a different function $f_2(\circ)$ of the modulating signal, whereas in direct "phase" modulation change in $\theta(t)$ would obey the same rule $f_1(\circ)$. Hence, both frequency and phase modulation are inseparable parts of angle modulation affecting the angle by two different functions of the modulating signal.

In the present chapter we would essentially discuss about amplitude modulation (AM) technique. The basic rule of modulation would be defined to model the method of AM mathematically. To maximize the efficiency of the transmission method, different variations of AM like DSB-SC (Double Side Band- Suppressed Carrier), SSB (Single Side Band) and VSB (Vestigial Side Band) are employed. After discussing about design of generic types of modulators and demodulators through choice of proper filters, design of AM radio receiver is also discussed in evolutionary manner.

4.2 Modulation and Its Measure: Global Definitions

4.2.1 Modulation

Modulation can be defined as a technique of modifying a high frequency single tone carrier by any of the three parameters of the modulating signal for the ease of transmission in terms of noise immunity, multiplexing, antenna length. The basic rule of modulation can be stated as: the change in parameter ($\Delta param$) of the carrier signal is directly proportional to the modulating signal ($m(t)$).

$$\Delta param \propto m(t) \tag{4.2}$$

4.2.1.1 Amplitude Modulation

Amplitude Modulation is defined as a technique of analog modulation where the amplitude of the carrier signal is modulated according to the instantaneous amplitude of the modulating signal. In other words change in amplitude (referring the Eq. 4.2), is proportional to the modulating signal $(m(t))$.

$$\begin{aligned} &\Delta\ amplitude \propto \quad m(t) \\ &\Rightarrow \Delta\ amplitude = m(t) \end{aligned} \tag{4.3}$$

by considering the proportionality constant 1.

4.2.2 Modulation Index

In the proceeding sections, we will understand that a single type of modulation can also be done in several variant. If we just concentrate on the method of modulation the specification would be incomplete.

Modulation index is the measure of extent of modulation done on the carrier signal by the modulating signal. This extent of modulation is obviously dependent on relative change of amplitude, frequency and phase on AM, FM and PM respectively.

4.2.2.1 Modulation Index for AM

Let's consider two carrier signals of same frequency and different amplitudes $[V_1, V_2;\ V_2 > V_1]$. If a modulating signal *m(t)* modulates both the carriers, it is easily understandable that the relative change of amplitude would be higher for low magnitude (V_1) career. We can therefore infer that the modulation index for AM (μ) is inversely proportional to the amplitude of carrier signal (V_c).

$$\mu \propto \frac{1}{V_c} \tag{4.4}$$

Now, if we consider two modulating signals of two different amplitudes available for two amplitude modulations on constant amplitude careers, the relative change of career would obviously be higher for high magnitude modulating signal. Therefore, modulation index is directly proportional to the amplitude of modulating signal (V_m).

$$\mu \propto V_m \tag{4.5}$$

By combining Eqs. 4.4 and 4.5 and taking proportionality constant 1, we can define the modulation index for AM as

$$\mu = \frac{V_m}{V_c} \tag{4.6}$$

Hence the depth of modulation is measured as $\mu \times 100\%$.

4.3 Math Model of AM

From Eq. 4.2, AM can be taken as a significant case where the amplitude of the carrier signal is taken as the parameter of modulation. Let's consider a *single tone*[3] modulating signal $m(t) = V_m \sin \omega_m t$ and single tone carrier signal $v_c(t) = V_c \sin \omega_c t$ where $\omega_c \gg \omega_m$. Generally we take single tone carrier for any modulation scheme but the type of modulation is determined by the choice of modulating signal. Here the type of modulation is ***single tone AM*** as the modulation signal considered is single tone and the process of modulation is amplitude modulation. *So*, from Eq. 4.3,

$$\Delta amplitude = m(t)$$

Un-modulated carrier is represented by $v_c(t) = V_c \sin \omega_c t$
Amplitude of the carrier before modulation $= V_c$
Modulated carrier would be

$$\varphi_{AM}(t) = A(t) \sin \omega_c t \tag{4.7}$$

Where, $A(t) = V_c + \Delta amplitude$

$$A(t) = V_c + m(t) = (V_c + V_m \sin \omega_m t) \tag{4.8}$$

From Eqs. 4.7 and 4.8,

$$\begin{aligned}\varphi_{AM}(t) &= (V_c + V_m \sin \omega_m t) \sin \omega_c t \\ &= V_c \sin \omega_c t + V_m \sin \omega_m t \sin \omega_c t \\ &= V_c \sin \omega_c t + \frac{V_m}{2} 2 \sin \omega_m t \sin \omega_c t \\ &= V_c \sin \omega_c t + \frac{V_m}{2} \{\cos(\omega_c - \omega_m)t - \cos(\omega_c + \omega_m)t\}\end{aligned}$$

Now, from Eq. 4.6, $\mu = \frac{V_m}{V_c}$ i.e., $V_m = \mu V_c$

$$\therefore \varphi_{AM}(t) = V_c \sin \omega_c t + \frac{\mu V_c}{2} \cos(\omega_c - \omega_m)t - \frac{\mu V_c}{2} \cos(\omega_c + \omega_m)t \tag{4.9}$$

As seen from Eq. 4.9, the amplitude modulated signal contains three frequency components as f_c, $(f_c + f_m)$ & $(f_c - f_m)$. Therefore, though the adjacent strip distance in the AM signal (Fig. 4.2) is constant, the signal does not contain a single frequency equal to the un-modulated or raw carrier frequency. As the two frequencies $(\omega_c - \omega_m)$ and $(\omega_c + \omega_m)$ are symmetrical around the carrier frequency ω_c, the single tone signals of the 2 said frequencies are named as lower and upper side frequency signals. Therefore, the ***bandwidth*** of the single tone AM is the difference between the maximum $(f_c + f_m)$ and minimum frequency $(f_c - f_m)$ present in the modulated signal, i.e., $\{(f_c + f_m) - (f_c - f_m)\} = 2f_m$.

[3] Only sinusoids (and cosinusoids) are single tone signal.

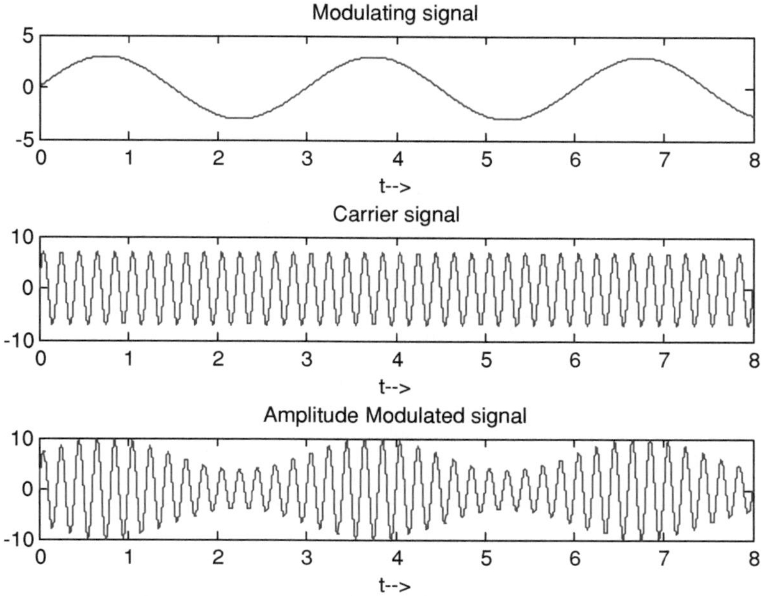

Fig. 4.2 Waveforms before and after amplitude modulation

After introducing general AM (composite tone aperiodic modulating signal), we'll show the frequency domain impulses would be converted to two finite width frequency bands to form lower side band (LSB) and upper side band (USB), respectively. As in the math model, it comprises of two side bands with carrier, the fundamental AM waveform can also be known as *double side band–with carrier* (*DSB-WC*) AM.

In the Fig. 4.2 the shape of changed amplitude of the carrier signal named as the *envelope*, is exactly following the shape of the modulating signal. In the demodulator side, if one can tract the peaks of the changed amplitude of the amplitude modulated carrier, that itself results recovery of the modulating signal. The technique is described with all its constraints in later section.

We can also observe that, in Fig. 4.2, the amplitude of the carrier is taken higher than the amplitude of the modulating signal. From Eq. 4.6, it signifies (**Case 1**) modulation index $\mu = \frac{V_m}{V_c} < 1$. This case is considered as *under-modulated AM*. In Fig. 4.3, other cases of μ, i.e., relation between V_m *and* V_c are considered. For the other two cases, i.e. (**Case 2**) $V_c = V_m$ or $\mu = 1$ and (**Case 3**) $V_c < V_m$ or $\mu > 1$ are called as *critical modulation* and *over-modulation*, respectively. In over-modulation, portions of the upper and lower envelopes change their polarities by crossing the limit of time axis. This polarity change results in 180° phase change of the modulated carrier as shown in the Fig. 4.3. This change of phase while crossing the time axis is known as "twist" of the modulated carrier.

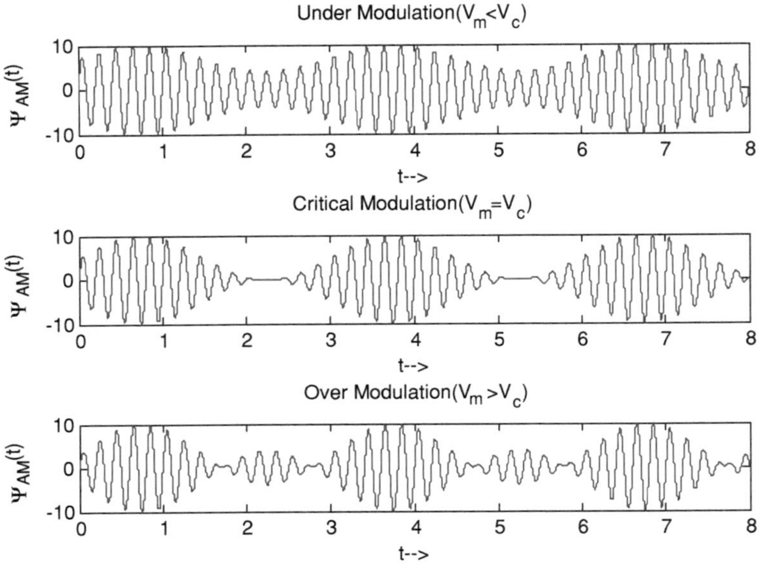

Fig. 4.3 Different cases of amplitude modulation with different modulation index

It is easily understandable that, detection of proper envelope is impossible in the case of over-modulation. That's why, for successful communication through AM, the modulation index is always taken as $\mu \leq 1$.

4.4 Transmission Power and Transmission Efficiency

As seen from the expression of single tone AM from Eq. 4.9, the amplitude modulated signal comprises of three single tone signals with peak amplitudes V_c, $\frac{\mu V_c}{2}$ and $\frac{\mu V_c}{2}$ respectively. As all of them are sinusoids, the r.m.s values need to be considered in all the three cases.
Total transmitted power for DSB-WC single-tone AM = carrier power + upper side frequency power + lower side frequency power.

$$P_{DSB-Wc} = P_c + P_{USB} + P_{LSB} \tag{4.10}$$

Now, considering the transmitted antenna resistance R,

$$P_c = \left(\frac{V_c}{\sqrt{2}}\right)^2 \frac{1}{R} = \frac{V_c^2}{2R} \tag{4.11}$$

$$P_{LSB} = P_{USB} = \left(\frac{\mu V_c}{2\sqrt{2}}\right)^2 \frac{1}{R} = \frac{\mu^2 V_c^2}{8R} \tag{4.12}$$

From Eqs. 4.10, 4.11 and 4.12,

$$P_{DSB-Wc} = \frac{V_c^2}{2R} + \frac{\mu^2 V_c^2}{8R} \times 2 = \frac{V_c^2}{2R}\left(1 + \frac{\mu^2}{2}\right) \tag{4.13}$$

From Eqs. 4.11 and 4.13,

$$P_{DSB-Wc} = P_c\left(1 + \frac{\mu^2}{2}\right) \tag{4.14}$$

As we know, the raw carrier component, present in the DSB-WC signal, does not contain the intended information to be transmitted. Therefore, the power taken to transmit the raw carrier can be considered as a "loss". To calculate the efficiency of DSB-WC AM transmission, we take the general definition of efficiency as follows.

$$Efficiency, \eta = \frac{Usable\ output\ power}{Total\ delivered\ power} \times 100\%$$

$$\Rightarrow \eta = \frac{P_{Sidebands}}{P_{DSB-WC}} \times 100\%$$

$$\Rightarrow \eta = \frac{P_c \frac{\mu^2}{2}}{P_c\left(1 + \frac{\mu^2}{2}\right)} \times 100\% = \frac{\mu^2}{(2 + \mu^2)} \times 100\% \tag{4.15}$$

For 100% modulation (critical modulation), i.e. $\mu = 1$,

$$\Rightarrow \eta = \frac{1}{3} \times 100\% = 33.33\% \tag{4.16}$$

This very low efficiency indicates high redundancy in transmission. For this redundancy, the raw carrier is responsible. Therefore we should try to remove the raw carrier before transmission retaining the modulated carrier only. This new type of AM is named as *double side band—suppressed carrier* (*DSB-SC*) amplitude modulation, described in the next section.

4.5 Double Side Band Suppressed Carrier (DSB-SC) Modulation

As understood from the previous discussion of transmission power and transmission efficiency, it is clear that the raw carrier component of the amplitude modulated signal is a burden to the signal itself. The efficiency of transmission can be improved significantly by removing the raw carrier component of the signal. The concept is really interesting. We are not removing the carrier itself; then there

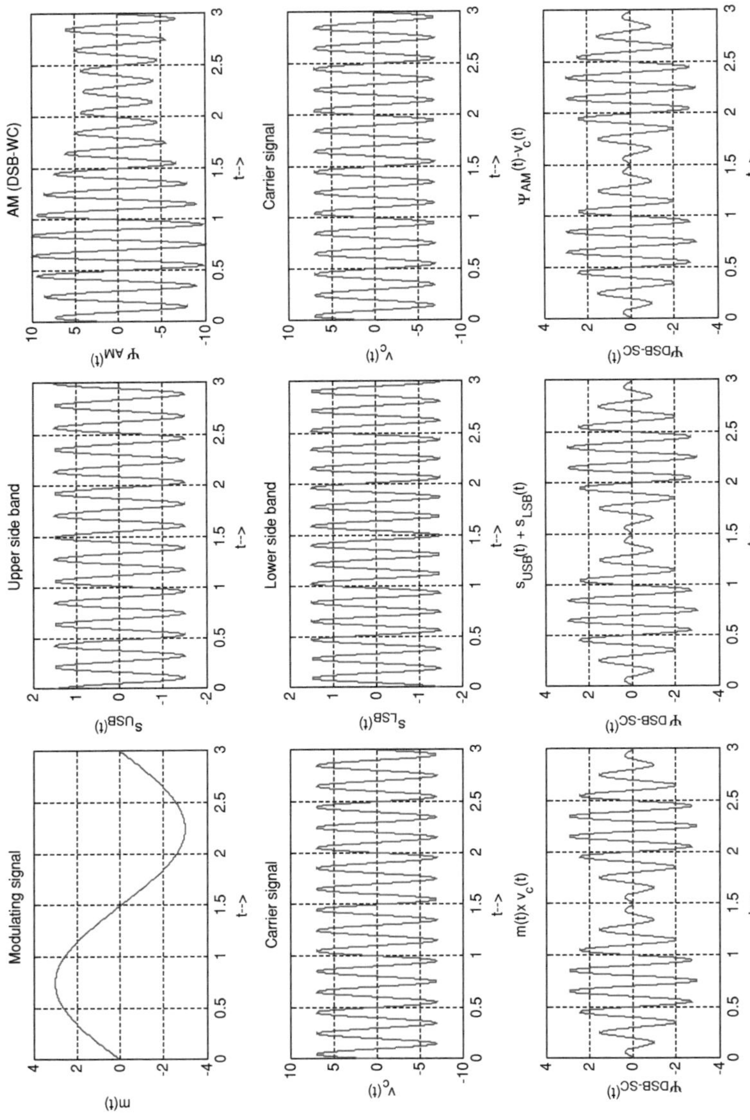

Fig. 4.4 Different ways of realizing DSB-SC

would be no modulation. By the methodology of modulation, the carrier signal gets modified or transformed or conditioned in terms of summation of three single tone signals. All those three single tone signals are carriers. Two of them form sidebands (modulated carrier) and the third one gets unchanged (raw or un-modulated carrier). Just by suppressing the raw carrier we get the purely

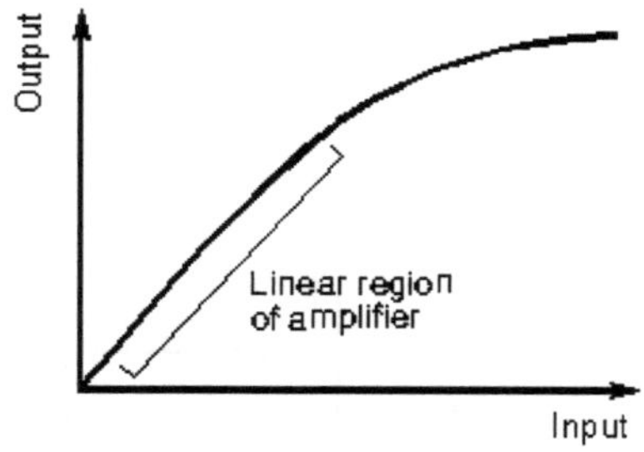

Fig. 4.5 Transfer characteristics of a non-linear amplifier

modulated carrier components. This new signal after suppressing the raw carrier component, in named as double side band suppressed carrier (DSB-SC) signal.

To interpret physically, three different methods of interpreting the same DSB-SC waveform is shown in Fig. 4.4. In the third column (c) of the Fig. 4.4, the raw carrier signal has been subtracted from the general single tone amplitude modulated signal to get DSB-SC AM signal. In the second column (b), the same DSB-SC waveform has been generated by just adding one lower frequency ($f_c - f_m$) and one higher frequency ($f_c + f_m$) carrier. Finally in the first column (a), just by multiplying the carrier with the modulating signal, identical DSB-SC waveform could be generated.

4.6 Balanced Modulator

4.6.1 Non-linear Amplifier

To understand the operation of balanced modulator it is necessary to understand at least the operating principle of non-linear amplifier. Non-linear amplifier (NLA) is defined as an amplifier in which a change in input does not produce a proportional change in output. It means the response of the non-linear amplifiers does not vary with respect to the excitation proportionally. In the lower order values of the input, the characteristics may seem to be a bit linear. But, as the dynamic range of the input increases, the slope of the transfer characteristic is decreased. We can have this type of characteristics by using simple diode or bipolar junction transistor as the active element of the transistor (Fig. 4.5). In case of linear amplifier, there is a direct proportionality between input and output. Therefore, it is having a simple gain (G) relationship between the ports of the amplifier as follows $(\text{Input port}:\ x(t),\ \text{Output port}:\ y(t))$.

$$\begin{aligned} & y(t) \propto x(t) \\ & \therefore\ y(t) = Gx(t) \end{aligned} \tag{4.17}$$

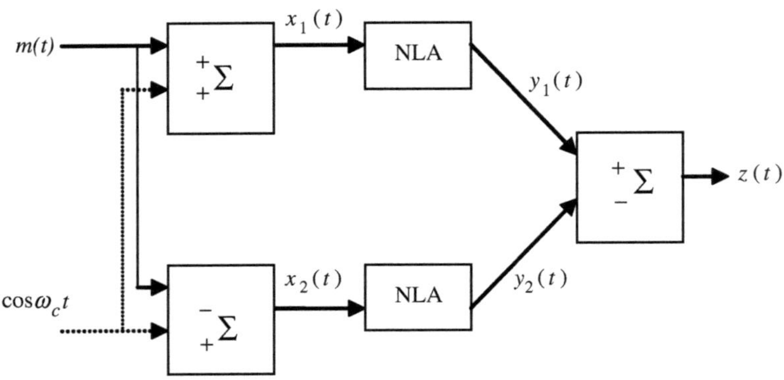

Fig. 4.6 Balanced modulator without output filtering

Whereas, in non-linear amplifiers, there are two gain components; one is for non-linear (a) and the other is for linear (b) amplification factor. Therefore, in NLA, the relationship between input and output ports is

$$y(t) = ax^2(t) + bx(t) \tag{4.18}$$

Considering the non-linear amplifier (NLA) as a component, here two different configurations of AM modulators are presented.

4.6.2 Configuration 1

Modulator is a complete device which generates modulated carrier from un-modulated carrier depending on the chosen primary parameter[4] of the information/base-band signal/the modulating signal. In the configuration 1 of balanced modulator as shown in Fig. 4.6, two intermediate signals $x_1(t)$ and $x_2(t)$ are generated simply by adding and subtracting the carrier and the modulating signal. After that, those two signals are being passed parallel through two non-linear amplifiers with unique amplification parameters. Finally the outputs of the two NLAs $(y_1(t)\&y_2(t))$ are again subtracted from each other to get $z(t)$. Now our primary objective should be to analyze $z(t)$ in time domain as well as in frequency domain if needed.

Expressions of the intermediate signals are as follows

$$x_1(t) = \cos\omega_c t + m(t) \tag{4.19}$$

$$x_2(t) = \cos\omega_c t - m(t) \tag{4.20}$$

[4] As discussed before, therefore three basic parameters of a signal, especially an electrical signal, namely amplitude, frequency and phase.

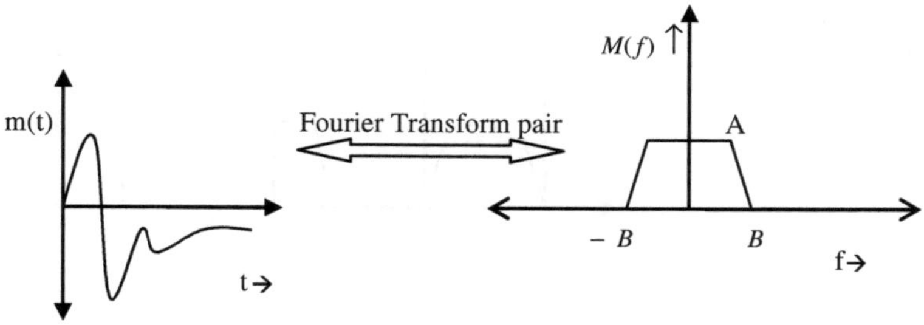

Fig. 4.7 Waveform (*left*) and spectrum (*right*) of modulating signal

$$y_1(t) = ax_1^2(t) + bx_2(t) \tag{4.21}$$

$$y_2(t) = ax_1^2(t) + bx_2(t) \tag{4.22}$$

Therefore the output of the configuration as shown in Fig. 4.6 is

$$\begin{aligned}
z(t) &= y_1(t) - y_2(t)\\
z(t) &= a\{x_1^2(t) - x_2^2(t)\} + b\{x_1(t) - x_2(t)\}\\
z(t) &= a\left\{(\cos\omega_c t + m(t))^2 - (\cos\omega_c t - m(t))^2\right\}\\
&\quad + b\{(\cos\omega_c t + m(t)) - (\cos\omega_c t - m(t))\}\\
z(t) &= 4am(t)\cos\omega_c t + 2bm(t)
\end{aligned} \tag{4.23}$$

We can see from the above equation that, the expression of *z*(*t*) does not lead us directly to any known result. The signal is not exactly expressible as any type of amplitude modulation. Therefore, to apply some signal conditioning technique is necessary. By applying Fourier transform over the signal *z*(*t*), we get

$$Z(f) = 2a[M(f + f_c) + M(f - f_c)] + 2bM(f) \tag{4.24}$$

To draw the amplitude spectrum of *Z*(*f*), we need to know the shape of the spectrum of the modulating signal, *M*(*f*). For the ease of understanding, we have considered that the information signal is band-limited to *B* Hz and the shape of the spectrum is trapezoidal, as shown in the Fig. 4.7. It is also shown that the amplitude of the spectrum is *A*.

Now, let's have a close look at the expression *Z*(*f*). The spectrum is having two major parts. The amplified (if $2b > 1$) or attenuated (if $2b < 1$) *M*(*f*) and amplified (if $2a > 1$) or attenuated (if $2a < 1$) spectrum of *M*(*f*) which is shifted in both the sides of *M*(*f*) by an amount f_c. In the Fig. 4.8, the spectrum of *Z*(*f*) is shown. The doubly shifted *M*(*f*) are of same height (2aA) and the amplified *M*(*f*) is of height 2bA. It easily observed that, the spectrum does not match any of the types of amplitude modulation. Now, our objective should be to make it a defined type of amplitude modulation by applying simple methods of signal conditioning. The spectrum *Z*(*f*) itself directs towards DSB-SC. As we know, in the modulated

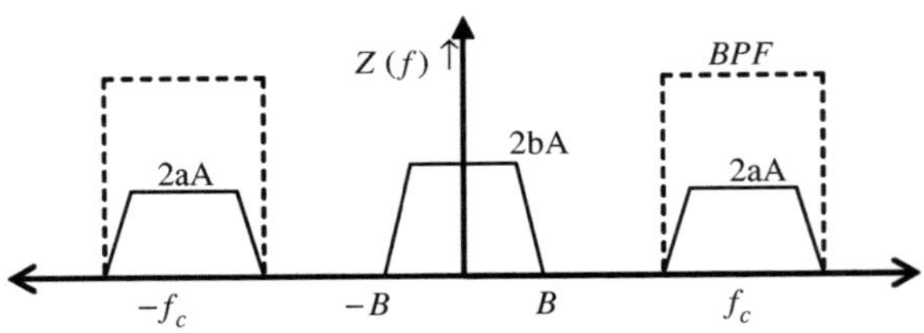

Fig. 4.8 Spectrum of balanced modulator output

spectrum base-band signal spectra cannot exist. If we remove the central spectrum *M(f)* from the entire spectra of *Z(f)* the resultant spectra will consist of the upper and lower sidebands with negative frequency validity. Therefore, to take the shifted spectra only, we must use one band pass filter of center frequency f_c and bandwidth $2B$ Hz (lower cut-off frequencies $(f_c + B)$ and $(f_c - B)$ Hz, respectively). We can conclude that, the complete configuration 1, as shown in Fig. 4.9 actually generates ***DSB-SC*** modulated signal. This is a defined AM signal.

4.6.3 Configuration 2

By a simple interchange between the inputs of the configuration 1 as shown in Fig. 4.6, we can get the next new configuration. This is now a new modulator with same circuit arrangement and same inputs. But the output would be a bit different, as the input terminals are interchanges as compared to the configuration 1. The new configuration is shown in the Fig. 4.10.

From the mathematical modeling of the configuration 1, we can directly get the output *z(t)* for configuration 2, as

$$z(t) = a\left\{(m(t) + \cos\omega_c t)^2 - (m(t) - \cos\omega_c t)^2\right\} \\ + b\{(m(t) + \cos\omega_c t) - (m(t) - \cos\omega_c t)\} \tag{4.25}$$

$$z(t) = 4am(t)\cos\omega_c t + 2b\cos\omega_c t \tag{4.26}$$

By applying Fourier transform over the signal *z(t)*, we get

$$Z(f) = 2a[M(f + f_c) + M(f - f_c)] + b[\delta(f + f_c) + \delta(f - f_c)] \tag{4.27}$$

Here, the spectrum comprises of shifted *M(f)* as well as shifted unit impulse functions. The amount of shift in both the directions is also same. Therefore, we can draw the spectrum as shown in Fig. 4.11.

The spectrum of the output of the modulator configuration 2 clearly shows that, the spectrum comprises of USB, LSB and raw carrier. Therefore, the output can easily be designated as ***DSB-WC*** modulated AM signal.

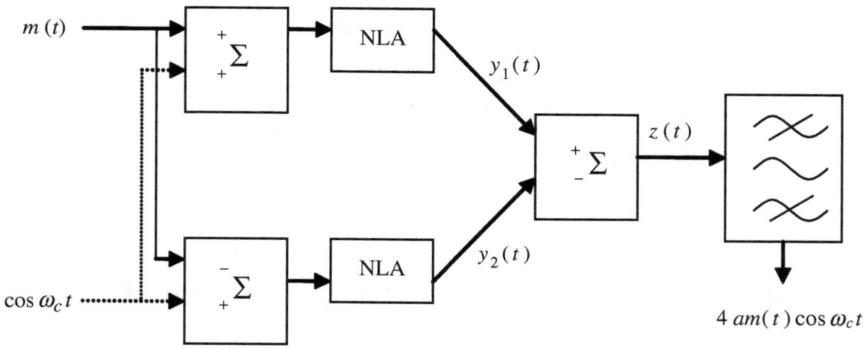

Fig. 4.9 Balanced modulator with output filtering

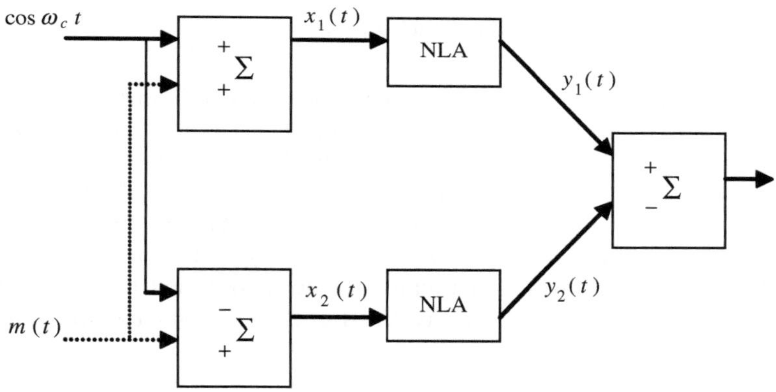

Fig. 4.10 Balance modulator configuration 2

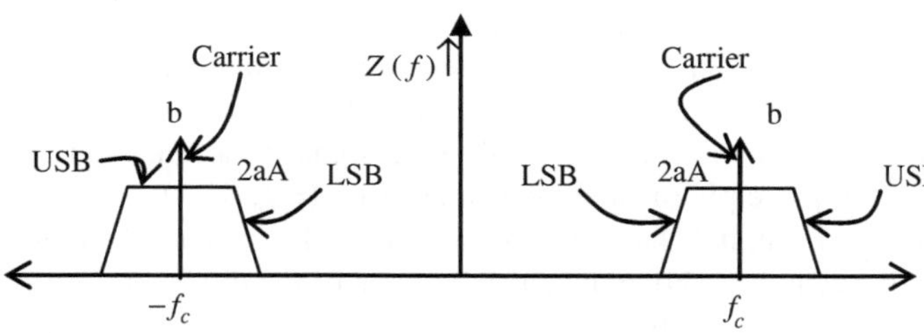

Fig. 4.11 Spectrum of balanced modulator output of configuration 2

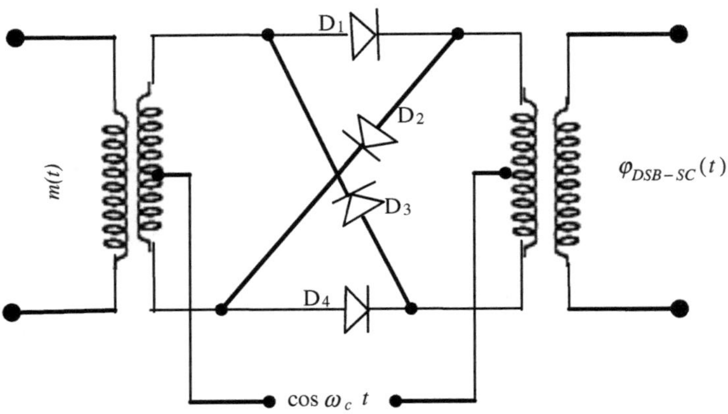

Fig. 4.12 Ring modulators

4.6.4 Why "Balanced" Modulator?

Looking at the responses of the two configurations shown in this section, we can conclude that the same modulator only with interchanged inputs can generate both DSB-SC and DSB-WC waveform. If we closely observe the expressions of inputs and the outputs, we can infer one interesting conclusion. In both the cases, only one input is present intact along with some new signal (the multiplied signal, which is neither input 1 not input 2). So, we can say, at the output, one input is present intact (may be with some amplitude deviation), and the other input is balanced. Therefore, this modulator is called as ***Single Balanced Modulator***.

4.7 Ring Modulator

The configuration shown in Fig. 4.12, represents a special type of modulator. The specialty is on the choice of carrier signal. The carrier used for modulation is a derived one from the given single tone carrier $\cos \omega_c t$. The configuration itself drives the physical phenomena of waveform generation for a specific type of signal conditioning where the physical carrier ultimately is being conditioned to a logical carrier [1]. The transformation from a single tone signal to a multi-tone periodic signal has made the principle of operation interesting (Fig. 4.12).

As we know, *cosine* signals have two polarities. At the instant when the carrier is in positive half-cycle, just above the cut-in voltage of the diodes in forward path i.e., D_1 and D_4 conduct. When the positive half-cycle is executed, other two diodes D_2 and D_3 do not conduct due to reverse voltage applied to their terminals. On the other hand, when negative half cycle of *cosine* signal is fed to the transformer, the two diodes D_2 and D_3 conduct, and D_1 and D_4 logically behave as open circuit for the aforesaid reason. Therefore, we get two different configurations at two different half-cycles, as shown in Fig. 4.13. Here, each of the half cycles of a cosine carrier

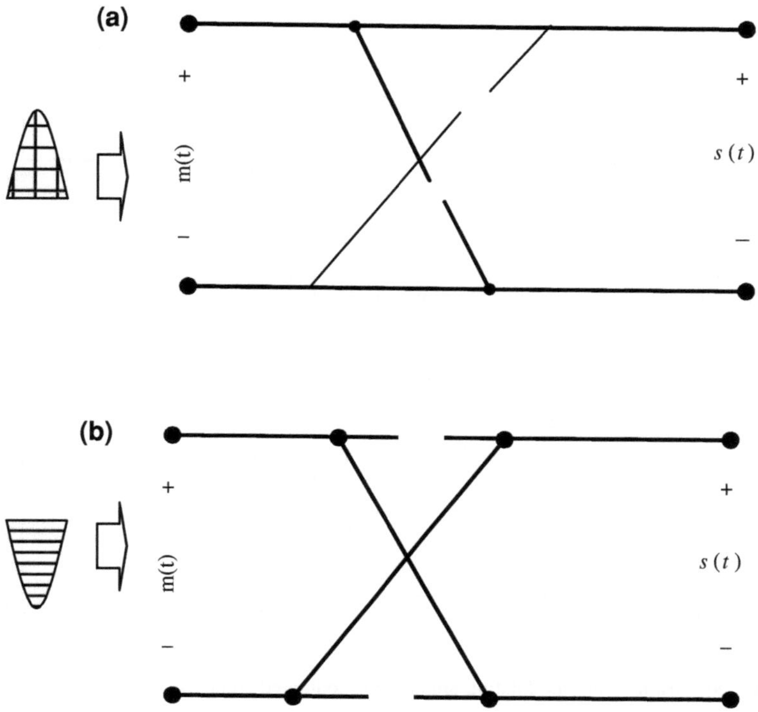

Fig. 4.13 Input–output relationship at **a** positive half-cycle, and **b** negative half-cycle of the *cosine* carrier

is considered to form two different configurations as shown in the figure above. The diodes open and short periodically with $\cos \omega_c t$, results in multiplying the input signal by $p(t)$, logically. At the duration of positive half cycle attained by the carrier, the output, $s(t)$ becomes exactly similar to the modulating signal input $m(t)$ (Fig. 4.13a). In the negative half cycle of the carrier (Fig. 4.13b), the output positive polarity is shorted to the negative polarity of the input modulating signal.[5] Therefore, we can understand that, the relationship between the input and the output is as follows:

$$\begin{aligned} s(t) &= \{m(t), -m(t), m(t), -m(t), \cdots\} \\ &= m(t) \times \{+1, -1, +1, -1, \cdots\} \end{aligned} \tag{4.28}$$

Obviously, the change of polarity would occur concurrently with the cosine carrier, as the carrier itself is responsible for the change in polarity. So, the output is, modulating signal alternatively multiplying +1 and −1 by carrier frequency. This

[5] The polarity here signifies the measured terminal to the reference terminal of the input and output.

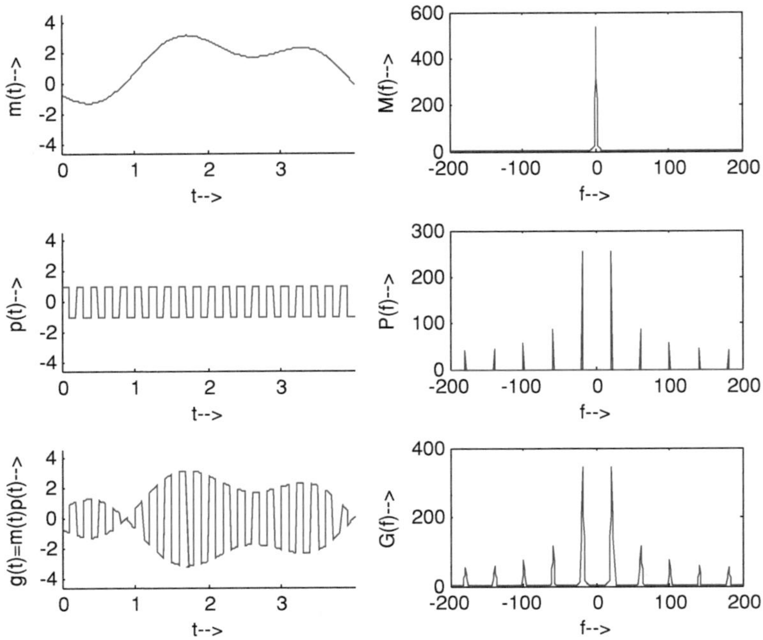

Fig. 4.14 Spectrum of the modulating signal and modulated signal by ring modulator

polarity change from +1 to -1 and vice versa forms a logical ring. Thus the modulator is called as ring modulator.

Therefore, we can look at the process of modulation as multiplication of the modulating signal with a square wave. Here one general continuous aperiodic signal, $m(t)$ is considered as modulating signal. The spectrum (Fig. 4.14) is therefore obtained by Fourier Transform. The periodic pulse train $p(t)$ (logically generated) is then multiplied to get the modulated signal $g(t)$. Spectrum of $p(t)$ is obtained by Continuous Time Fourier Series (CTFS). As, the multiplication in time is dual to convolution in frequency domain, multiplication between the signals $p(t)$ and $m(t)$ corresponds to convolution between the spectra $M(f)$ and $P(f)$. Therefore, the resultant signal $G(f)$ would be repeated spectrum $M(f)$. The spectrum centered at f_c can now be filtered in by band-pass filter centered at f_c with bandwidth 2B Hz, when the bandwidth of the modulating signal is B Hz.

As discussed, the expression of the modulated signal would be

$$g(t) = m(t) \times p(t) \tag{4.29}$$

Where,
The Trigonometric Fourier Series Expression of the pulse train (Refer Chap. 2),

$$p(t) = \frac{4}{\pi}\left(\sin\omega_c t + \frac{1}{3}\sin 3\omega_c t + \frac{1}{5}\sin 5\omega_c t + \cdots\right) \tag{4.30}$$

From the above two equations,

$$g(t) = \frac{4}{\pi}\left(m(t)\sin\omega_c t + \frac{1}{3}m(t)\sin 3\omega_c t + \frac{1}{5}m(t)\sin 5\omega_c t + \cdots\right) \quad (4.31)$$

$$G(f) = \frac{2}{\pi}\begin{bmatrix} \{M(f+f_c) + M(f-f_c)\} + \frac{1}{3}\{M(f+3f_c) + M(f-3f_c)\} \\ +\frac{1}{5}\{M(f+5f_c) + M(f-5f_c)\} + \cdots \end{bmatrix} \quad (4.32)$$

After passing trough BPF centered at f_c, we get

$$g_{BPF}(t) = \frac{4}{\pi}m(t)\sin\omega_c t$$

i.e.,

$$G(f) = \frac{2}{\pi}[\{M(f+f_c) + M(f-f_c)\}] \quad (4.33)$$

One interesting thing to notice is that, the output signal consists of a completely different signal than the modulating and cosine carrier signal inputs and even the logical pulse carrier input. From the classical discussion of balanced modulator, we can infer that here both of the inputs are balanced. Thus, the ring modulator is also designated as *Double Balanced Modulator.*

4.8 Phasor Diagram

In the definition of "signal", one important property is reflected which says the signal must be a physically measurable quantity. As discussed in Chap. 3, phasor is actually the vector form of representation of signal. The single tone AM signal has three different frequency components f_c, $(f_c + f_m)$, and $(f_c - f_m)$ as seen from Eq. 4.9.

As seen from the mathematical expression of AM, derived in Eq. 4.9,

$$\varphi_{AM}(t) = (1 + V_m \sin\omega_m t)\cos\omega_c t = (1 + \mu\sin\omega_m t)\cos\omega_c t \quad (4.34)$$

We have taken $\cos\omega_c t$ as carrier signal for ease of representation with unit magnitude. The observations from the phasor representation would be unchanged if the carrier is considered as $V_c \sin\omega_c t$. The above equation then can be expanded as

$$\varphi_{AM}(t) = \cos(\omega_c t) + \frac{\mu}{2} \times \{\sin(\omega_c + \omega_m)t - \sin(\omega_c - \omega_m)t\} \quad (4.35)$$

In vector form,

$$\vec{\varphi}_{AM} = \vec{v}_c + \vec{R}_{SB} \quad (4.36)$$

As we know, for successful transmission, the modulation index of AM is generally taken as $0 < \mu \leq 1$, preferably $0 < \mu < 1$. Therefore, the sideband phasors would

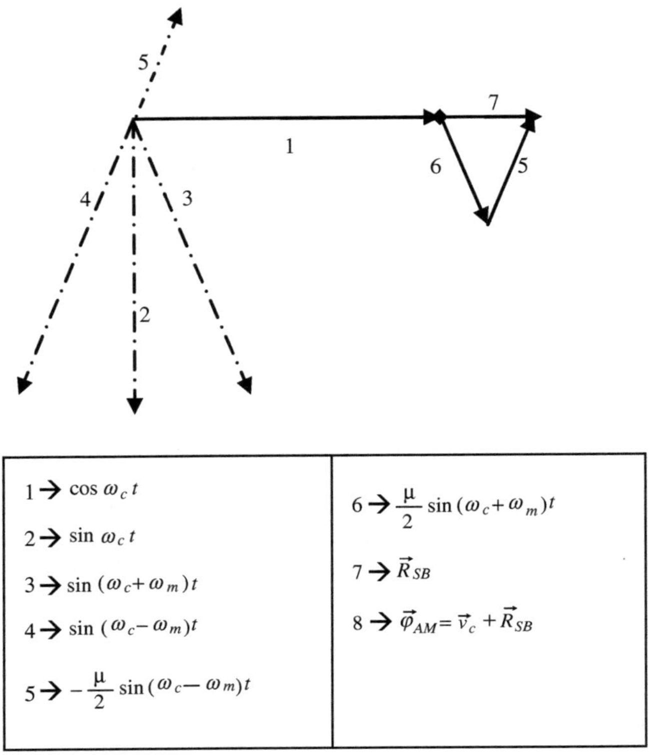

Fig. 4.15 Phasor representation of amplitude modulated signal

have magnitude less than half the magnitude of the carrier signal. As seen from the Eq. 4.35, the sidebands are rotated quadrature phasor with respect to the raw carrier phasor. The phasor (or vector) representation of AM is shown in Fig. 4.15.

The carrier phasor is first rotated clockwise to get the *sine* carrier (Quadrature carrier) component. Next the *sine* carrier forms the two sideband phasors rotating anticlockwise and clockwise by same angle. Next, the phasor/vector $-\frac{\mu}{2}\sin(\omega_c-\omega_m)t$ is formed by reducing the magnitude and inverting the direction of $\sin(\omega_c-\omega_m)t$. Similarly the upper sideband phasor is formed by reducing the magnitude of the $\sin(\omega_c+\omega_m)t$ phasor and placed at the terminating position of carrier phasor. The lower sideband phasor (shown as '5' in the Fig. 4.15) is placed at the end of the upper sideband phasor. Vector triangle law is used here for realizing the complete AM phasor. Firstly, $\frac{\mu}{2}\sin(\omega_c+\omega_m)t$ and $-\frac{\mu}{2}\sin(\omega_c-\omega_m)t$ are added to form the consolidated sideband phasor $\vec{R}_{SB}$. Eventually $\vec{R}_{SB}$ is found to be aligned with the carrier phasor. Finally, the carrier phasor $(\vec{v}_c)$ is added to the sideband phasor $(\vec{R}_{SB})$ to form complete AM phasor, $\vec{\varphi}_{AM}$.

4.8.1 Observed Properties of AM

From the phasor representation of AM as shown in Fig. 4.15, we can note a few interesting properties of AM. The observed properties are listed below:

Observation 1: The modulated carrier phasor $\vec{\varphi}_{AM}$ and the unmodulated carrier phasor $\vec{v}_c$ are perfectly aligned. This is the evidence of '*amplitude modulation*'.
Observation 2: The sideband phasor and carrier phasor are in phase.
Observation 3: As because the sideband phasor and the carrier phasor are in phase, just by adding the magnitude of the unmodulated carrier and the sideband we can get the magnitude of the modulated carrier. It signifies significant amplitude modulation and no angle modulation.

4.9 Envelope Detector

The basic assumption of any type of modulation is the frequency of the carrier signal is much higher the frequency of the modulating signal. By considering the case of safe and successful AM transmission, i.e., amplitude modulation with modulation index less than 100% the waveform of under-modulated AM waveform is obtained. It can be seen from Figs. 4.2 and 4.3 that the ***envelope*** (borderline of the extent of modification of carrier amplitude) of the AM wave follows the pattern of the modulating signal. Thus the desired demodulation can be accomplished by extracting the envelope of the resultant AM wave [4].

The principle and configuration of demodulator (or detector) is known as envelope detector. A typical version of the configuration is shown in the Fig. 4.16a. The detector consists of a diode and a resistor–capacitor filter. The operation of the detector can be understood as follows. On the positive half cycle of the input signal the diode is forward biased and the capacitor C charges up **rapidly** to the peak of the input modulated signal, as shown in Fig. 4.16b. When the input signal falls below the value the diode becomes reverse biased and the capacitor C discharges through the load resistance R_L. The discharge process continues until the next positive half cycle. The process repeats at each cycle of the modulated carrier signal. We assume the diode to be near ideal in the sense that it presents zero impedance to the current in the forward biased region. We have also considered that the AM wave has been applied to the envelope detector from a voltage source having a series resistance R_S. The charging time constant $\tau_{ch\arg e} = R_S C$ must be very small compared to the carrier time period (fundamental time period measured by adjacent strip distance), that is

$$\tau_{ch\arg e} = R_S C \ll \frac{1}{f_c} \qquad (4.37)$$

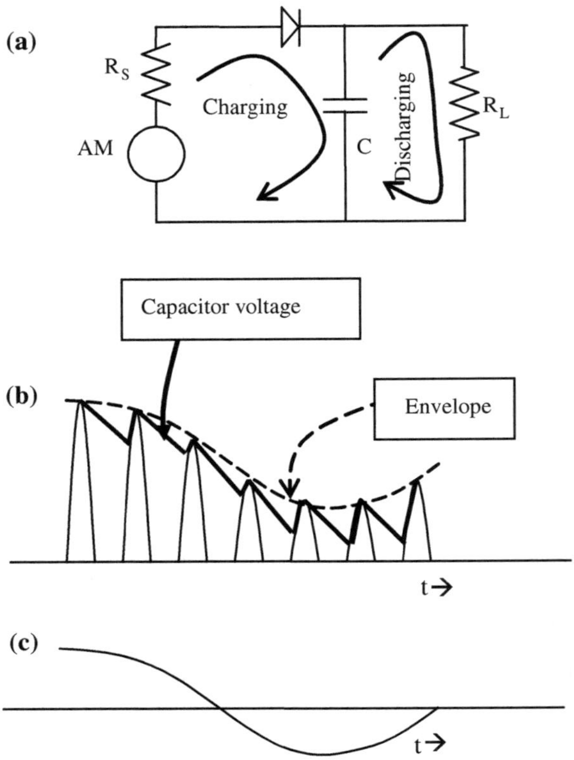

Fig. 4.16 Envelope detector (**a**) circuit configuration, (**b**) AM wave (half wave rectified), and (**c**) envelope detector output

For this configuration, the capacitor charges rapidly and hereby follows the applied voltage when the diode is conducting. On the other hand, the discharging time constant $\tau_{discharge} = R_L C$ must be long enough to ensure that the capacitor discharges slowly through the resistor R_L between positive peaks of the modulated carrier signal, but not so long that the capacitor voltage will not discharge at the maximum rate of change of modulating signal (of bandwidth B Hz), that is

$$\frac{1}{f_c} \ll R_L C \ll \frac{1}{B} \tag{4.38}$$

As a result, the capacitor voltage which is the detector output, is very nearly same as the envelope of the AM wave as shown in Fig. 4.16c. In the detected output some ripples may be present which can be finally removed by an additional low pass filter (LPF).

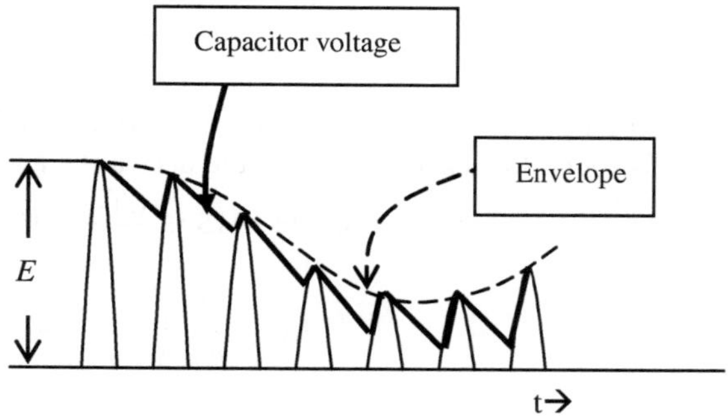

Fig. 4.17 Capacitor discharge in an envelope detector

Example 4.1 Determine the upper limit of RC to ensure that the capacitor voltage follows the envelope in single tone AM demodulation.

According to the Fig. 4.17, the capacitor discharges from the peak value E starting at reference time $t = 0$. The voltage v_c across the capacitor is given by

$$v_c = Ee^{-\frac{t}{RC}} \tag{4.39}$$

As the time constant $(\tau = RC)$ of the discharging circuit is chosen much larger than the interval between the two successive cycles of the carrier $(RC \gg 1/f_c)$, the capacitor voltage v_c discharges exponentially for a short time compared to its time constant. By considering the 1st two terms of Taylor's series we can approximate the exponent $Ee^{-\frac{t}{RC}}$ by a straight line as

$$v_c \approx E\left(1 - \frac{t}{RC}\right) \tag{4.40}$$

The slope of the discharge is $-\frac{E}{RC}$. In order for the capacitor to follow the envelope *E*(*t*), the magnitude of the slope of the RC discharge must be greater than the magnitude of the slope of the envelope *E*(*t*). Hence,

$$\left|\frac{dv_c}{dt}\right| = \frac{E}{RC} \geq \left|\frac{dE}{dt}\right| \tag{4.41}$$

But the envelope *E*(*t*) of a tone-modulated carrier is

$$E(t) = V_c\{1 + \mu \cos \omega_m t\}$$
$$\Rightarrow \frac{dE(t)}{dt} = -V_c \mu \omega_m \sin \omega_m t$$

Therefore, from Eq. 4.41,

$$\frac{V_c\{1 + \mu \cos \omega_m t\}}{RC} \geq -V_c \mu \omega_m \sin \omega_m t; \forall t$$

$$\Rightarrow RC \leq \frac{1 + \mu \cos \omega_m t}{\mu \omega_m \sin \omega_m t}; \forall t \tag{4.42}$$

The worst possible case occurs when the right hand side (RHS) is the minimum. This is found (as usual, by taking the derivative and setting it to zero) to be when $\cos \omega_m t = -\mu$. For this case the RHS is $\frac{\sqrt{1-\mu^2}}{\mu\omega_m}$. Hence,

$$RC \leq \frac{1}{\omega_m}\left(\frac{\sqrt{1-\mu^2}}{\mu\omega_m}\right) \tag{4.43}$$

4.10 Quadrature Amplitude Modulation

Both in the DSB-SC and DSB-WC modulation schemes of amplitude modulation, the requirement of channel bandwidth is twice the bandwidth (B Hz) of the modulating/baseband signal. In the present section we shall discuss about a unique method of efficient bandwidth utilization.

Say two modulating or baseband signals $m_1(t)$ and $m_2(t)$ each of bandwidth B Hz are to be communicated through one channel. If general method of DSB-SC is used, the total bandwidth required is 2B + 2B = 4B. Here the concept of orthogonal phasors needs to be considered, to achieve higher efficiency. The two modulations can be done by two carriers of same frequency and quadrature phase. The scheme is shown in the Fig. 4.18.

4.10.1 QAM Transmitter

The two baseband signals $m_1(t)$ and $m_2(t)$ are DSB-SC modulated by carrier signals of same frequency and quadrature in phase, as shown in Fig. 4.18a. As understood from the previous discussions, DSB-SC shifts the spectrum by an amount equal to the carrier frequency. Therefore, after modulation both the spectrum would be positioned to same frequency f_c, here. But, as sine and cosine are orthogonal single tone signals, the plane of the shifted spectrum would be quadrature in phase as seen in Fig. 4.18b. Just after adding these two DSB-SCs, we can transmit the two modulated signals $m_1(t)\cos\omega_c t$ and $m_2(t)\sin\omega_c t$ without any interference. The "no interference" state is ensured by the orthogonal properties of the carriers used for the respective modulations. This technique is known as *Quadrature Amplitude Modulation* (*QAM*) and given by

$$\varphi_{QAM}(t) = m_1(t)\cos\omega_c t + m_2(t)\sin\omega_c t \tag{4.44}$$

By this way, the same bandwidth 2B Hz is utilized by both the baseband signals. Therefore the utilization of bandwidth is maximized.

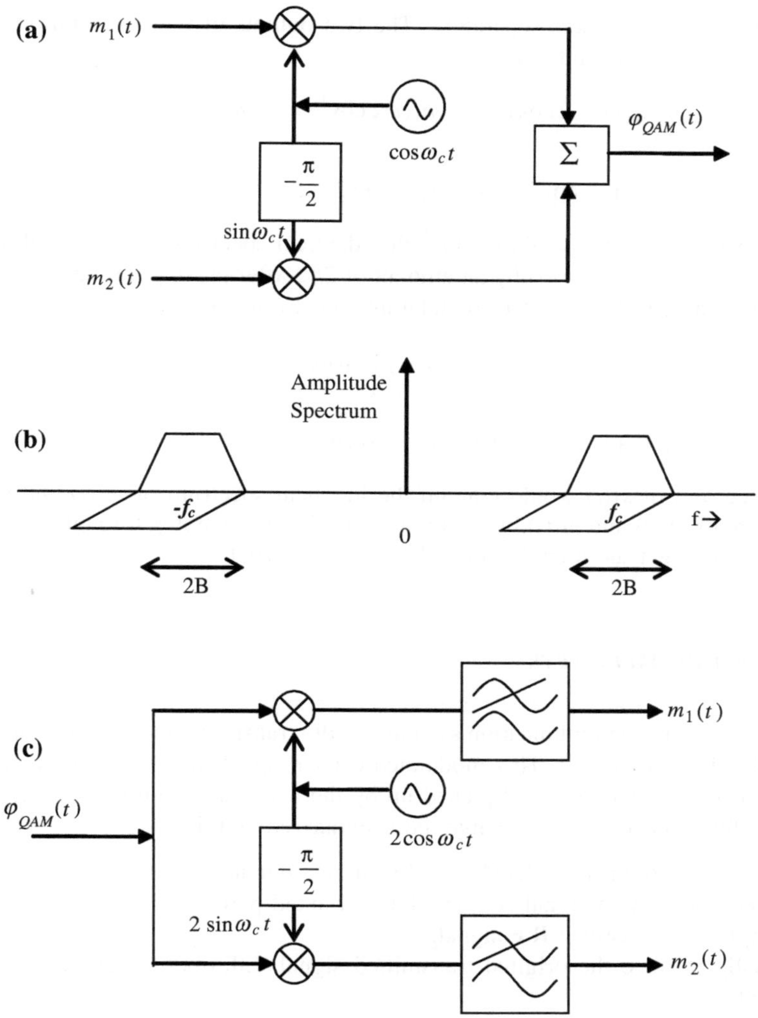

Fig. 4.18 Quadrature amplitude modulation (*QAM*) (**a**) transmitter, (**b**) amplitude spectrum, and (**c**) receiver

4.10.2 QAM Receiver

The composite DSB-SC signal is now to be separated and demodulated. The end response of the QAM receiver must be the two baseband signals $m_1(t)$ and $m_2(t)$. The process of coherent detection is used in this case. The locally generated carrier signal is used to regenerate the quadrature phase carrier signal here again at the receiver side. The entire QAM signal gets multiplied with the in phase and quadrature phase carriers separately. The two outputs of the two multipliers are given below as $s_1(t)$ and $s_2(t)$.

$$s_1(t) = \varphi_{QAM}(t) \times 2\cos\omega_c t = m_1(t)2\cos^2\omega_c t + m_2(t)2\sin\omega_c t\cos\omega_c t$$

or

$$s_1(t) = m_1(t)(1 + \cos 2\omega_c t) + m_2(t)\sin 2\omega_c t \tag{4.45}$$

The spectrum of $s_1(t)$ would contain the identical spectrum of $m_1(t)$ and shifted spectrum of $m_1(t)$ and $m_2(t)$ by an amount of $2f_c$. After passing through a low pass filter, we can get the intended modulating signal $m_1(t)$.Similarly,

$$s_2(t) = \varphi_{QAM}(t) \times 2\sin\omega_c t = m_1(t)2\sin\omega_c t\cos\omega_c t + m_2(t)2\sin^2\omega_c t$$

or,

$$s_1(t) = m_1(t)\sin 2\omega_c t + m_2(t)(1 - \cos 2\omega_c t) \tag{4.46}$$

The spectrum of $s_1(t)$ would contain the identical spectrum of $m_2(t)$ and shifted spectrum of $m_2(t)$ and $m_1(t)$ by an amount of $2f_c$. After passing through a low pass filter, we can get the intended modulating signal $m_2(t)$.

4.11 Radio Receivers

As we know, in a communication system, a radio transmitter radiates or transmits a modulated carrier signal. This modulated carrier signal travels down the channel *i.e.* transmission medium and picked up by the antenna of radio receiver.

The functions of a receiver may be summarized as follows:

1. Collect incoming modulated signal using an antenna.
2. Select the desired signal & reject the unwanted parts.
3. Amplify the received R.F signal.
4. Finally retrieve the original transmitted signal (information) from the modulated one.

In the present section, the fundamental AM receiver, known as tuned radio frequency receiver is described. Analyzing some design drawback one advanced version of receiver is then explained as super-heterodyne receiver.

4.11.1 Tuned Radio Frequency (TRF) Receiver

Tuned radio frequency receiver [3] is the simplest form of a 'logical' receiver. Figure 4.19 shows the block diagram of a tuned radio frequency receiver.

There are plenty of modulated and un-modulated signals coming to the receiver terminal. In the first stage of RF amplification, the receiver selectively amplifies

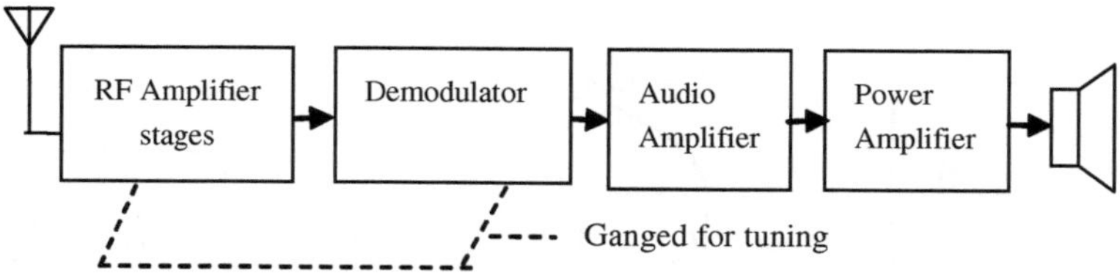

Fig. 4.19 Tuned radio frequency receiver

the incoming frequency and rejects the unwanted signal from it. As the range of commercial AM broadcasting is 536–1,640 kHz, the RF amplifier needs to change its resonance or tuning frequency within the said range. This amplified incoming modulated signal is then applied to the demodulator to get modulating or baseband signal (i.e. Audio Signal) back. In AM demodulation specifically, the knowledge of carrier signal frequency used at the time of modulation is important. Therefore, for the sake of un-erroneous detection of the modulating signal, the same tuner (used in RF amplifier stages) is ganged to the demodulator also. Due to natural decay, the signal strength is expected to be reduced below the range of sensitivity. Therefore another amplifier working in audio frequency range (audio amplifier) is employed to get amplified modulating signal. Finally it is passed through a power amplifier before it is fed finally to a loudspeaker. A loudspeaker is a transducer which converts the electrical signal into sound signal.

There are some basic advantages and drawbacks of using this kind of receiver.

4.11.1.1 Advantages of TRF Receiver

1. Operates at lower broadcast frequencies (536–1,640 kHz).
2. Simple to design.

4.11.1.2 Limitations/Drawbacks of TRF Receiver

1. *Tendency to oscillate at higher frequencies*

We know from the Barkhausen Criterion that oscillation occurs due to positive feedback and the system become unstable. In TRF receiver, due to high gain (say 40,000) is achieved by the small amplifier stage at a single frequency, a small portion $\left(1/40,000\right)$ is leaked back to the input of RF stage which causes positive feedback and oscillation making the system unstable.

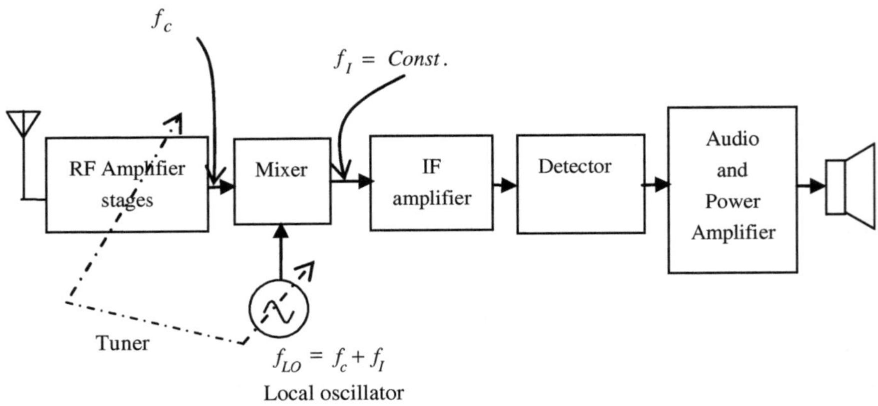

Fig. 4.20 Super-heterodyne receiver

2. *Poor Selectivity*

At high frequencies the ability of the TRF receiver to distinguish the desirable signal from the undesirable signal is affected adversely.

3. *Bandwidth Variation over the tuning range*

Consider a tuned circuit having bandwidth of 10 kHz at a frequency of 535 kHz. Thus the quality factor (Q) would be $f/\Delta f = 535/10$ i.e. 53.5. Hence at the other end of the broadcast band say have a frequency of 1,640 kHz. Thus Q factor of the coil increased by a factor 1,640/535 to 164. Practically various frequency dependent looses prevent such high increase and is unlikely to exceed 120, giving a bandwidth of 1,640/120 = 13.8 kHz. Hence the receiver will pick up adjacent channel frequencies along with the desired one.

4.11.2 Super Heterodyne Receiver

All the drawbacks of TRF receiver are taken care of in super-heterodyne receiver. It is the most widely used receiver in television, radar receiver etc. Figure 4.20 shows the block diagram of super-heterodyne receiver. The word *heterodyne* stands for mixing. In this Receiver the incoming RF signal frequency is combined with a local oscillator frequency through a mixer circuit is converted to a fixed frequency is known a **Intermediate Frequency** (Fig. 4.20).

In a superheterodyne Receiver, a constant frequency difference is maintained between the local oscillator signal frequency and incoming RF signals frequency. The intermediate frequency amplifier provides most of the sensitivity and selectivity of the receiver. After the I.F. amplifier the signal is applied to the decoder or demodulator unit which extracts the original modulating signal. The purpose of

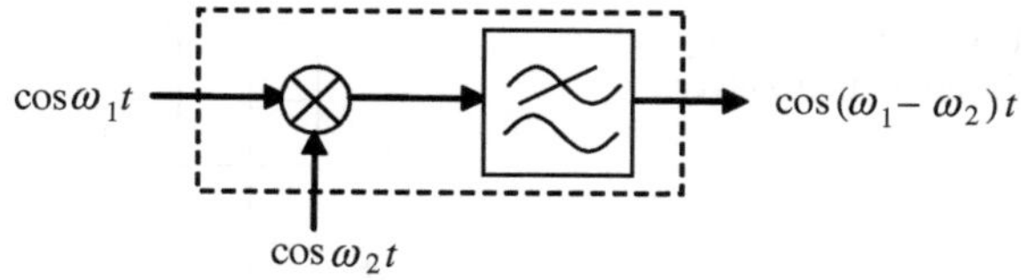

Fig. 4.21 Internal configuration of mixer

Automatic Gain Control (AGC) is to vary the overall gain of a receiver with changing strength of received signal for keeping output signal constant. The modulating signal is then amplified by the audio amplifier and power amplifier to get a particular voltage level and power level so that it may activate the speaker. The loudspeaker is a transducer which converts the electrical signal into sound signal.

4.11.2.1 Mixer

The firstly introduced commercial AM radio receiver, TRF had a major drawback of insufficient adjacent channel rejection. If we just try to find the fundamental reason of the drawback, we'll ultimately reach to inability of Q-adjustment of the heart of the entire radio receiver configuration, named as the *demodulator*. We are forcing the demodulator to work in such a large frequency range (536–1,640 kHz) and therefore it results in variation of bandwidth with respect to central or tuned carrier frequency.

By any means if we manage to exclude the demodulator from the process of tuning, then the problem of insufficient adjacent channel rejection can be taken care of. The only way to achieve this state is by maintaining the incoming carrier frequency to the demodulator constant. Let's call this constant frequency as intermediate frequency f_{IF}. In the improved receiver configuration (super-heterodyne), we shall have to then convert the antenna received carrier frequency (536–1,640 kHz) to a constant frequency. This needs a frequency conversion sage. The block responsible for this frequency conversion is known as mixer. The internal configuration of mixer stage is shown in Fig. 4.21. As shown, there is a multiplier stage followed by a low pass filter. By considering the two inputs of the mixer as $\cos\omega_1 t$ and $\cos\omega_2 t$, the output of the multiplier is

$$x(t) = \cos\omega_1 t\cos\omega_2 t = \frac{1}{2}\times 2\cos\omega_1 t\cos\omega_2 t$$

or,

$$x(t) = \frac{1}{2}\times\{\cos(\omega_1+\omega_2)t + \cos(\omega_1-\omega_2)t\} \tag{4.47}$$

It is obvious that after passing through the low pass filter we'll get the difference frequency signal $\frac{1}{2}\cos(\omega_1-\omega_2)t$ at the output.

To ensure the output signal frequency of the mixer constant even for varying input carrier signal, we have to tune the RF amplifier stages along with the frequency of the other input of the mixer. The other input signal of the mixer is generated locally. Therefore the signal generator is known as local oscillator as seen in the super-heterodyne receiver configuration as given in Fig. 4.20.

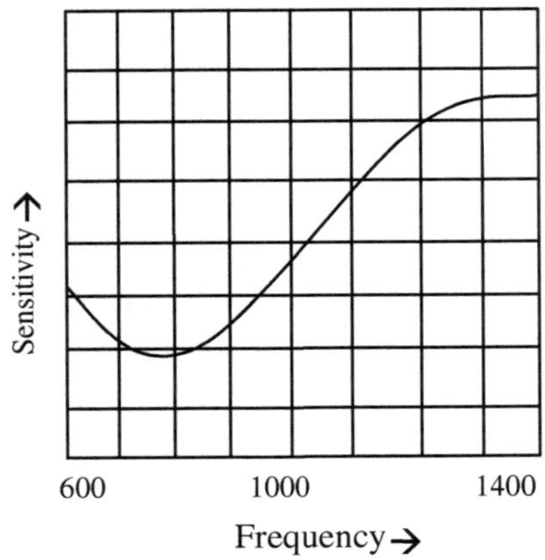

Fig. 4.22 Sensitivity curve for domestic receiver

4.11.2.2 Advantages of Super-Heterodyne Receiver

1. Bandwidth is constant over the entire operating range.
2. It has a high sensitivity and selectivity.
3. It exhibits comfortable adjacent channel rejection.

4.11.3 Receiver Characteristics

As superheterodyne receiver is the most commonly used type of AM receiver hence all characteristic discussed here would be with respect to superheterodyne receiver.

4.11.3.1 Sensitivity

Sensitivity of a radio receiver is the ability to amplify weak signals. It is often measured in volts but sometimes in microvolts or in decibels below 1 V.

The most important factor which determines sensitivity syperheterodyne receive is the gain of the IF amplifier(s) and RF amplifier. The noise figure also plays a vital role in determining the sensitivity. Figure 4.22 shows the sensitivity plot for a good domestic receiver.

4.11.3.2 Selectivity

Selectivity of a radio receiver is the ability to reject unwanted signals. It varies with the receiving frequency if ordinary tuned circuits are used in IF section and becomes even worse for higher receiving frequency.

In general selectivity of a superheterodyne receiver Selectivity of a radio receiver is the ability to reject unwanted signals. It varies with the receiving frequency if ordinary tuned circuits are used in IF section. The RF amplifier also

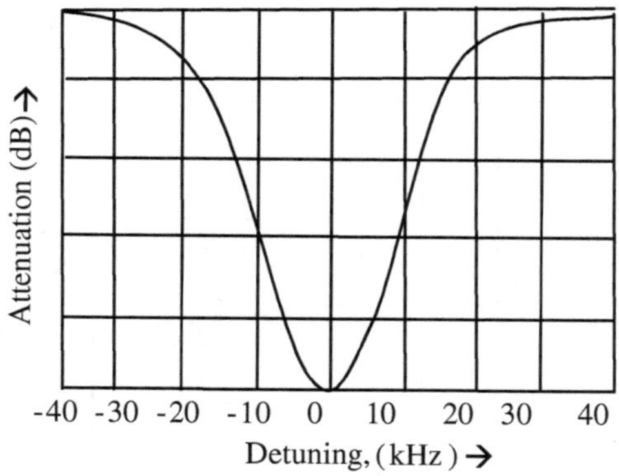

Fig. 4.23 A typical sensitivity curve for a good quality domestic receiver

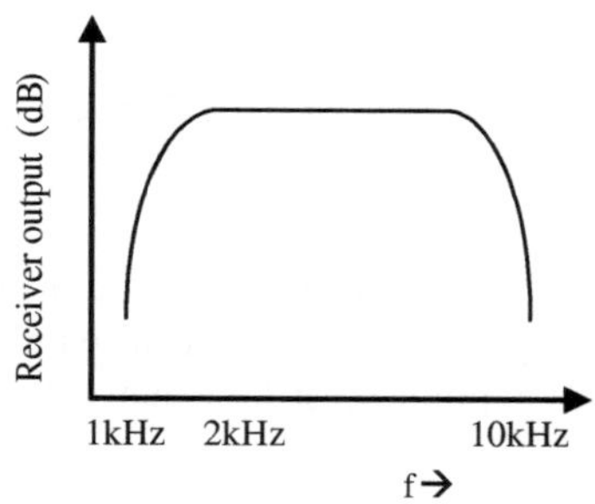

Fig. 4.24 Fidelity characteristics of a receiver

plays significant role in selectivity determination. The Fig. 4.23 shows the attenuation offered by the receiver is higher to the frequencies adjacent the tuned frequency and almost zero for the tuned frequency. It is the main determining factor for adjacent channel rejection of a receiver.

4.11.3.3 Fidelity

Fidelity [2] of a receiver is the ability to reproduce all the modulating frequencies equally. It mainly depends on the frequency response of the AF amplifier. Fidelity curve for a receiver is shown in Fig. 4.24. An ideal receiver should have flat frequency response over the entire range of audio frequencies, but practically it becomes decreasing on the higher and lower frequency side.

4.11.3.4 Double Spotting or Adjacent Channel Selectivity

It is the phenomena of a radio receiver picking up the same shortwave station at two nearby points on the receiver dial. The main reason for occurence of the double spotting phenomena is the poor front-end selectivity (i.e. inadequate image frequency rejection). This phenomena may be have adverse effect because due to

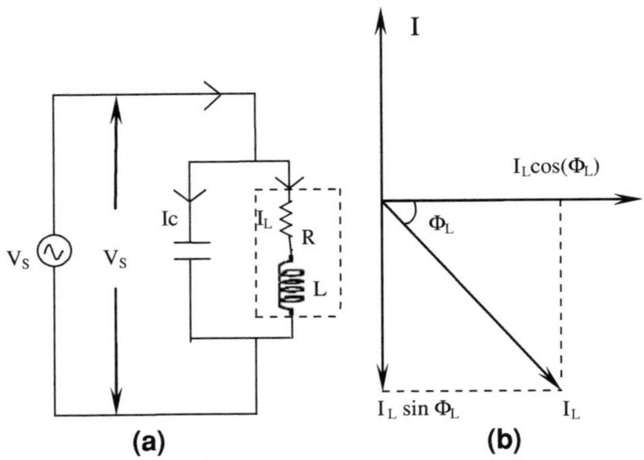

Fig. 4.25 **a** Tuned circuit. **b** Corresponding phasor diagram

reception of a strong station, a nearby weak signal may be masked. But Double Spotting can also be used to calculate the IF of an unknown receiver, since the spurious point must lie exactly $2f_i$ below the correct frequency.

4.11.4 Tuned Circuit

A tuned circuit [2] is commonly known as parallel resonance circuit as shown in the Fig. 4.25a. We will consider the practical case of a coil in parallel with a capacitor. We know when the frequency of the applied source become equal to the natural frequency of the LC circuit, electrical resonance occurs. In other words a circuit is said to be in electrical resonance when the reactive component of the line current becomes zero. The frequency at which this happens is known as resonant frequency (Fig. 4.25).

$$\text{Net reactive or wattles component } = I_C - I_L \sin \Phi_L \tag{4.48}$$

$$\text{As at resonance its value is zero so, } I_C = I_L \sin \Phi_L \tag{4.49}$$

$$\text{Now, } I_L = V/Z_L \tag{4.50}$$

$$\sin \Phi_L = X_L \tag{4.51}$$

$$\text{and } I_C = {}^{V}\!/_{X_C} \tag{4.52}$$

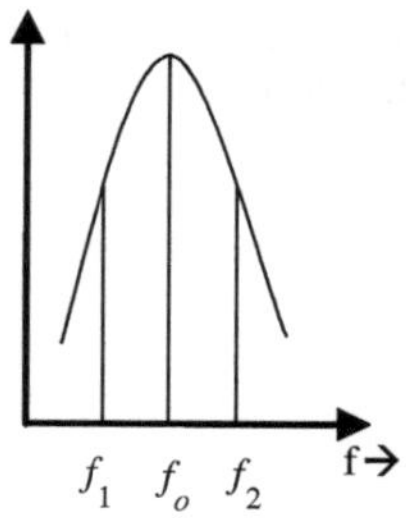

Fig. 4.26 Bandwidth measurement from a Q-curve

Hence the condition of resonance becomes,

$$V/Z_L \times X_L/Z_L = V/X_C$$

or

$$X_L \times X_C = Z_L \quad (4.53)$$

Now,

$$X_L = \omega L \text{ and } X_C = 1/\omega C$$

or,

$$\frac{L}{C} = R^2 + X_L^2 = R^2 + (2\pi f_O L)^2,$$

where f_O is the resonating frequency

or,

$$(2\pi f_O L)^2 = \frac{L}{C} - R^2$$

or,

$$2\pi f_O = \sqrt{\frac{1}{LC} - \frac{R^2}{L^2}}$$

or

$$f_O = \frac{1}{2\pi}\sqrt{\frac{1}{LC} - \frac{R^2}{L^2}} \quad (4.54)$$

In practice the resistance of the inductor is very small.

So, the tuning frequency $f_O = \frac{1}{2\pi}\sqrt{\frac{1}{LC}}$ (4.55)

4.11.4.1 Characteristic of Tuned Circuit

1. *Quality Factor*

It is defined as the ratio of the current circulating between its two branches to the line drawn from the supply or simply as the current magnification. Hence the Q-factor $= I_C/I$. Now,

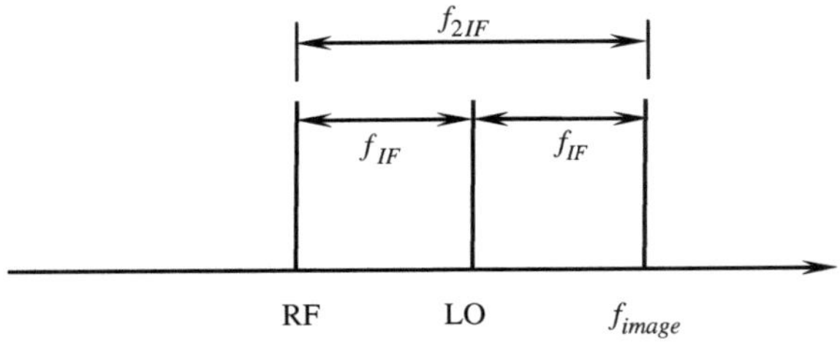

Fig. 4.27 Concept of image frequency

$$I_C = V/X_C = V\Big/ 1/\omega C = \omega CV$$

So the Q-factor is

$$\omega L/R = 2\pi f_o/R \tag{4.56}$$

It shows that current falls rapidly if we move off the resonance (for low value of R or high value of Q). This means that such a circuit will have better frequency discrimination. Thus it can be said that Q is the measure of ability to select a band of frequency (Fig. 4.26).

2. *Bandwidth*

Bandwidth represents the frequency range at which the circuit offers low impedance path. So, Bandwidth = BW $= f_2 - f_1$. In terms of Q-factor Bandwidth may be written as,

$$\text{BW} = f_o/Q \tag{4.57}$$

4.11.5 Image Frequency

The principle of operation of the super-heterodyne receiver depends on the use of heterodyning or frequency mixing. As understood from the previous discussions, the mixer is used in a super-heterodyne receiver to create the constant intermediate frequency (IF). But a mixer creates both sum and difference frequencies. By corollary, there are two RF frequencies that will produce the exact same IF. The unwanted signal frequency is called the **Image frequency.** The unwanted frequency is called the *image* of the wanted frequency, because it is the "mirror image" of the desired frequency reflected f_{LO} (Fig. 4.27).

We can see from the Fig. 4.27 that

$$f_{image} = f_{LO} + f_{IF} \tag{4.58}$$

Substituting,

$$f_{LO} = f_{IF} + f_{RF}$$

$$f_{image} = f_{RF} + 2f_{IF} \tag{4.59}$$

Therefore, an image frequency is an undesired input frequency equal to the station frequency plus twice the intermediate frequency. The image frequency results in two stations being received at the same time, thus producing interference.

4.11.5.1 Solution to Image Frequency Problem

The main idea of solving the image frequency problem is by employing the higher IF and putting the image farther in spectrum. But stability of the IF amplifier gets poorer. If the pre-selector (the primary band-pass filter for all the noise removal except the band-pass random noise) does not allow the image frequency then there is no image problem. To ensure this, higher Q factor needs to be employed.

The ability of a receiver to reject interfering signals at the image frequency is measured by the image rejection ratio. The image rejection ratio, or ***image frequency rejection ratio***, is the ratio of the output of the receiver from a signal at the received frequency, to its output for an equal-strength signal at the image frequency. This parameter is generally measured in dB.
Mathematically,

$$IFRR = \sqrt{1 + Q^2 \rho^2} \tag{4.60}$$

Where,

$$\rho = \left(f_{image}/f_{RF} \right) - \left(f_{RF}/f_{image} \right) \tag{4.61}$$

$$IFRR(dB) = 10 \log(IFRR) \tag{4.62}$$

4.12 MATLAB Codes

4.12.1 AM

```
% AM
% Apurba Das (22/08/11)
clc;
clear all;
close all;

t=0:0.001:5;
T=3;

Tc=0.2;
fm=1/T;
fc=1/Tc;
Vm=3;        %change and watch
Vc=7;        % change and watch
vm=Vm.*sin(2*pi*fm*t);
subplot(311)
plot(t,vm)
title('Modulating signal');
xlabel('t-->')
vc=Vc.*sin(2*pi*fc*t);
subplot(312)
plot(t,vc)
title('Carrier signal');
xlabel('t-->')
am=(Vc+vm).*sin(2*pi*fc*t);
subplot(313)
plot(t,am)
title('Amplitude Modulated signal');
xlabel('t-->')
```

4.12.2 DSB-SC

```
% AM
% Apurba Das (22/08/11)
clc;
clear all;
close all;

t=0:0.00001:3;
T=3;
Tc=0.2;
fm=1/T;
fc=1/Tc;
Vm=3;
Vc=7;
vm=Vm.*sin(2*pi*fm*t);
figure(1)
subplot(331)
plot(t,vm)
ylabel('m(t)');
xlabel('t-->');
title('Modulating signal');
grid;
vc=Vc.*sin(2*pi*fc*t);
subplot(334)
plot(t,vc)
ylabel('v_c(t)');
xlabel('t-->')
title('Carrier signal');
grid;
dsb_sc=vm.*sin(2*pi*fc*t);
subplot(337)
plot(t,dsb_sc)
title('m(t)x v_c(t)');
xlabel('t-->')
ylabel('\psi_{DSB-SC}(t)');
grid;
usb=0.5*Vm*(cos(2*pi*(fc-fm)*t));
subplot(332)
plot(t,usb);
ylabel('s_{USB}(t)');
xlabel('t-->')
title('Upper side band');
grid;
```

```
% dsbsc=0.5*Vm*(cos(2*pi*(fc-fm)*t)-
cos(2*pi*(fc+fm)*t));
lsb=-0.5*Vm*(cos(2*pi*(fc+fm)*t));
subplot(335)
plot(t,lsb);
ylabel('s_{LSB}(t)');
xlabel('t-->')
title('Lower side band');
grid;
dsbsc=usb+lsb;
subplot(338)
plot(t,dsbsc)
title('s_{USB}(t) + s_{LSB}(t)');
xlabel('t-->')
ylabel('\psi_{DSB-SC}(t)');
grid;
am=(Vc+vm).*sin(2*pi*fc*t);
subplot(333)
plot(t,am);
title('AM (DSB-WC)');
xlabel('t-->')
ylabel('\psi_{AM}(t)');
grid;
subplot(336)
plot(t,vc)
title('Carrier signal');
xlabel('t-->')
ylabel('v_c(t)');
grid;

dd=am-vc;
subplot(339)
plot(t,dd)
title('\psi_{AM}(t)-v_c(t)');
xlabel('t-->')
ylabel('\psi_{DSB-SC}(t)');
grid;
```

4.12.3 Ring Modulator

```
% Ring modulator
% Apurba Das (Signal Conditioning) 20/04/2011

clc;
clear all;
t=0:0.01:(4-0.01);
T=10;
f=1/T;
a=2*sin(2*pi*f*t)-1.5*cos(4*pi*f*t)...
    +cos(10*pi*f*t-30)-cos(2*pi*f*t-40);
subplot(321)
axis([0 (4-0.01) -4.5 4.5]); hold on;
plot(t,a);
xlabel('t-->');
ylabel('m(t)-->');
pulses=[ones(1,10) -ones(1,10)];
pul=repmat(pulses,1,20);
subplot(323)
axis([0 (4-0.01) -4.5 4.5]); hold on;
plot(t,pul);
xlabel('t-->');
ylabel('p(t)-->');

subplot(325)
a1=pul.*a;
axis([0 (4-0.01) -4.5 4.5]); hold on;
plot(t,a1);
xlabel('t-->');
ylabel('g(t)=m(t)p(t)-->');

%%%%%%%%%%%%%%%%%% FFT %%%%%%%%%%%%%%%%
s1=abs(fftshift((fft(a))));
subplot(322)
pt=20;
f=-199:200;
```

```
% f=-99:100;
% axis([-4 (4-0.01) -3.5 3.5]); hold on;
% s1=s1(-99:100);
plot(f,s1)
xlabel('f-->');
ylabel('M(f)-->');

s2=fftshift(abs(fft(pul)));
subplot(324)
f=-199:200;
plot(f,s2)
xlabel('f-->');
ylabel('P(f)-->');
s3=fftshift(abs(fft(a1)));
subplot(326)
f=-199:200;
plot(f,s3)
xlabel('f-->');
ylabel('G(f)-->');

figure,
f=-99:100;
plot(f,s3(100:299))
xlabel('f-->');
ylabel('G(f)-->');
```

References

1. Lathi, B.P.: Modern Digital and Analog Communication Systems, 3rd edn. Oxford University Press, NY (2005)
2. Thereja, B.L., Thereja, A.K.: Textbook of Electrical Technology. S. Chand and Co. Ltd., India (2000)
3. Kennedy, G.: Electronic Communication Systems, 4th edn. Tata McGraw Hill, India (1993)
4. Chakroborti, P.: Analog Communication Systems. Dhanpat Rai & Co, India (2005)

Chapter 5
Angle Modulation Technology

5.1 Introduction

From the understanding of the primary properties of an electrical signal, we understood that, we can either modulate amplitude or the phase of the signal. In phase again we have two tunable parts namely frequency and initial phase. The algorithm of modulation in this case may be of two types. We may modulate the frequency of carrier with respect to the modulating signal, or we may directly modulate the entire phase of the carrier depending upon the modulating signal. In both the cases, the resultant signal will form an appreciable amount of angle with the phasor of raw carrier component. Whichever may be the case, ultimately the angle will be modulated. Therefore, frequency modulation (FM) and phase modulation (PM) are the indispensable part of angle modulation.

Two major disturbances, occurred in communication or signal transfer technology, are noise and distortions. In broader sense we can differentiate noise and distortion from the place of generation point of view. The disturbances occurred in the transmitter and the receiver stages are generally called as "distortions" whereas, disturbance on the signal, while that is being traversed trough the channel is named as "noise". It can be well understood that, any type of distortion or noise is severe for amplitude of a signal than that of the frequency. It would be easier for even an impulsive noise to distort instantaneous amplitude of any signal but not so easy to distort frequency, as definition of frequency does not permit the term "instantaneous".

5.2 Concept of Instantaneous Frequency

Any type of modulation in consideration must obey the same rule of dependency. In AM, the instantaneous amplitude of the carrier was modified or modulated with respect to the modulating signal, whereas in FM also we need to consider the instantaneous frequency. But, the concept of instantaneous frequency is not so straight forward like instantaneous amplitude, which is defined as amplitude at a

A. Das, *Signal Conditioning*, Signals and Communication Technology,
DOI: 10.1007/978-3-642-28818-0_5, © Springer-Verlag Berlin Heidelberg 2012

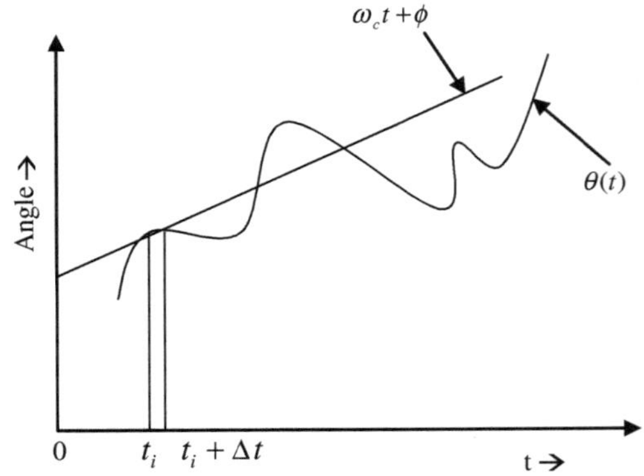

Fig. 5.1 Concept of instantaneous frequency

particular instant of time. Frequency can not be measured at an instant, as frequency needs to be calculated over at least one time period. Therefore, we need to understand the concept of instantaneous frequency in a different way.

Let's consider two signals $s_1(t) = A\cos\theta(t)$ and $s_2(t) = A\cos(\omega_c t + \phi)$. Now, if we plot the angles introduced by the two signals above with respect to time, we can get the graph as shown in Fig. 5.1.

The angles $\theta(t)$ and $(\omega_c t + \phi)$ both are function of time. The first one is varying more or less randomly and the second one linearly with slope ω_c and intercept ϕ. It is observed that, the straight line formed by the angle $(\omega_c t + \phi)$ becomes the tangent of the angle $\theta(t)$ in a small duration of time $[t_i, t_i + \Delta t]$, or in other words, within the aforesaid interval of time $\theta(t) = (\omega_c t + \phi)$. Therefore, within the mentioned interval, frequency of first signal $s_1(t) = A\cos\theta(t)$ is ω_c. It is to be noticed that, ω_c is indeed the slope of that curve at that duration Δt or at instant t_i taking $\Delta t \to 0$. Therefore, the instantaneous frequency is defined as

$$\omega_i = \frac{d\theta_i(t)}{dt} \tag{5.1}$$

and, instantaneous angle,

$$\theta_i = \int_{-\infty}^{t} \omega_i(t)dt \tag{5.2}$$

5.3 Mathematical Model

We can modulate the angle or phase of a signal in two ways. As the angle of a signal is essentially a function of the frequency, we can modulate frequency (FM) or directly modulate the entire angle/phase (PM) to get angle modulation. Frequency Modulation can be defined as the technique of modulation where the change in frequency is directly proportional to the modulating signal.

If we consider the carrier signal $v_c(t) = A\cos\omega_c t$ and modulating signal $m(t)$, the frequency of the carrier before modulation is ω_c and frequency after modulation is $\omega_c + \Delta\omega$.

$$\Delta\omega \,\alpha\, m(t)$$

$$\Delta\omega = k_f m(t),$$

where k_f is the constant of frequency modulation.
Therefore, the instantaneous frequency of carrier after modulation is

$$\omega_i = \omega_c + \Delta\omega = \omega_c + k_f m(t) \tag{5.3}$$

The angle derived from the instantaneous frequency is

$$\begin{aligned}\theta_i &= \int_{-\infty}^{t} \omega_i(t)dt = \int_{-\infty}^{t} (\omega_c + k_f m(t))dt \\ &= \int_{-\infty}^{t} \omega_c dt + \int_{-\infty}^{t} k_f m(t)dt \\ &= \omega_c t + k_f \int_{-\infty}^{t} m(t)dt \end{aligned} \tag{5.4}$$

Therefore, the Frequency Modulated signal is now expressed as

$$\begin{aligned}\varphi_{FM}(t) &= A\cos\theta_i(t) \\ &= A\cos\left(\omega_c t + k_f \int_{-\infty}^{t} m(t)dt\right)\end{aligned} \tag{5.5}$$

Similarly, if we employ Phase Modulation, the change in angle is directly proportional to the modulating signal, here.

If we consider the carrier signal $v_c(t) = A\cos\omega_c t$ and modulating signal $m(t)$, the frequency of the carrier before modulation is $\omega_c t$ and frequency after modulation is $\omega_c t + \Delta\theta$.

$$\Delta\theta \,\alpha\, m(t)$$

$$\Delta\theta = k_p m(t),$$

where k_p is the constant of phase modulation.
Therefore, the instantaneous angle/phase of carrier after modulation is

$$\theta_i = \omega_c t + \Delta\theta = \omega_c t + k_p m(t) \tag{5.6}$$

The instantaneous frequency derived from the angle using Eq. 5.1 is

$$\begin{aligned}\omega_i &= \frac{d\theta_i(t)}{dt} \\ &= \frac{d}{dt}\left(\omega_c t + k_p m(t)\right) \\ &= \omega_c + k_p \dot{m}(t)\end{aligned} \tag{5.7}$$

Therefore, the Phase Modulated signal is now expressed as

$$\begin{aligned}\varphi_{PM}(t) &= A\cos\theta_i(t) \\ &= A\cos\left(\omega_c t + k_p m(t)\right)\end{aligned} \tag{5.8}$$

5.4 FM and PM are Interchangeable

From the basic discussion of angle modulation we can infer that, we can change the entire angle (or phase) of carrier signal either by modifying the frequency or by modifying the phase. That's why PM and FM are the two sub-classes of the broader class of angle modulation. From the mathematical models derived in the previous section, it can also be shown that FM and PM are interchangeable and thereafter inseparable.

Let's consider one situation when we are having one FM modulator, but we need a phase modulated carrier. For the sake of simplicity, here we have considered the constant of frequency and phase modulation identical as equal to one unique constant of angle modulation, *k*. When we apply *M*(*t*) as modulating signal into one FM modulator, the carrier $A\cos\omega_c t$ would be modulated using the Eq. 5.4, i.e., the output of the said block would be $A\cos\left(\omega_c t + k\int_{-\infty}^{t} M(t)dt\right)$. Now, let's consider *M*(*t*) as a pre-processed modulating signal instead of original modulating signal, as $M(t) = dm(t)/dt$, when *m*(*t*) is the original modulating signal, then the output of the FM block would be $A\cos(\omega_c t + km(t))$, which is indeed the mathematical model of PM as derived in Eq. 5.8. Similarly, if another situation is in consideration where, we have one PM modulator, but we need a frequency modulated carrier. When we apply *N*(*t*) as modulating signal into one PM modulator, the carrier $A\cos\omega_c t$ would be modulated using the Eq. 5.8, i.e., the output of the said block would be $A\cos(\omega_c t + kN(t))$. Now, let's consider *N*(*t*) as a pre-processed modulating signal instead of original modulating signal, as $N(t) = \int_{\infty}^{t} m(t)dt$, when *m*(*t*) is the original modulating signal, then the output of the PM block would be $A\cos\left(\omega_c t + k\int_{-\infty}^{t} m(t)dt\right)$, which is indeed the mathematical model of FM as derived in Eq. 5.4. From the discussion above and the illustration shown in Figs. 5.2 and 5.3 it can be inferred that, "FM and PM are interchangeable and can only be treated as inseparable methods under the purview of Angle modulation technology".

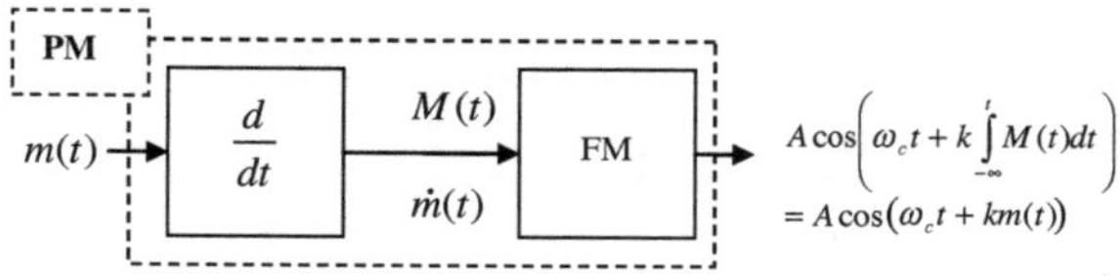

Fig. 5.2 FM modulator along with a differentiator used as PM modulator

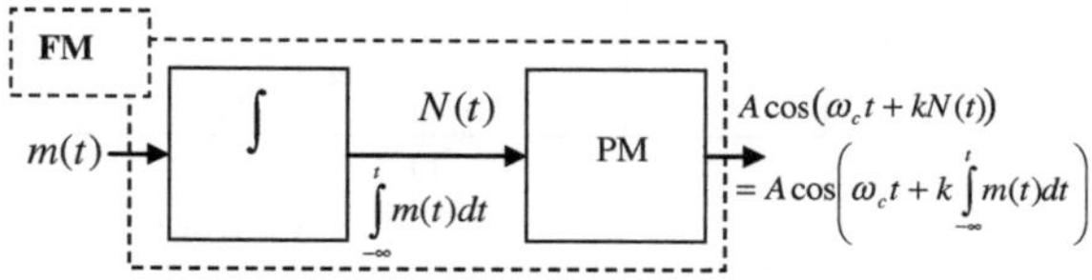

Fig. 5.3 PM modulator along with an integrator used as FM modulator

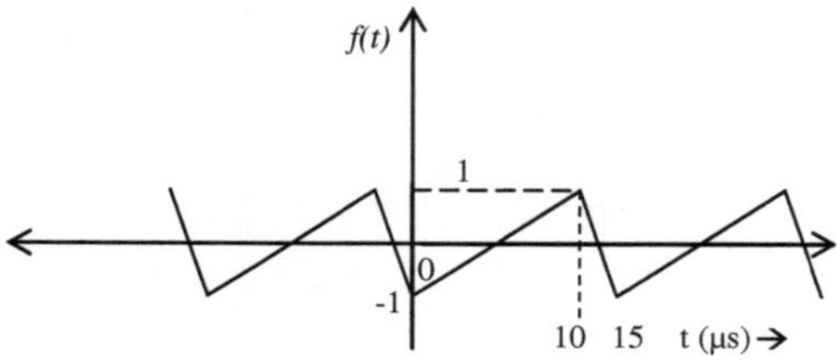

Fig. 5.4 A sawtooth modulating signal

5.4.1 Example 1

Draw the FM and PM waveform for the modulating signal given in Fig. 5.4 and indicate the maximum and minimum frequency attained by the modulated carrier.

Given, Carrier signal: $\cos 2\pi t$

$$k_f = 1.5\pi$$

$$k_p = 2\pi$$

As the modulating signal is a saw tooth waveform, it can be considered as piecewise linear, i.e., has some positive ramp and negative ramp region as well. From the definition of FM, it is easily understood that, the frequency modulated carrier will attain a maximum frequency and a minimum frequency per time period of $f(t)$. Between the two peaks of $f(t)$, the FM waveform will gradually change its frequency from maximum to minimum and vice versa (Fig. 5.5).

Time period of the modulating signal is 15 μs. The signal $f(t)$ can be defined mathematically as

$$f(t) = \begin{cases} \frac{1}{5}t - 1; & 0 \le t \le 10 \\ -\frac{2}{5}t + 5; & 10 < t \le 15 \end{cases} \tag{5.9}$$

5.4.1.1 FM

The maximum frequency will be attained by the frequency modulated signal at time 10 and 25 μs. Similarly the minimum frequency would be attained by the FM signal at time instants 15 and 30 μs. The instantaneous frequency is

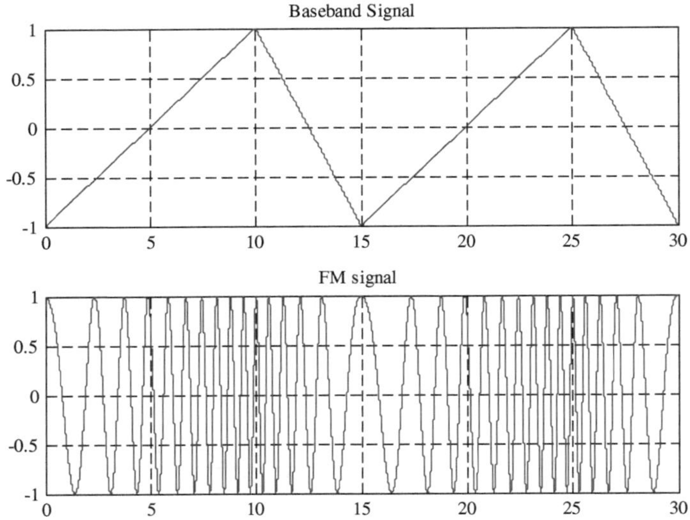

Fig. 5.5 Frequency modulated signal

$$\omega_i = \omega_c + k_f f(t)$$

$$f_i = f_c + \frac{k_f}{2\pi} f(t) = 1 + 0.75 f(t)$$

Here only *f*(*t*) is variable. So to get the extreme instantaneous frequencies, f(t) maximum and minimum values would be responsible as follows:

$$f_{\max} = 1 + 0.75|f(t)|_{\max} = 1.75\text{kHz}$$

$$f_{\min} = 1 + 0.75|f(t)|_{\min} = 1 - 0.75 = 0.25\text{kHz}$$

5.4.1.2 PM

Now, from the discussion of interchangeability of FM and PM, it is understood that, PM signal can be obtained by doing simple FM on the differentiated modulating signal *f*(*t*). As the modulating signal is a Shaw tooth type signal, it can be considered as a piecewise linear signal. One piece of the period is with positive slope and other with negative slope. Obviously, the slope will be reflected as constant vale to the $\frac{df(t)}{dt}$ signal. So the $\frac{df(t)}{dt}$ would be a rectangular wave as shown in Fig. 5.6. As the derived signal is having two distinct values, the modulated signal would have also two distinct values of frequencies. Therefore, maximum frequency and minimum frequency would be attained in a range of time not in an instant as seen in the case of FM.

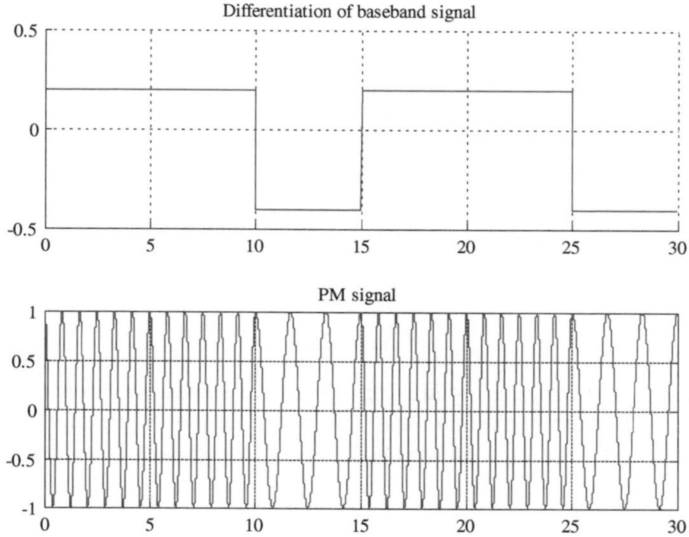

Fig. 5.6 Phase modulated signal

The maximum frequency attained by the phase modulated signal within the time range (0–10) and (15–25) μs. Similarly, the minimum frequency attained by the phase modulated signal within the time range (10–15) and (25–30) μs.

The instantaneous frequency is

$$\omega_i = \omega_c + k_p \frac{df(t)}{dt}$$

$$f_i = f_c + \frac{k_p}{2\pi}\frac{df(t)}{dt} = 1 + \frac{df(t)}{dt}$$

To get the extreme instantaneous frequencies, $\frac{df(t)}{dt}$ maximum and minimum values would be responsible as follows:

$$f_{\max} = 1 + \left|\dot{f}(t)\right|_{\max} = 1 + 0.2 = 1.2kHz$$

$$f_{\min} = 1 + \left|\dot{f}(t)\right|_{\min} = 1 - 0.4 = 0.6kHz$$

5.4.2 Example 2

Draw the FM and PM waveform for the modulating signal given in Fig. 5.7 and indicate the maximum and minimum frequency attained by the modulated carrier.

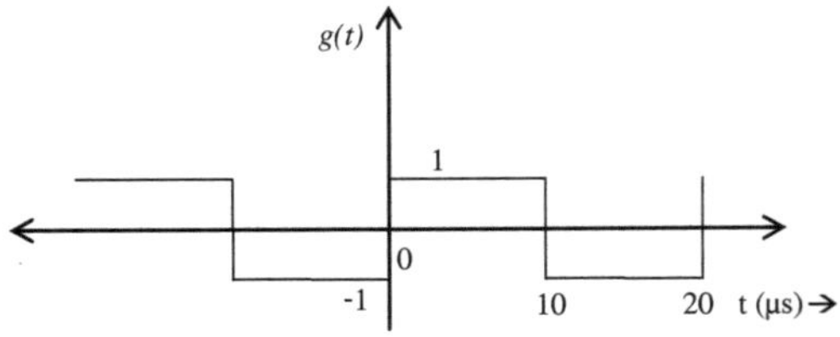

Fig. 5.7 A square type modulating signal

Given, Carrier signal: $\cos \pi t$ and $\sin \pi t$ for FM and PM respectively.

$$k_f = 1.5\pi$$

$$k_p = 2\pi$$

In the present problem, the modulating signal is a square waveform. So, the FM waveform will attain maximum and minimum frequency throughout the T_{ON} [where $g(t) = +1$] and T_{OFF} [where $g(t) = -1$] respectively (Fig. 5.8).

5.4.2.1 FM

Between the two polarities of $g(t)$, the FM waveform will *drastically* change its frequency from maximum to minimum (Fig. 5.8).

The maximum frequency attained by the frequency modulated signal within the range of time (0–10) and (20–30) μs. Similarly the minimum frequency would be attained by the FM signal in the time range (10–20) and (30–40) μs. The instantaneous frequency is

$$\omega_i = \omega_c + k_f f(t)$$
$$f_i = f_c + \frac{k_f}{2\pi} f(t) = 1 + 0.75 f(t)$$

Here only $f(t)$ is variable. So to get the extreme instantaneous frequencies, f(t) maximum and minimum values would be responsible as follows:

$$f_{\max} = 1 + 0.75 |f(t)|_{\max} = 1.75\text{kHz}$$
$$f_{\min} = 1 + 0.75 |f(t)|_{\min} = 1 - 0.75 = 0.25\text{kHz}$$

5.4.2.2 PM

As the principle modulating signal is itself a square waveform, analog differentiation cannot be done here. So, discrete difference operation results in discrete impulses as shown in Fig. 5.9.

It is interesting to observed that, 180° phase shift does occur in the phase modulated (PM) signal. This has observed only for square or rectangular waveforms. As the symbol representation in digital communication is mostly rectangular type, in digital PM, i.e., PSK, this phase shift of 180° always happens [1].

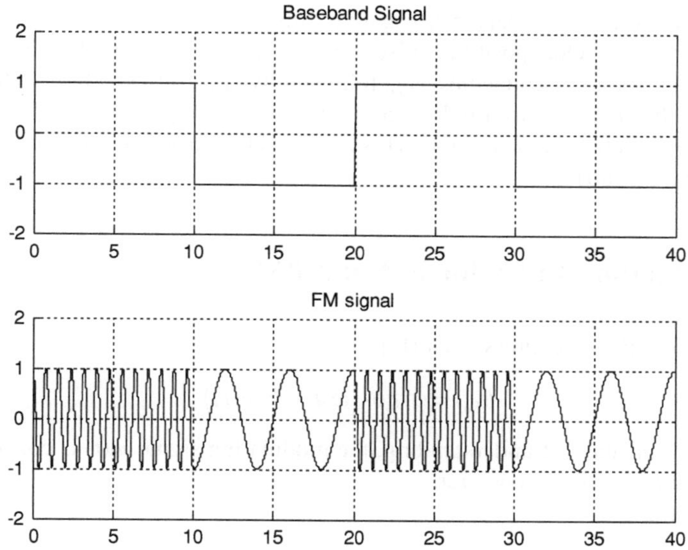

Fig. 5.8 Frequency modulated signal

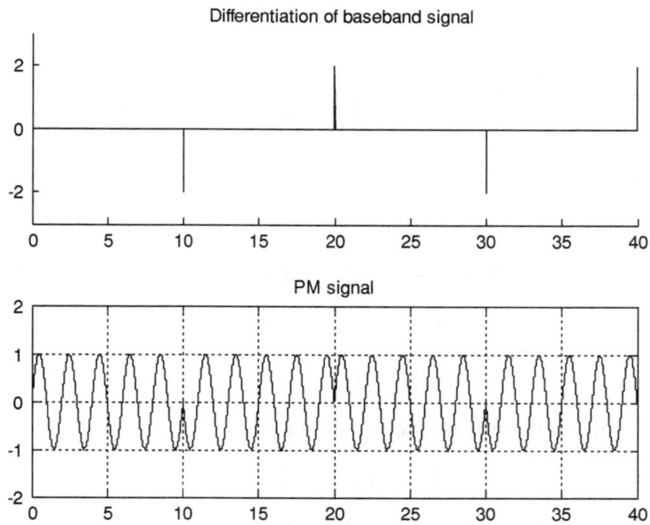

Fig. 5.9 Phase modulated signal

As there is no sustain variation of amplitude, apparently the derived signal shows zero amplitude every time expect the impulses. So the frequency would be essentially identical to the carrier signal. An impulse as considered as a zero width pulse here for understanding. A positive pulse of very narrow width responses to a high frequency carrier whereas a negative pulse of narrow width responses low carrier frequency. One positive pulse surrounded by zero amplitude results in combination of three frequencies as f_c followed by f_{max} followed by f_c. This three-frequency transition is occurred in zero time for an impulse. The similar thing happens for negative impulses, too. This is indeed the reason of 180° phase shift of the single tone carrier.

5.5 Modulation Index for FM and PM

A single tone FM is expressed as (Eq. 5.5)

$$\phi_{FM} = A\cos\left(\omega_c t + k_f m(t)\right)$$

Now, for the ease of understanding, the modulating signal $m(t)$ is considered as a single tone signal, $V_m \cos \omega_m t$.

Therefore,

$$\phi_{FM} = A\cos\left(\omega_c t + k_f V_m \sin \omega_m t\right) \tag{5.10}$$

From the above equation, the total angle is

$$\theta(t) = \left(\omega_c t + k_f V_m \sin \omega_m t\right) \tag{5.11}$$

$$\therefore \frac{d\theta(t)}{dt} = \left(\omega_c + k_f V_m \omega_m \cos \omega_m t\right) = \omega_i \tag{5.12}$$

But, we know that,

$$\omega_i = \omega_c + \Delta\omega \tag{5.13}$$

Comparing Eqs. (5.12) and (5.13), the angular frequency deviation

$$\Delta\omega = k_f V_m \omega_m \cos \omega_m t$$

$$\Rightarrow \Delta f = k_f V_m f_m \cos \omega_m t$$

Therefore, the maximum frequency deviation,

$$\delta_f = \Delta f_{\max} = k_f V_m f_m \tag{5.14}$$

From the basic definition of modulation (Chap. 1), here also we can derive the measure or extent of modulation done on the carrier by the modulating signal, as follows.

In FM, it is quite obvious that modulation index (β) would be directly proportional to maximum frequency deviation (δ_f). Actually the "deviation of frequency" concept is in-built into the definition of FM. On the other hand, if the frequency of modulating signal is lower, the relative effect of frequency deviation on FM would be higher. So, we can frame modulation index (β) for FM as,

$$\beta \propto \delta_f \tag{5.15}$$

$$\beta \propto \frac{1}{f_m} \tag{5.16}$$

Combining the above two equations, and considering unit proportionality constant, the modulating index

$$\beta = \frac{\delta_f}{f_m} \tag{5.17}$$

Now, expressing Eq. (5.10) in terms of modulating index,

$$\begin{aligned}\phi_{FM} &= A\cos(\omega_c t + k_f V_m \sin\omega_m t)\\ &= A\cos\left(\omega_c t + \frac{k_f V_m f_m}{f_m}\sin\omega_m t\right)\\ &= A\cos\left(\omega_c t + \frac{\delta_f}{f_m}\sin\omega_m t\right)\\ &= A\cos(\omega_c t + \beta\sin\omega_m t)\end{aligned} \tag{5.18}$$

Similarly, by considering the same equation (Eq. 5.10) for PM (changing the constant of modulation from k_f to k_p), we can deduce the modulation index for PM.

From Eq. (5.11) it is clear that, the phase deviation ($\Delta\theta$) is

$$\Delta\theta = k_p V_m \sin\omega_m t$$

Therefore, maximum phase deviation is

$$\Delta\theta_{\max} = k_p V_m \tag{5.19}$$

PM is not at all affected by frequency of modulating signal or other such factors. Therefore, Modulation index (μ) for PM is exactly same as the maximum frequency deviation for PM.

$$\mu = \Delta\theta_{\max} \tag{5.20}$$

Now expressing PM in terms of its modulation index,

$$\begin{aligned}\phi_{PM} &= A\cos(\omega_c t + k_p V_m \sin\omega_m t)\\ &= A\cos(\omega_c t + \Delta\theta \times \sin\omega_m t)\\ &= A\cos(\omega_c t + \mu\sin\omega_m t)\end{aligned} \tag{5.21}$$

5.6 Bandwidth of FM

A single tone FM is expressed as

$$\varphi_{FM} = A\cos(\omega_c t + \beta\sin\omega_m t) \tag{5.22}$$

The expression can be considered as real part of the exponential given by,

$$\hat{\varphi}_{FM} = Ae^{j(\omega_c t+\beta \sin \omega_m t)} = Ae^{j\omega_c t}e^{j\beta \sin \omega_m t} \tag{5.23}$$

In the above expression, the second exponential is a periodic function of period $1/f_{\mathrm{m}}$ and may be expressed in the form of a two way (exponential) Fourier series as:

$$e^{j\beta \sin \omega_m t} = \sum_{n=-\infty}^{\infty} D_n e^{jn\omega_m t} \text{ for } -\frac{1}{2f_m} \leq t \leq \frac{1}{2f_m} \tag{5.24}$$

Here the Fourier coefficient is given by

$$D_n = f_m \int_{-\frac{1}{2f_m}}^{\frac{1}{2f_m}} e^{j\beta \sin \omega_m t} e^{-jn\omega_m t} dt$$

Substituting $x = \omega_m t$ we get

$$D_n = f_m \int_{-\pi}^{\pi} e^{j(\beta \sin x - nx)} dx \tag{5.25}$$

The above expression is the ***n*th *order Bessel function of the first kind*** with argument β. This function is represented as $J_n(\beta)$.

Therefore,

$$D_n = J_n(\beta) \tag{5.26}$$

Now expressing Eq. 5.24 in terms of the Bessel function,

$$e^{j\beta \sin \omega_m t} = \sum_{n=-\infty}^{\infty} J_n(\beta) e^{jn\omega_m t} \tag{5.27}$$

Substituting the above expression in Eq. 5.23 we get

$$\begin{aligned} \hat{\varphi}_{FM} &= Ae^{j\omega_c t} \sum_{n=-\infty}^{\infty} J_n(\beta) e^{jn\omega_m t} \\ &= A \sum_{n=-\infty}^{\infty} J_n(\beta) e^{j(\omega_c + n\omega_m)t} \end{aligned} \tag{5.28}$$

As described in Eqs. 5.23 and 5.24, the real part of $\hat{\varphi}_{FM}$ is the frequency modulated (FM) signal, i.e.,

$$\varphi_{FM}(t) = A \sum_{n=-\infty}^{\infty} J_n(\beta) \cos(\omega_c + n\omega_m)t \tag{5.29}$$

Thus, the single tone FM signal has been expressed in terms of series of Bessel function.

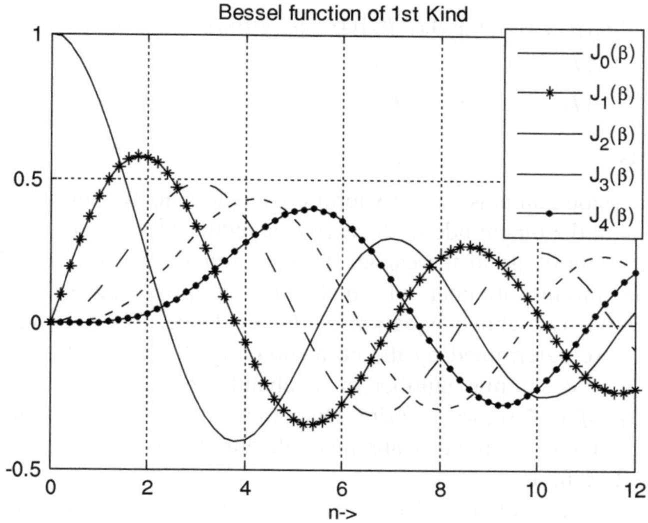

Fig. 5.10 Bessel functions of first kind

Now, the Bessel function can be represented using power series as

$$J_n(\beta) = \sum_{m=0}^{\infty} \frac{(-1)^m \left(\frac{\beta}{2}\right)^{n+2m}}{m!(n+m)!} \tag{5.30}$$

In Fig. 5.10, a parametric graph has been presented with varying n and taking β as a parameter.

Some important properties of the Bessel function are as follows:

(1)

$$J_n(\beta) = J_{-n}(\beta) \text{ for even } n$$
$$J_n(\beta) = -J_{-n}(\beta) \text{ for odd } n$$

(2) For a small value of β

$$J_0(\beta) \cong 1$$
$$J_1(\beta) \cong \frac{\beta}{2}$$
$$J_n(\beta) \cong 0 \quad \mathit{for} \quad n > 1$$

(3)

$$\sum_{n=-\infty}^{\infty} J_n^2(\beta) \cong 1$$

Making use of the first property, the Eq. 5.29 can be written as

$$\begin{aligned}\varphi_{FM}(t) =& AJ_0(\beta)\cos\omega_c t + AJ_1(\beta)\{\cos(\omega_c+\omega_m)t - \cos(\omega_c+\omega_m)t\} \\ &+ AJ_2(\beta)\{\cos(\omega_c+2\omega_m)t + \cos(\omega_c+2\omega_m)t\} \\ &+ AJ_3(\beta)\{\cos(\omega_c+3\omega_m)t - \cos(\omega_c+3\omega_m)t\} + \cdots\end{aligned} \tag{5.31}$$

Observations

(1) The expression contains a carrier term $\cos\omega_c t$ having magnitude $AJ_0(\beta)$. This signifies that the magnitude of the carrier is reduced by a factor of $J_0(\beta)$. Thus is FM, the amplitude of the carrier does not remain constant as AM.

(2) There an infinite number of sidebands are produced in FM, at least theoretically. The above equation is the evidence. The magnitudes of the sidebands are determined by the corresponding Bessel function $J_n(\beta)$. Thus, the presence of infinite number of sidebands components makes ***the ideal bandwidth of FM signal infinite***. But practically, the distant sidebands are ignored as their magnitudes are insignificant. It makes the practical bandwidth of FM finite.

(3) It is easily understandable that, the number of significant sidebands is determined by the value of modulation index β. Therefore, the bandwidth of FM is dependent on the modulation index.

(4) From the second property of Bessel function it is clear that, for lower value of β, i.e.,$\beta < 0.5$, only $J_0(\beta)$ and $J_1(\beta)$ have significant values and higher terms are negligible. Thus for small values of β, FM has two sidebands along with the carrier term. This FM is characterized as *Narrow Band Frequency Modulation* (NBFM). Taking the property two as limiting condition, the Eq. 5.31 would converge to the expression of NBFM (Eq. 5.34).

(5) As the FM signal does not have varying envelope, i.e., amplitude of modulated and unmodulated carrier are equal, the power of the FM signal also will be same as unmodulated carrier power.

5.7 Phasor Diagram

As discussed in Chap. 4, phasor is actually the vector form of representation of signal. The single tone narrow band frequency modulated (NBFM) signal is having three different frequency components like single tone AM. The three frequency components, f_c, $(f_c + f_m)$ and $(f_c - f_m)$ are common to both AM and NBFM. Only the understanding of the modulation techniques through phasor diagram can distinguish these two modulation schemes. Phasor representation also justifies the modulation techniques to be ‘amplitude’ and ‘angle’ modulation respectively.

Let’s start with the mathematical expression of FM, derived in Eq. 5.31. The same equation is applicable for NBFM if the modulation index β is a small fraction, i.e., $\beta << 1$.

$$\varphi_{NBFM}(t) = \cos(\omega_c t + \beta \sin \omega_m t)$$

Here, the amplitude of the cosine carrier is considered as unity.

The above equation now can be expanded as

$$\varphi_{NBFM}(t) = \cos(\omega_c t)\cos(\beta \sin \omega_m t) - \sin(\omega_c t)\sin(\beta \sin \omega_m t) \tag{5.32}$$

We know,

$$\underset{\theta \to 0}{Lt} \cos \theta = 1 \text{ and}$$

$$\underset{\theta \to 0}{Lt} \sin \theta = \theta$$

For NBFM, $\beta \ll 1$,

Therefore Eq. 5.32 can be simplified as

$$\varphi_{NBFM}(t) = \cos(\omega_c t) - \beta \sin \omega_m t \sin(\omega_c t) \tag{5.33}$$

$$\begin{aligned} &= \cos(\omega_c t) - \frac{\beta}{2} \times 2 \sin \omega_m t \sin(\omega_c t) \\ &= \cos(\omega_c t) - \frac{\beta}{2} \times \{\cos(\omega_c - \omega_m)t - \cos(\omega_c + \omega_m)t\} \\ &= \cos(\omega_c t) + \frac{\beta}{2}\cos(\omega_c + \omega_m)t - \frac{\beta}{2} \times \cos(\omega_c - \omega_m)t \end{aligned} \tag{5.34}$$

In vector form,

$$\vec{\varphi}_{NBFM} = \vec{v}_c + \vec{R}_{SB} \tag{5.35}$$

This equation is now used for constructing the phasor diagram. The carrier phasor is rotated anticlockwise and clockwise by same angle to form two sideband phasors $\cos(\omega_c + \omega_m)t$ and $\cos(\omega_c - \omega_m)t$ respectively. Next, the phasor/vector $-\frac{\beta}{2}\cos(\omega_c - \omega_m)t$ is formed by reducing the magnitude and inverting the direction of $\cos(\omega_c - \omega_m)t$. The magnitude is reduced as $\beta \ll 1$and the direction has been inverted for the −ve sign of the vector. Similarly the upper sideband phasor is formed by reducing the magnitude of the $\cos(\omega_c + \omega_m)t$ phasor and placed at the terminating position of carrier phasor. The lower sideband phasor (shown as '4' in Fig. 5.11) is placed at the end of the upper sideband phasor.

Vector triangle law is used here for realizing the NBFM phasor. Firstly, $\frac{\beta}{2}\cos(\omega_c + \omega_m)t$ and $-\frac{\beta}{2}\cos(\omega_c - \omega_m)t$ are added to form the consolidated sideband phasor $\vec{R}_{SB}$. Eventually $\vec{R}_{SB}$ has made a perfect right angle with the carrier phasor. Finally, the carrier phasor $((\vec{v}_c))$ is added to the sideband phasor $(\vec{R}_{SB})$ to form complete NBFM phasor, $\vec{\varphi}_{NBFM}$.

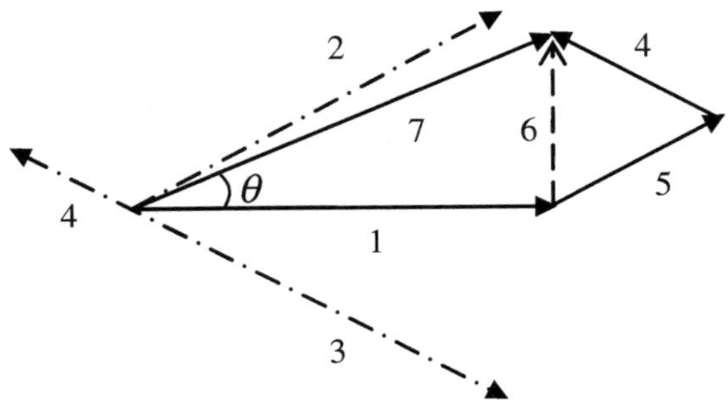

$1 \rightarrow \cos\omega_c t$	$5 \rightarrow \frac{\beta}{2}\cos(\omega_c+\omega_m)t$
$2 \rightarrow \cos(\omega_c+\omega_m)t$	$6 \rightarrow \vec{R}_{SB}$
$3 \rightarrow \cos(\omega_c-\omega_m)t$	$7 \rightarrow \vec{\varphi}_{NBFM} = \vec{v}_c + \vec{R}_{SB}$
$4 \rightarrow -\frac{\beta}{2}\cos(\omega_c-\omega_m)t$	

Fig. 5.11 Phasor representation of frequency modulated signal

5.7.1 Observed Properties of NBFM

From the phasor diagram of NBFM as shown in Fig. 5.11, we can note a few interesting things about NBFM, most of which are applicable for generic FM also. The observed properties are listed below:

- **Observation 1:** The modulated carrier phasor $\vec{\varphi}_{NBFM}$ has made an angle θ with the unmodulated carrier phasor $\vec{v}_c$. This is the evidence of 'angle modulation'.
- **Observation 2:** The sideband phasor and carrier phasor are orthogonal to each other. From the nature of the triangle formed by the phasors '4–6' it can easily be proved that the angle between phasors '1' and '6' is exactly 90°.
- **Observation 3:** When we measure the amplitude of the modulated carrier, the measurement should be done as a measure of component with respect to the unmodulated carrier. Here, when the magnitude of the modulated carrier phasor is measured from the top, as aligned with the horizontal axis, the magnitude would be exactly same as unmodulated carrier. Therefore, it can be inferred that, no amplitude modification/modulation has been occurred here.

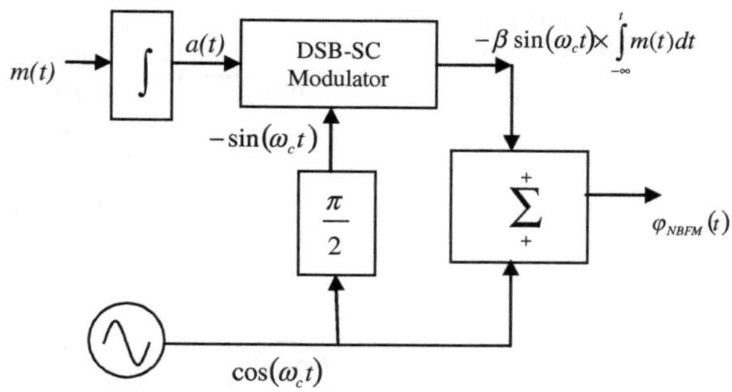

Fig. 5.12 NBFM generation using a DSB-SC modulator (indirect method)

5.8 NBFM and NBPM Generation: Indirect Method

As derived in the previous section, Narrow Band Frequency Modulated waveform is expressed as

$$\varphi_{NBFM}(t) = \cos(\omega_c t) - \beta \sin \omega_m t \sin \omega_c t$$

The above equation was derived for single tone modulation. But, the same expression can also be realized when we consider multi-tone or general signal as modulating signal, as

$$\varphi_{NBFM}(t) = \cos(\omega_c t) - \beta \sin(\omega_c t) \times \int_{-\infty}^{t} m(t)dt \tag{5.36}$$

It is clearly observed that, the NBFM modulated signal has one raw carrier signal and a 'mixed'[1] signal. From the knowledge of DSB-SC modulation, we can easily understand the 'mixed' term as DSB-SC signal. Therefore, the second term of the above equation can be generated by any DSB-SC modulator like balanced modulator. As we know, *sine* phasor always lags 90° behind *cos* phasor, we have shifted the phase of raw carrier, $\cos(\omega_c t)$ by an amount of 90° in positive direction (clockwise) to get $-\sin(\omega_c t)$ instead of $\sin(\omega_c t)$. The modulating signal *m*(*t*) is passed through an integrator (Fig. 5.12) to get $a(t) = \int_{-\infty}^{t} m(t)dt$. Here a DSB-SC modulator is employed to get the multiplied signal (a signal with two sidebands) expressed as $-\beta \sin(\omega_c t) \times \int_{-\infty}^{t} m(t)dt$. At the final end, the DSB-SC output is added to the raw carrier $\cos(\omega_c t)$ to get the NBFM modulated signal.

---

[1] MIXER stage generates summation of two signals with additive and subtractive frequencies after low pass filtering we get only the subtracted signal. Here 'mixed' signal signifies signal obtained before low pass filter.

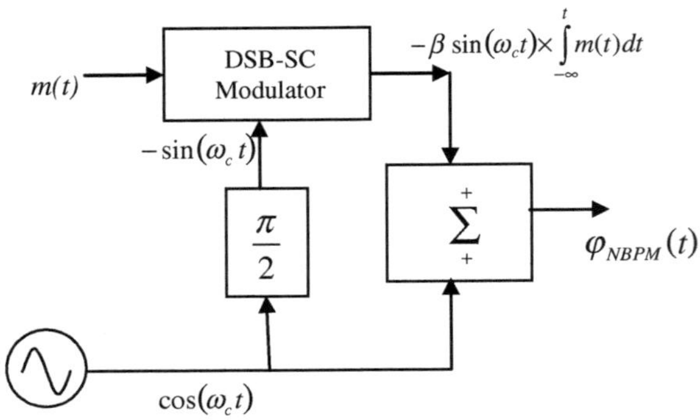

Fig. 5.13 NBPM generation using a DSB-SC modulator (indirect method)

From the concept of interchangeability between FM and PM, as discussed in Sect. 5.4, we can easily design the NBPM modulator (indirect method). As seen from the Sect. 5.4, just by adding one differentiator at the input port of modulating signal, one FM modulator can be converted to PM modulator. In Fig. 5.12, the input of the DSB-SC modulator would be $m(t)$ instead of $a(t) = \int_{-\infty}^{t} m(t)dt$, to realize NBPM modulator. From the Eq. 5.36, we can also express NBPM when general signal is used as modulating signal, as

$$\varphi_{NBPM}(t) = \cos(\omega_c t) - \beta \sin(\omega_c t) \times m(t) \tag{5.37a}$$

Thus to add one differentiator signifies removing the integrator as both of the operations are linear. Therefore NBPM generator can be depicted as in Fig. 5.13.

5.9 Wide Band FM Generation: Indirect Method of Armstrong

Though the bandwidth of FM is infinite ideally, in practice wide bandwidth of FM is considered. The wideband FM (WBFM) has high finite bandwidth. As the major difference between NBFM and WBFM is the bandwidth issue, we can first design a NBFM modulator using a DSB-SC modulator (indirect method) as discussed in the previous section. Next we can use suitable *frequency multiplier* stages to increase the bandwidth. In the present discussion of indirect method of Armstrong for WBFM generation we'll take the help of an example for the ease of understanding.

Let us formulate a problem of WBFM generation. One NBFM generator is given which generates signal with bandwidth 25 Hz. In the NBFM generator, we have used a crystal oscillator of frequency 200 kHz for carrier frequency generation. Now, our objective is to generate a WBFM signal with carrier frequency 91.2 MHz and bandwidth 75 kHz. While designing the WBFM generator, we may have to handle the trade off between the accuracy of bandwidth achievement and that of the carrier frequency. Here we have to take a decision of priority. As the carrier frequency is a designation of a particular channel, we have to ensure getting the carrier frequency without any compromise. Whereas to get the bandwidth exactly, as specified in the problem description is not so important. Therefore, for the sake of the design constrains we can compromise the bandwidth to some extent but not the carrier frequency.

Looking after the change in bandwidth from NBFM to WBFM, we have the bandwidth ratio $\frac{\Delta f_{out}}{\Delta f_{in}} = \frac{75kHz}{25Hz} = \frac{75{,}000}{25} = 3{,}000 = 3 \times 2^3 \times 5^3$. The choice in multiplication states are determined by the expression of the total multiplication by powers of prime numbers. From the above decomposition, we can see the number of states required is three (3, 2 and 5). To reduce the design complexity, we can take the closest approximation of the multiplier states by reducing the required states to two (3 and 2) as $\frac{\Delta f_{out}}{\Delta f_{in}} = 3 \times 2^{10} = 3 \times 1{,}024 = 3{,}072$. Now, the entire frequency multiplication can be divided in two parts as $3{,}072 = 64 \times 48$. After getting the NBFM signal, by the procedure stated in the previous section, we'll apply the first stage of frequency multiplication i.e., $(\times 64)$. After the multiplier, we get a modified FM signal with bandwidth $\Delta f = 25 \times 64\text{Hz} = 1.6\text{kHz}$ and the carrier frequency becomes $f_{C2} = 200 \times 64\text{kHz} = 12.8\text{MHz}$. Now, we have a remaining amplification factor of $(\times 48)$. Finally we have to have carrier frequency of 91.2 MHz. Just before the final stage, the frequency multiplier stage is to be attached. Therefore, the carrier frequency at the input of the second stage of multiplier $(\times 48)$ must be $f_{C3} = \frac{91.2MHz}{48} = 1.9MHz$.

Thus we have to solve a new problem of frequency conversion. The intermediate carrier frequency f_{C3} should be converted from f_{C2} by using a simple frequency converter or mixer. We know that, the frequency of the output signal of the mixer stage is always the difference between the frequencies of the input signals. So, the frequency of the signal generated by the local oscillator (associated to the mixer, as discussed in the chapter of super heterodyne receiver) should be $f_{C2} - f_{C3} = (12.8 - 1.9)\text{MHz} = 10.9\text{MHz}$. The bandwidth will not be affected by the mixer stage as bandwidth also signifies a difference between two specific frequencies. By mixer, we are actually shifting the entire range of frequency which does not result in any difference frequency. In the final stage of WBFM generation, the bandwidth obtained would be $1.6\text{kHz} \times 48 = 76.8\text{kHz}$ which is a bit higher than the specified value of $\Delta f = 75\text{kHz}$. But the obtained value of bandwidth is within the permissible range depending on the present problem specification. The entire method of WBFM generation using indirect method of Armstrong is presented in Fig. 5.14.

5.10 Direct Method of FM Generation: Using VCO

To understand the direct method of FM modulator design, it is important to understand the operating principle of voltage controlled oscillator (VCO). As the name suggests, the frequency (ω_i) of the output signal of the VCO is controlled by the voltage $(v_i(t))$ of the input signal. The relationship is linear as follows.

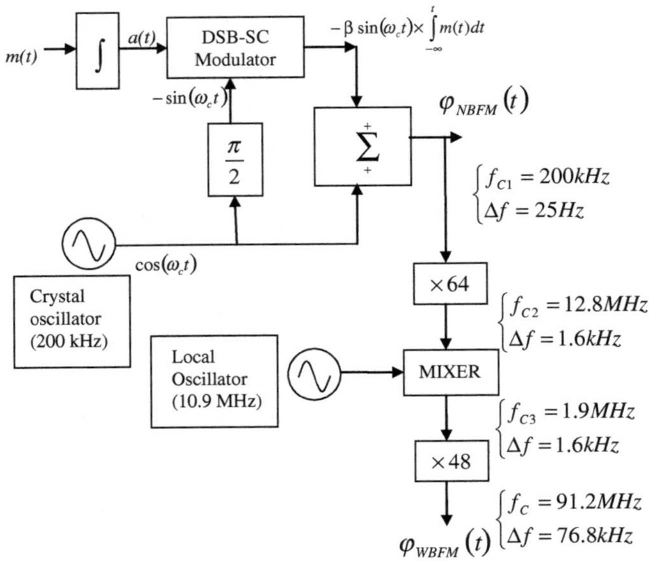

Fig. 5.14 WBFM generation using indirect method of Armstrong

$$\omega_i(t) = \omega_0 + Kv_i(t) \tag{5.37b}$$

From the VCO characteristics shown in Fig. 5.15, it is observed that even if no input voltage is applied to the VCO, the oscillator generates a signal with a quotient frequency, ω_0. As the frequency can be defined as the frequency of oscillation when no control voltage applied, it is known as *free-running frequency*.

Now let's consider a case when a modulating signal is applied to the input of the VCO. The free-running frequency is here considered as the frequency before modulation or raw carrier frequency. Therefore the change in frequency is given by $Kv_i(t)$ i.e., directly proportional to $v_i(t)$. The proportionality constant K is called as the 'constant of frequency modulation' and defined as the slope of the VCO characteristic curve. But in the present application of direct method of FM modulator design, we consider K as constant of frequency modulation K_f. Thus it is now clearly understood that as the input voltage (here the modulating signal) increases, the frequency of oscillation of VCO increases; output signal frequency becomes ω_0 when the instantaneous magnitude of the base-band signal becomes zero; the oscillation frequency decreases by decreasing input voltage. This is indeed the FM waveform.

Fig. 5.15 VCO characteristics

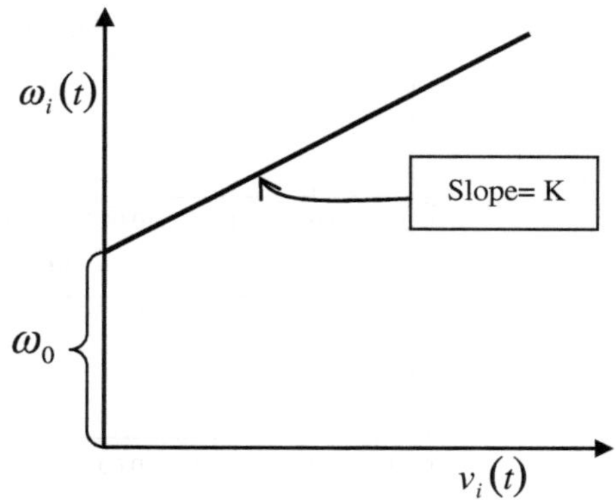

5.11 Indirect Method of FM Demodulation

As understood from the previous discussions, indirect method of FM generation signifies generation of FM signal with the help of AM modulators. Similarly, in indirect method of FM demodulation also we would prefer to use demodulator of any another modulation technique, like AM.

5.11.1 Slope Detector

To use the AM demodulator, we need to reform FM signal so that the high frequency career signal is grouped and bounded into an envelope, which would be following the pattern of the modulating or baseband signal used at the time of modulation. To determine the required margin or modified version of varying instantaneous amplitude modulated signal, we passed the FM signal through a differentiator as indicated in the following equations [2], [3].

From expression of FM (Eq. 5.5),

$$\begin{aligned}
\varphi_{FM}(t) &= A\cos\left(\omega_c t + k_f \int_{-\infty}^{t} m(t)dt\right) \\
\therefore \frac{d}{dt}\varphi_{FM}(t) &= \left|\left\{A\omega_c + k_f m(t)\right\}\right| \sin\left(\omega_c t + k_f \int_{-\infty}^{t} m(t)dt\right) \\
\Rightarrow \dot{\varphi}_{FM}(t) &= \left|\left\{B + Cm(t)\right\}\right| \sin\left(\omega_c t + k_f \int_{-\infty}^{t} m(t)dt\right) \\
\Rightarrow \dot{\varphi}_{FM}(t) &= Amp(t)\sin\left(\omega_c t + k_f \int_{-\infty}^{t} m(t)dt\right)
\end{aligned} \tag{5.38}$$

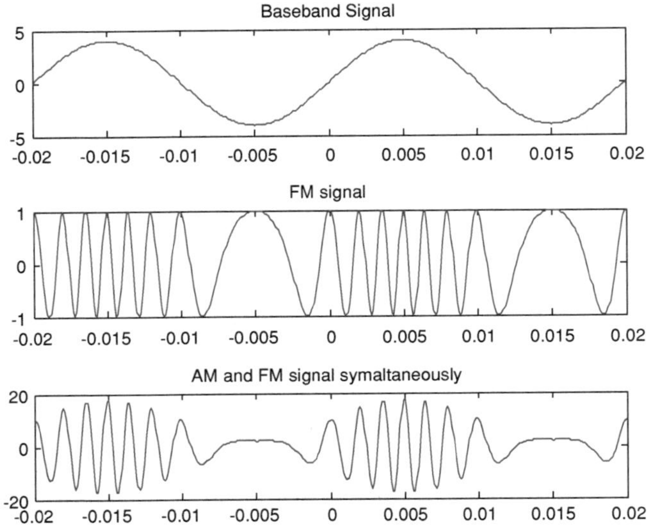

Fig. 5.16 Envelope is emphasized by slope detection (differentiation) of FM signal

In Eq. 5.38, the time varying amplitude *Amp*(*t*) can be considered as a linear function of modulating signal *m*(*t*). If the amplitude before modulation were $A\omega_c = B$ then the change in amplitude is $Cm(t)$. As the change of amplitude is directly proportional to the modulating signal, it is satisfying the definition of AM. Hence, the differentiator output would be a modulated signal depicted as a combination of AM and FM. There would be continuous amplitude and frequency variation simultaneously as shown in Fig. 5.16.

As there is a strong correspondence between the envelope and the shape of the modulating signal, by applying the envelope detector, thus we can extract the modulating signal. By differentiation, we essentially detect the slope. So, this demodulator is named as 'slope detector'. In Fig. 5.17 we have shown the amplitude response of the differentiated FM signal. The response of the system is like a narrow band BPF. Differentiation essentially returns the slope of a dynamics. The positive slope of the amplitude response curve is region of interest here. From the detected slope, amplitude at low frequency and high frequency are assigned to A_{max} and A_{min} respectively, as shown in Fig. 5.16.

5.11.2 Dual Slope Detector

There are two basic problems in single slope detection technique described in the previous subsection as,

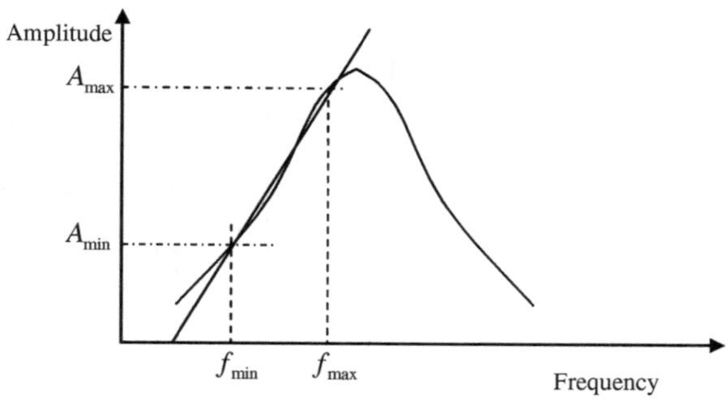

Fig. 5.17 Frequency response characteristics of single slope detector

(1) The detector also responds to the spurious amplitude variation of the input FM.
(2) The range of the linear slope is quite small.

A limiter preceding the tuned circuit or a differentiator takes care of the former while the extended linearity can be achieved in the present model of dual slope detection as shown in Fig. 5.18. The balanced discriminator as shown in Fig. 5.18 contains two resonant circuits; one tuned above f_c (just below f_{max}) and other is tuned below f_c (just above f_{min}). As the two resonant circuits are coupled in opposite direction, the resultant frequency response (frequency-to-voltage characteristics) would be the difference of those two frequency response curves as shown in Fig. 5.18b. As the DC component is automatically cancelled, the amplitude/voltage at frequency f_c becomes zero, as desired. A hard limiter (zero crossing detector) however, is applied as preprocessor to avoid the spurious amplitude variation of the practical incoming FM signal. It is customary to mention that here the difference of amplitude corresponding to the highest and lowest frequencies becomes higher with respect to the previous approach, resulting the range of the linear slope high. As a consequence the detection of the slope would be much easier in the sense of suitable discrimination between peak levels. In this present configuration, we have actually used two different resonant curves which ultimately results in an indifferent contiguous linear slope. So this configuration is called as *dual slope detector*.

5.12 Stereophonic FM

Stereophonic transmission can be considered as a special type of multiplexed channel transmission [5]. The main objective is to communicate two different channels in synchronized and integrated manner, without crosstalk. If we consider the case of audio signal, in transmitter side there two separate transducers (microphone) are required for two-channel information. Let's assume mic1 is assigned for female voice and mic2 for male voice. In stereophonic communication, the aim is to design the communication system such that in receiver side also

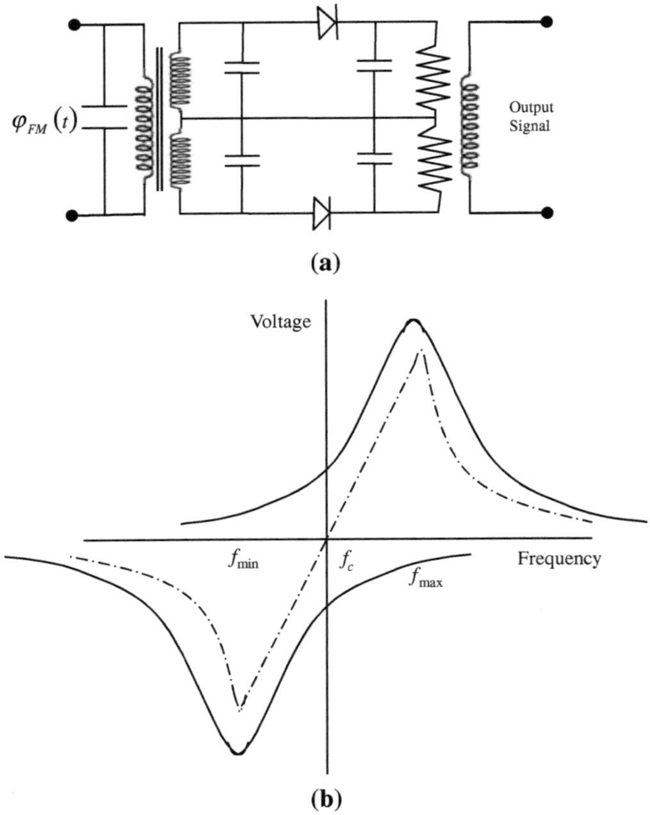

Fig. 5.18 **a** Balance discriminator. **b** Frequency response characteristics of resonant circuits

two segregated transducers can respond for mic1 and mic2 signals respectively, in synchronized manner.

As the two different information captured by two different microphones (transducers) needs to be transmitted through one channel, the process of multiplexing [2], [4] needs to be employed over here. In the present design we use the scheme namely Frequency Division Multiplexing (FDM) [2]. The left [L(t)] and right [R(t)] microphone captures signals (audio signal) of same bandwidth of 15 kHz. Firstly, we added and subtracted the L(t) and R(t) signals to get (L + R) and (L − R) respectively. It is customary to mention that simple addition and subtraction operations do not change the frequency occurrences and hence the bandwidth of the signal. Therefore, bandwidth of (L + R) and (L − R) would be 15 kHz again.

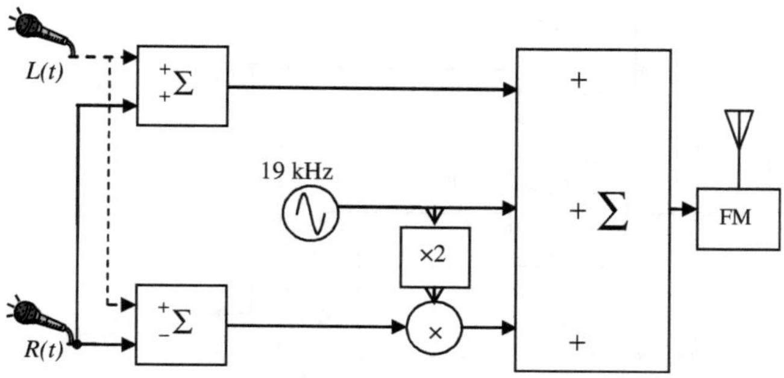

Fig. 5.19 Stereophonic FM transmitter

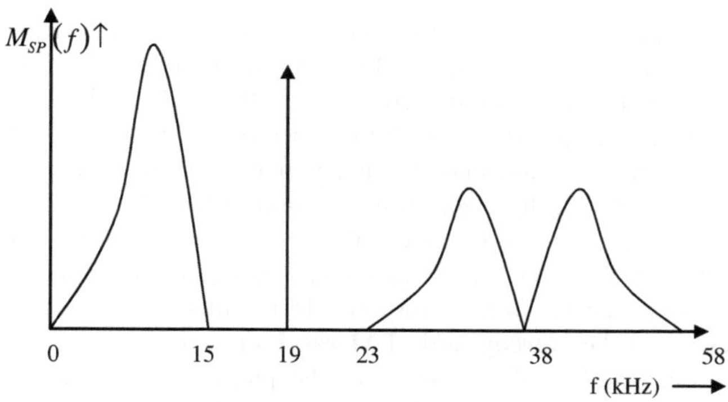

Fig. 5.20 Spectrum of stereophonic baseband signal: FDM

To avoid overlapping between these two spectra, we need to change the position index of either of the two signals. As understood, the magnitude of the added spectrum must be higher than the amplitude of the subtracted spectrum. Therefore, the (L−R) spectrum has been chosen for frequency shifting. For this purpose we have taken a single tone signal (cosinusoid) of frequency 19 kHz as the pilot carrier signal. Easily a single tone signal of frequency 38 kHz can be generated by passing the pilot carrier through a frequency doubler (multiplication by 2). This single tone signal of frequency 38 kHz is used for shifting the spectrum of (L − R). Finally the signals $(L+R), (L-R)\cos 2\omega_p t, \cos \omega_p t$are added to get the multiplexed (FDM) baseband signal. In the last stage, any frequency modulator circuit can be used to generate stereophonic FM as shown in Fig. 5.19.

The choice of the pilot carrier is really essential in the design of stereophonic FM transmitter (Fig. 5.19) and receiver (Fig. 5.21). As seen from the spectrum of the FDM signal (Fig. 5.20), by considering the negative frequency

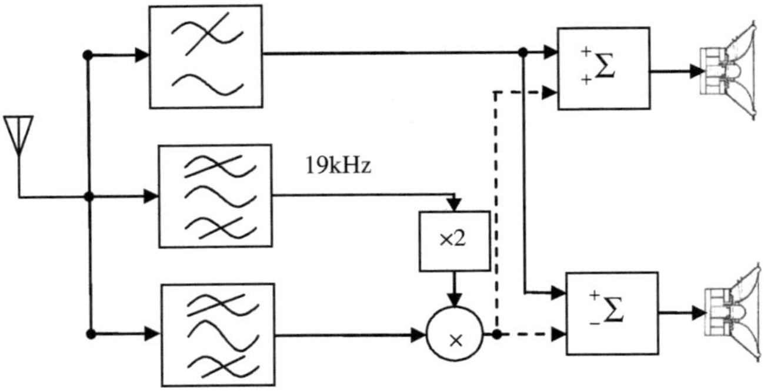

Fig. 5.21 Stereophonic FM receiver

concept, the bandwidth of the shifted (L − R) spectrum becomes double the original bandwidth of 15 kHz, i.e., 30 kHz. As the multiplier single tone signal frequency is 38 kHz, the central frequency of the shifted (L − R) would be 38 kHz. However, the spread of the shifted spectrum would be 30 kHz centered at 38 kHz, i.e., the start and end frequency marker would be 23 and 53 kHz respectively. It is now really interesting to observe that the width of the guard bands spaced between (L + R) and pilot carrier and spaced between pilot carrier and shifted (L − R) are all identical and equal to 4 kHz. The suitable choice of the pilot carrier only is responsible for this equispaced guard bands.

While designing the **stereophonic FM receiver**, one frequency discriminator realized by three parallel filters can serve the purpose of demodulation. In the design as shown in Fig. 5.21, we have used one low pass filter with cut off frequency 15 kHz, one narrow band pass filter centered at 19 kHz and one band pass filter centered at 38 kHz, of bandwidth 30 kHz to filter in (L + R), pilot carrier and shifted (L − R) respectively. The pilot carrier again passed through a local frequency doubler for coherent detection of the (L − R) signal as discussed in Chap. 4. In the final stage the (L + R) and (L − R) again are added and subtracted to retrieve the $L(t)$ and $R(t)$ to excite two different loud speakers or two earphones.

5.13 Matlab Programs

5.13.1 Bessel Function of First Kind

```
% FM -> Bessel's function representation
% Apurba Das (01/04/2011)
clc;
clear all;
close all;
z = (0:0.2:12)';

% a=besselj(0:3,z);
a0=besselj(0,z);
plot(z,a0,'k');hold on;
a1=besselj(1,z);
plot(z,a1,'k*-')
a2=besselj(2,z);
plot(z,a2,'k--')
a3=besselj(3,z);
plot(z,a3,'k:')
a4=besselj(4,z);
plot(z,a4,'k.-')
leg-
end('J_0(\beta)','J_1(\beta)','J_2(\beta)','J_3(\beta)'
,'J_4(\beta)')
xlabel('n->')
title('Bessel function of 1st Kind')
grid;
%%
z = (0:0.25:5)';
% z=0.2;

A=besselj(0:5,z);
T=[z,A(:,1),A(:,2)];
%
% N=10;
% Ta-
ble=[z(1:2:N),a0(1:.4:N),a1(1:.4:N),a2(1:.4:N),a3(1:.4:
N),a4(1:.4:N)];
```

5.13.2 FM and PM Signal Generation

```
% FM and PM signal generation
% Apurba Das (01/04/2011)
clc;
clear all;
close all;
%%
ts=0.0001;
t=-.02:0.0001:0.02;
m_sig=4*sin(2*pi*20*t);
%%
kf=160*pi;
m_intg=kf*ts*cumsum(m_sig);
s_fm=cos(2*pi*400*t+m_intg);
s_pm=cos(2*pi*400*t+pi*m_sig);
subplot(311)
plot(t,m_sig);
title('Baseband Signal');
subplot(312)
plot(t,s_fm)
title('FM signal')
subplot(313)
plot(t,s_pm)
title('PM signal')
```

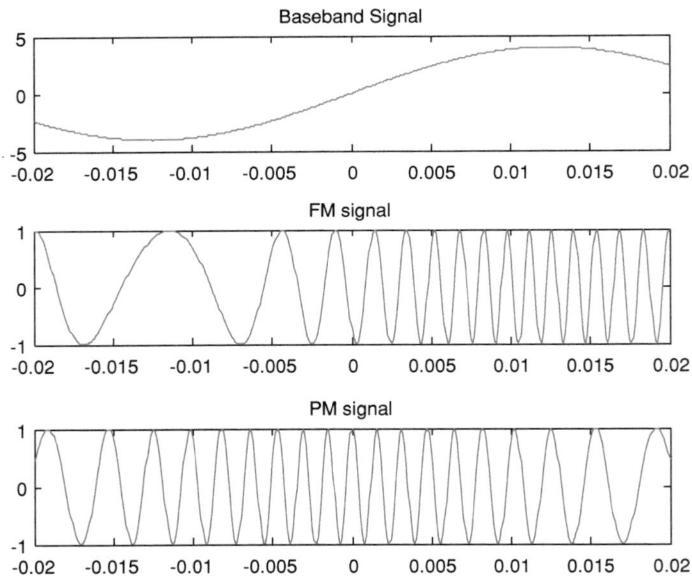

References

1. Das, A.: Digital Communication Principle and System Modelling. Springer, Berlin (2010)
2. Das, A.: Line Communication System, 2nd edn. New Age International Publishers, New Delhi (2010)
3. Lathi, B.P.: Modern Digital and Analog Communication Systems, 3rd edn. Oxford University Press, New York (2005)
4. Carlson, A.B., Crilly, P.B., Rutledge, J.C.: Communication Systems, 4th edn. McGraw Hill International, Boston (2002)
5. Taub, H., Schilling, D.L.: Principles of Communication Systems, 2nd edn. Tata McGraw-Hill, New Delhi (2005)

Part II
Discrete Signal Conditioning: 1D & 2D

Chapter 6
Discrete Time Transformations: DTFS and DTFT

6.1 Introduction

In the subsequent chapters we are going to discuss about the discrete signal conditioning including related transformations both in single dimension (1D) and two dimensions (2D). The present chapter would emphasize on the fundamental frequency domain representations in discrete time domain. Fourier series and transform would be illustrated for discrete time signals, here. They will be named as Discrete Time Fourier Series (DTFS) and Discrete Time Fourier Transform (DTFT), for sampled periodic and non-periodic signals (or sequences) respectively.

To understand the properties and representations of discrete time signals and their spectra, we have gone through some pre-requisites in the primary sections of the present chapter. The process of sampling, properties of its spectra, aliasing, concept of analog and digital frequencies and their mathematical relationships are discussed briefly. Another important pre-requisite is already described in the introductory chapter (Chap. 1) with title "convolution".

The two derived transformations DTFS and DTFT are then introduced in a non-conventional way. In the descriptions, I have stressed upon more on the physical interpretation than mathematical derivations.

6.2 Concept of Sampling

The word 'sample' means a small portion of physical element/phenomena which singularly or collectively has all the characteristics of the original physical element/ phenomena. As discussed in Chap. 1, this is the first step towards analog to digital signal conversion [2]. Let's consider the signal m(t) of bandwidth B Hz. When the signal is multiplied by a train of impulses, the resultant signal is obviously the sampled version of m(t). The impulses are equi-spaced by an amount of time T_S. After sampling, the signal g(t) becomes discrete in time domain (Fig. 6.1).

A. Das, *Signal Conditioning*, Signals and Communication Technology, DOI: 10.1007/978-3-642-28818-0_6, © Springer-Verlag Berlin Heidelberg 2012

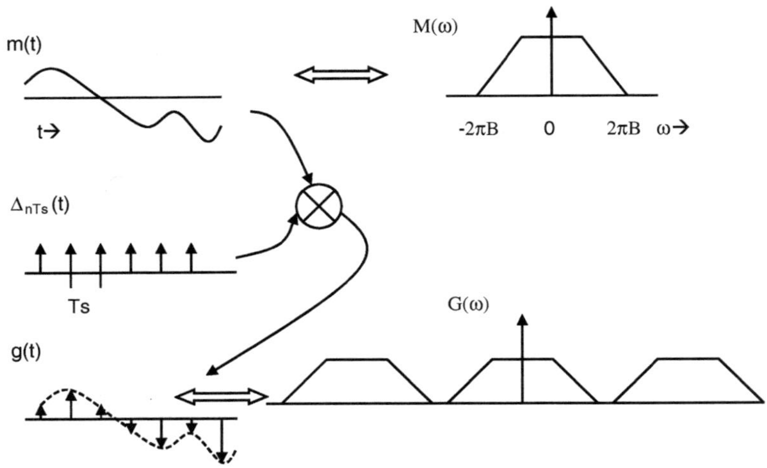

Fig. 6.1 The process of sampling to get discrete time signal

To establish the relationship [1] between the analog signal and discrete (sampled) signal, let's take the example of sampling of a sinusoid S(t), sampled at a rate of f_S, i.e., sampling interval T_S.

The analog signal is given by,

$$S(t) = A\sin(\Omega t + \phi) \tag{6.1}$$

where,
A is the amplitude,
Ω is the analog angular frequency $= 2\pi F$
F is the analog frequency,
ϕ is the initial phase

After sampling, we are taking the signal at t = 0, T_S, $2T_S$.,……NT_S. Or, t = nT_S, (n = 0, 1, 2, 3,….N), (N+1) = number of samples.

Therefore, from Eq. 6.1,

$$\begin{aligned} S(nT_S) &= A\sin(2\pi F n T_S + \phi) \\ \Rightarrow S(n) &= A\sin\left(2\pi \times \frac{F}{F_S} + \phi\right), \text{ as } T_S = 1/F_S \\ \Rightarrow S(n) &= A\sin(2\pi \times f + \phi) \\ \Rightarrow S(n) &= A\sin(\omega n + \phi) \end{aligned} \tag{6.2}$$

taking $f = \dfrac{F}{F_S}$

From Eqs. 6.1 and 6.2, the correspondence of sampling is shown. The sequence formed as S(n) and the input signal S(t) to the sampler, are correspondent.

Looking closely regarding the units of the different variations of frequencies, generated as a bi-product of the process of sampling, we have the following derived units.

Unit of F = cycles/sec
Unit of F_S = samples/sec
Therefore,
Unit of digital frequency, f = cycles/sample
Unit of analog angular frequency, Ω, i.e, 2πF = rad.cycles/sec
Unit of digital angular frequency ω = rad/samples.

Here, if one decrease the spacing between the samples (T_S), i.e., increase the rate of sampling F_S, the reconstruction would be easier at the receiver side. But, the bandwidth of transmission would be increased thereafter, it will affect the processing time of the sampled signal. So, there is a trade-off between noise and transmission bandwidth. However, it's really important to know the lower limit of the choice of the sampling frequency for successful reconstruction.

6.2.1 Sampling Theorem

An analog signal can be reconstructed from its sampled values un-erroneously, if the sampling frequency is at least twice the bandwidth of the analog signal.

Say an analog signal m(t) has three different frequency components f_1, f_2,and f_3, or combination of all of them, where $f_1 < f_2 < f_3$ then the bandwidth of that m(t) signal must be f_3. For ease of calculation, let's consider the bandwidth of that signal is B Hz, in general.

Therefore we can say the spectrum M(ω) is band limited by 2πB in ω scale, and band limited by B Hz in f scale. Now, sampled signal is nothing but the signal obtained by multiplying m(t) by unit impulse train $\Delta_{nTs}(t)$. From the figure the sampled signal is g(t) = m(t) × Δ_{nTs} (t). Time period of the impulse train is T_s. therefore frequency is $f_s = 1/T_s$. Let's now expand the Δ_{nTs} (t) signal in Fourier series so that we can study the spectrum of g(t) signal. As $\Delta_{nT_S}(t)$ is even function of time, By Fourier series,

$$\Delta_{nT_S}(t) = a_0 + \sum_{n=1}^{\infty} (a_n \cos n\omega_s t + b_n \sin n\omega_s t) \tag{6.3}$$

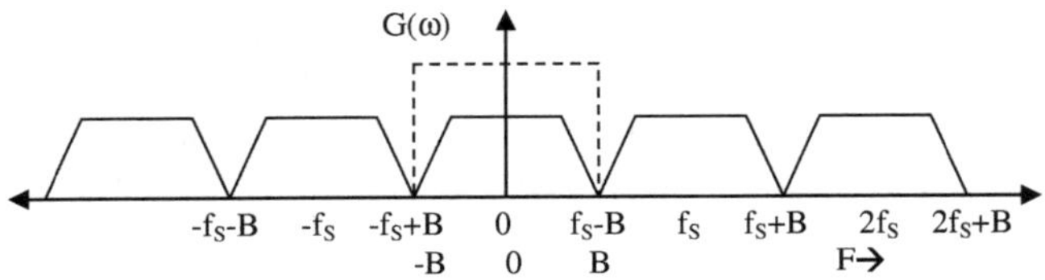

Fig. 6.2 Sampled spectra for $f_S = B$

Here,

$$a_0 = \frac{1}{T_S}\int_{T_S} \Delta_{nT_S}(t)dt = \frac{1}{T_S}$$
$$a_n = \frac{2}{T_S}\int_{T_S} \Delta_{nT_S}(t)\omega_0 t\, dt = \frac{2}{T_S}$$
$$b_n = 0, \mathit{for\ even function}$$

Here, $\int_{T_S}$ signifies integration within upper limit to lower limit difference T_S. Therefore,

$$\Delta_{nT_S}(t) = \frac{1}{T_S}(1 + 2\cos\omega_s t + 2\cos 2\omega_s t + 2\cos 3\omega_s t + \ldots\ldots)$$

$$\begin{aligned}
&\Rightarrow g(t) = m(t)\times\Delta_{nT_S}(t)\\
&= \frac{1}{T_S}(m(t) + 2m(t)\cos\omega_s t + 2m(t)\cos 2\omega_s t + 2m(t)\cos 3\omega_s t + \ldots\ldots)\\
&= \frac{1}{T_S}\left(m(t) + 2m(t)\times\frac{e^{j\omega_s t}+e^{-j\omega_s t}}{2} + 2m(t)\times\frac{e^{2j\omega_s t}+e^{-2j\omega_s t}}{2} + 2m(t)\times\frac{e^{3j\omega_s t}+e^{-3j\omega_s t}}{2} + \ldots\ldots\right)\\
&= \frac{1}{T_S}\left(m(t) + m(t)\left(e^{j\omega_s t}+e^{-j\omega_s t}\right) + m(t)\left(e^{2j\omega_s t}+e^{-2j\omega_s t}\right) + m(t)\left(e^{3j\omega_s t}+e^{-3j\omega_s t}\right) + \ldots\ldots\right)
\end{aligned} \tag{6.4}$$

From the first term of the equation above, it is very clear that spectrum of g(t) will be similar to M(ω) except the amplitude. Amplitude will be $1/T_s$ times of that of G(ω). Second term indicates M(ω) shifted in both sides by the amount of ω_s (using negative frequency concept). Similarly third term indicates M(ω) shifted in both sides by the amount of $2\omega_s$ and so on.

Now if we can extract the central spectrum by low pass filter from the assembly of spectrum (Fig. 6.2), then we can easily reconstruct the m(t) signal.

Case I From the figure above, a specific f_s i.e. ω_s is taken into account. It is very clear that central spectrum can be extracted using LPF. Therefore it is *success* of reconstruction.

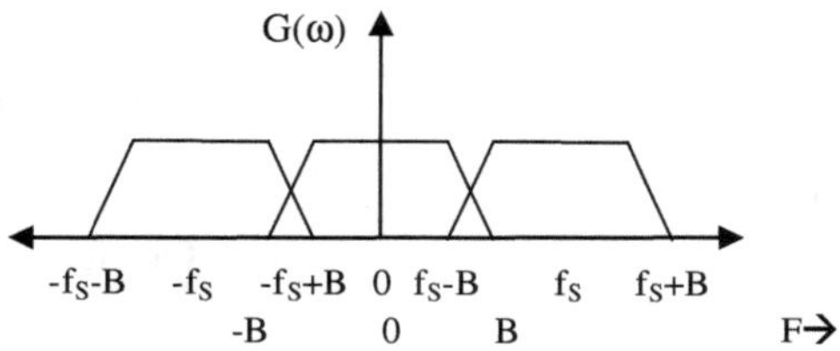

Fig. 6.3 Sampled spectra for $f_S < B$

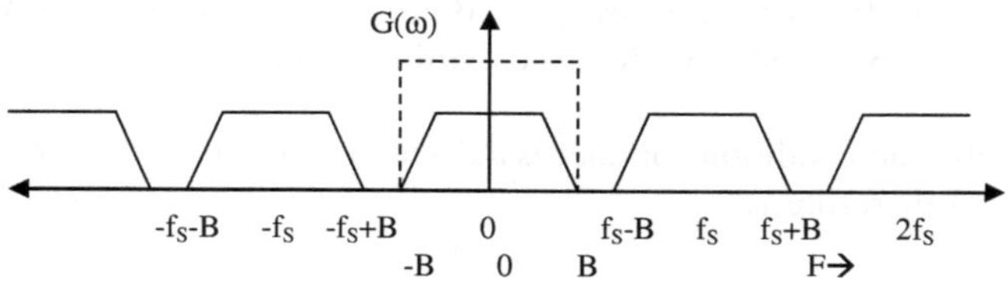

Fig. 6.4 Sampled spectra for $f_S > B$

Here

$$\omega_s - 2\pi B = 2\pi B$$

or,

$$\omega_s = 4\pi B$$

or,

$$f_s = 2B$$

Case II From the figure above other f_s i.e. ω_s is taken into account. It is very clear that central spectrum cannot be extracted using LPF; because, there is an overlap region of two spectra. The error due to this overlap region is called as *aliasing error*. Therefore it is *failure* of reconstruction (Fig. 6.3).
Here

$$\omega_s - 2\pi B < 2\pi B$$

or,

$$\omega_s < 4\pi B$$

or,

$$f_s < 2B$$

Case III From the figure above specific f_s i.e. ω_s is taken into account. It is very clear that central spectrum can be extracted using LPF more easily than case I. Therefore it is *success* of reconstruction (Fig. 6.4).
Here

$$\omega_s - 2\pi B > 2\pi B$$

or,

$$\omega_s > 4\pi B$$

or,

$$\text{Or, } f_s > 2B$$

Therefore collectively we can say $f_s > 2B$ i.e., sampling theorem is proved. The minimum frequency of sampling for successful reconstruction is Nyquist sampling rate ($f_s = 2B$)

Collectively, the condition for successful reconstruction of the message signal can be derived therefore as,

$$f_S \geq 2B \tag{6.5}$$

6.3 Aliasing

Say, two sequences are there s1(n) and s2(n). If the digital angular frequency difference between these two sequences is integral multiple of 360°, the sequences would be identical.

Say,

$$\begin{aligned} s1(n) &= A\sin(\omega_1 n + \phi) \\ s2(n) &= A\sin(\omega_2 n + \phi) \end{aligned} \tag{6.6}$$

and, $\omega_1 = \omega_2 + 2\pi$

Then,

$$s1(n) = A\sin(\omega_2 n + \phi + 2\pi n) = A\sin(\omega_2 n + \phi) \tag{6.7}$$

It is a severe error. The above equation signifies, after sampling the uniqueness of the signal destroys. Two different sinusoids map to a single sequence after sampling. This error is called as aliasing error and these two sequences are called as alias of other (Fig. 6.5).

To overcome the problem of aliasing, the choice of ω should be such that, any ω must reside within $\pm\pi$. Then only the difference between any two ω's be restricted within the range of 2π as shown in the Fig. 6.6.

Therefore, for zero aliasing, the digital angular frequency should be as follows

$$\begin{aligned} &|\omega| \leq \pi \\ &\Rightarrow 2\pi\frac{F}{F_S} \leq \pi \\ &\Rightarrow F \leq \frac{F_S}{2} \\ &\Rightarrow F_S \geq 2F \end{aligned} \tag{6.8}$$

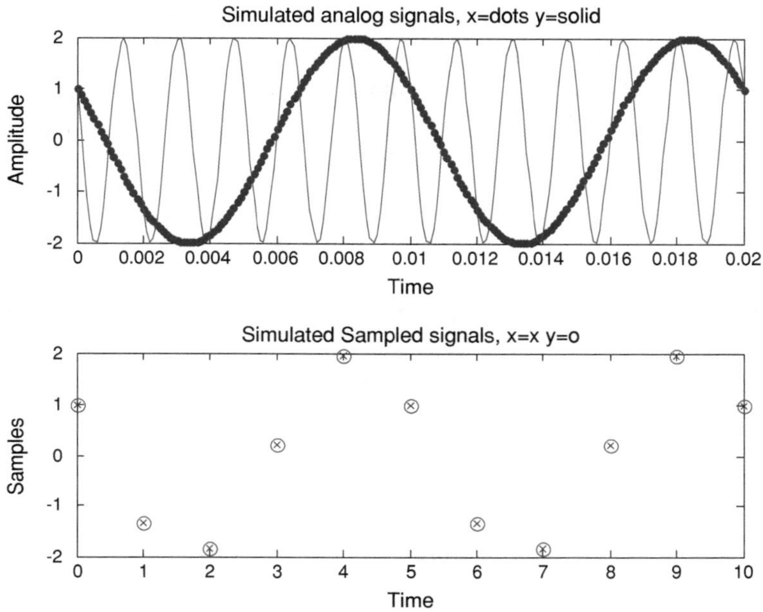

Fig. 6.5 Illustration of the sampling theorem

Fig. 6.6 Solution to the error of aliasing

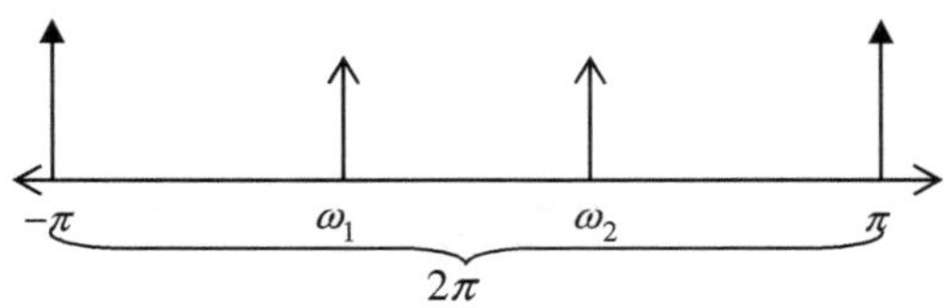

Equation 6.8 is indeed the mathematical model for statement of the sampling theorem, and the process of sampled sinusoid signal correspondence is another way to prove the sampling theorem [1], [2].

6.4 Discrete Time Fourier Series

As understood from the discussion of Continuous Time Fourier Series (CTFS) in Chap. 2, we have seen the following properties of the time and frequency domain representation

I. **Time domain:** Periodic and continuous
II. **Frequency domain:** Non-periodic and discrete

The method of sampling is nothing but multiplying a given signal with an impulse train. Again the theorem of convolution [2] suggests, if multiplication is done between two time domain signals, it is equivalent to convolution between

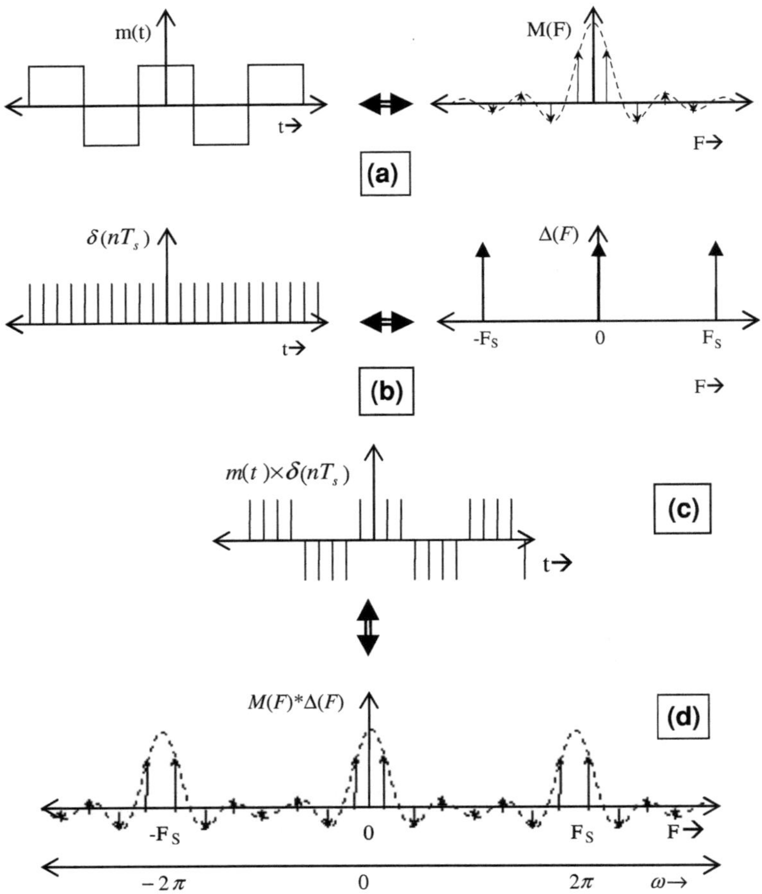

Fig. 6.7 **a** Continuous time periodic signal: waveform and spectrum. **b** Impulse train with equal interval $T_s$and its spectrum. **c** Sampled m(t): $m(t) \times \delta(nT_s)$. **d** DTFS (spectrum of sampled m(t)): $M(F) * \Delta(F)$

their respective spectra. From Chap. 2, Example 2.2 we derived that the spectrum of the continuous square wave is discrete *sinc* function. In the Fig. 6.7 we have illustrated the idea of the discrete domain transformation DTFS for periodic signal.

In Fig. 6.7a a DC free square wave m(t) is shown both in time (waveform) and frequency (spectrum) domain. For equidistance and uniform $(t = nT_S)$ sampling of the signal m(t), we have multiplied the same with the impulse train $\delta(nT_s)$. The Fourier series expansion of the impulse train is again an impulse train with periodicity and height $F_S = \frac{1}{T_S}$ as discussed in the previous sections.

To know the shape and pattern of the sampled spectra we need to execute convolution between two spectra $M(F)$ and $\Delta(F)$. The operation "convolution" comprises of (1) Folding, (2) Shifting and (3) Measuring overlap. The discrete *sinc* spectrum is even function of frequency. Therefore, due to single mirror symmetry the folding operation results the same shape and pattern of the spectrum. The entire spectrum would find overlap with each of the units (impulses) of $\Delta(F)$. Impulse would behave as unitary window is such case where the entire spectrum (*sinc*) would be replicated at the position of the impulse spectrum. Thus the discrete *sinc* spectrum would be repeated like discrete periodic spectra with periodicity $F = F_S$. In the Fig. 6.7d, two frequency axes are shown. According to analog frequency, the periodicity is $F = F_S$. From the definition of digital angular frequency (ω) as derived in Sect. 6.2, we know $\omega = 2\pi \frac{F}{F_S}$. So, the periodicity can also be expressed in digital angular frequency (ω) domain as $\omega = 2\pi \frac{F}{F_S}\Big|_{F=F_S} = 2\pi \frac{F_S}{F_S} = 2\pi$. This periodicity concept is quite in agreement with the concept of aliasing as shown in Fig. 6.6.

The spectrum shown in Fig. 6.7d is the Discrete Time Fourier Series (DTFS) representation and its pattern must be *discrete and periodic*.

6.5 Discrete Time Fourier Transform

As understood from the discussion of Continuous Time Fourier Transform (CTFT) in Chap. 3, we have seen the following properties of the time and frequency domain representation

III. **Time domain:** Non-periodic and continuous
IV. **Frequency domain:** Non-periodic and continuous

As the method of sampling described in the previous section, it can be inferred as multiplying a given signal with an impulse train. Hence from the theorem of convolution, the frequency domain representation of the sampled non-periodic signal would be convolution between the spectra of impulse train and the non-periodic continuous time signal.

In Fig. 6.8a a causal exponentially decaying signal m(t) is shown both in time (waveform) and frequency (spectrum) domain. For equidistance and uniform ($t = nT_S$) sampling the signal m(t), we have multiplied the same with the impulse train $\delta(nT_s)$. The Fourier series expansion of the impulse train is again an impulse train with periodicity and height $F_S = \frac{1}{T_S}$ as discussed in the previous sections.

Here also, to know the shape and pattern of the sampled spectra, we need to execute convolution between two spectra $M(F)$ and $\Delta(F)$. The operation "convolution" comprises of (1) Folding, (2) Shifting and (3) Measuring overlap. The continuous *Gaussian* like spectrum is even function of frequency. Therefore, due to single mirror symmetry the folding operation results the same shape and pattern of the spectrum. The entire spectrum would find overlap with each of the units (impulses) of $\Delta(F)$. Impulse would behave as unitary window in such case where the entire spectrum (*Gaussian*) would be replicated at the position of the impulse spectrum. Thus the spectrum would be repeated with periodicity (in frequency domain) $F = F_S$. In the Fig. 6.8d, two frequency axes are shown. According to

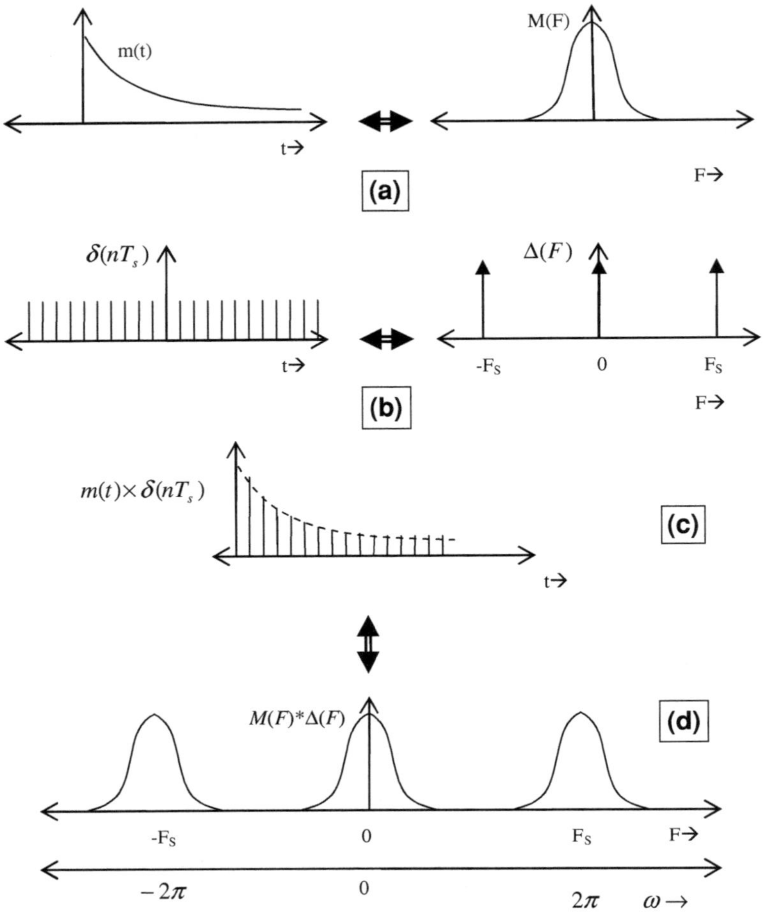

Fig. 6.8 **a** Continuous time non-periodic signal: waveform and spectrum. **b** Impulse train with equal interval T_s and its spectrum. **c** Sampled m(t): $m(t) \times \delta(nT_s)$. **d** DTFT (spectrum of sampled m(t)): $M(F) * \Delta(F)$

analog frequency, the periodicity is $F = F_S$. From the definition of digital angular frequency (ω) as derived in Sect. 6.2, we know $\omega = 2\pi \frac{F}{F_S}$. So, the periodicity can also be expressed in digital angular frequency (ω) domain as $\omega = 2\pi \frac{F}{F_S}\Big|_{F=F_S} = 2\pi \frac{F_S}{F_S} = 2\pi$. This periodicity concept is quite in agreement with the concept of aliasing as shown in Fig. 6.6.

The spectrum shown in Fig. 6.8d is the Discrete Time Fourier Transform (DTFT) representation and its pattern must be *continuous and periodic*.

In practical transactions, we generally handle the non-periodic signals as baseband for different types of signal processing including signal conditioning.

Table 6.1 Relationship between time and frequency domain representation of periodic and non-periodic signals

Time domain		Frequency domain		Operation
Continuous(C)/ discrete (D)	Periodic (P)/non-periodic (N)	Continuous(C)/ discrete (D)	Periodic (P)/non-periodic (N)	
C	P	D	N	CTFS
C	N	C	N	CTFT
D	P	D	P	DTFS
D	N	C	P	DTFT

Perfect periodic signal is communicated just by pre-defined commands as they can be regenerated in the destination side, very easily. Next, if we talk about digital signal conditioning, digital communication or any type of digital signal processing (DSP), we need to handle always finite and definite time and frequency samples. But even in the case of discrete time Fourier transform (DTFT), we get continuous spectrum. Between any two chosen frequency values there are infinite number of possible frequency samples. This spectrum is generated from a discrete time signal but the frequency domain representation is not discrete at all. Therefore to make it usable by digital systems, we would go for frequency domain sampling which is discussed in the subsequent two chapters of Discrete Fourier Transform (DFT) and Fast Fourier Transform (FFT).

In the Table 6.1 a chart is shown which can help students to establish logic of transformed domain representation both in continuous and discrete domain.

Table 6.1 reveals some interesting facts. The observed relationships are listed below.

I. For continuous time signal the spectrum is always non-periodic. The operation is either CTFS or CTFT.
II. For discrete time signals the spectrum is always periodic. The operation must be either DTFS or DTFT.
III. From I and II it is clear that there is a direct relationship between the first column of time domain representation and second column of frequency domain representation. We can infer, "By sampling a signal in time domain, the spectrum of the signal becomes periodic".
IV. For periodic signals the spectrum is always discrete. The operation must be Fourier series (CTFS or DTFS) not Fourier transform.
V. For non-periodic signals the spectrum is always continuous. The operation must be Fourier transform (CTFT or DTFT) not Fourier series.
VI. Hence the observations IV and V suggest that there is a direct relationship between second column of time domain representation and first column of frequency domain representation.

6.6 MATLAB Programs

6.6.1 Aliasing

```
% Aliasing
% Programed by Apurba Das (Jan,'10)

clear all;
close all;
clc;
freq=100;phase=pi/3;
mag=2;fs=500;Ts=1/fs;
k=1; % number of repetations
num_points=200; % No. of Points used=200 (smooth)
num_samples=11; % How many samples to simulate reading
                % 11 puts "sampled" under "analog"
step=2/(freq*num_points);   % get a nice step size
t=0:step:2*(1/freq);
n=0:num_samples-1;
% x and y are simulated analog functions
x= mag*cos(2*pi*freq*t+phase);
y= mag*cos(2*pi*(freq+k*fs)*t+phase);
% x2 and y2 are simulated sampled version of x and y
x2(n+1)=mag*cos(2*pi*freq*n*Ts+phase);
y2(n+1)=mag*cos(2*pi*(freq+k*fs)*n*Ts+phase);
% Plot the analog signal
subplot(2,1,1);
plot(t,x,'r.-');hold on;plot(t,y,'b-');
title('Simulated analog signals, x=dots y=solid');
xlabel('Time');ylabel('Amplitude');
% Plot "sampled" signals
subplot(2,1,2);
plot(n,x2,'rx');hold on;plot(n,y2,'bo');
title('Simulated Sampled signals, x=x y=o');
xlabel('Time');ylabel('Samples');
figure;
nn=0:0.002:0.02;
plot(t,x,'r.');hold on;plot(t,y,'b-');stem(nn,x2,'kv');
```

References

1. Das, A.: Line Communication System—A Telecommunication Switching Approach, 2nd edn. New Age International Publishers, New Delhi (2010a)
2. Das, A.: Digital Communication—Principles and System Modeling. Springer-Verlag, London (2010b)

Chapter 7
Discrete Fourier Transform

7.1 Introduction

The concept of signal conditioning holds true for synthesis domain (time or space domain) and analysis domain/transformed domain (frequency domain). When we are considering the discrete signal processing concept, the signal must be discrete both in time (or space) domain and in frequency domain. But from the study of Discrete Time Fourier Transform (DTFT) in the previous chapter, we are now convinced that, the discrete time signal is transformed into continuous frequency signal in analysis domain. From the basic concept of sampling and convolution discussed in the introductory chapter of signal conditioning (Chap. 1), we have successfully interpreted the reason of being the transformed domain signal (DTFT) continuous and periodic[1] with periodicity 2π. Now, our objective should be to condition/ process the continuous periodic spectrum such that the spectrum can be used by discrete signal processors like computers and hand-held devices like mobile phones, directly. In the present chapter, we have first introduced the classical algorithm of Discrete Fourier Transform (DFT). A complete flow from analog continuous time signal to discrete spectrum analysis is represented. Next, a clockwise rotated phasor (twiddle factor) is presented to simplify the DFT algorithm. Some interesting observations related to the DFT operation both in electrical signal (1D) and image (2D) are also presented with fair amount of elaboration and quest, at the rear end of the chapter.

7.2 The DFT Algorithm

Discrete signal processors (or digital signal processors) cannot handle the term 'infinity'. As discussed in the previous chapter of Discrete Time Fourier Transform (DTFT), we are now convinced that, the discrete time signal is transformed into

[1] DTFT always generates frequency-repetitive signal with F_{repeat} interval 2π.

A. Das, *Signal Conditioning*, Signals and Communication Technology,
DOI: 10.1007/978-3-642-28818-0_7, © Springer-Verlag Berlin Heidelberg 2012

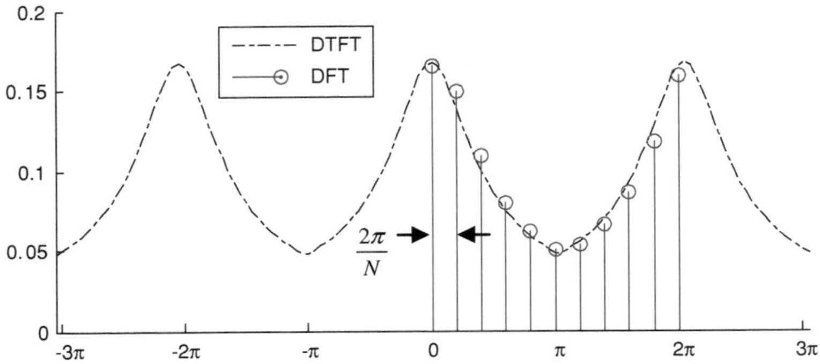

Fig. 7.1 Sampling DTFT spectrum in frequency domain in the range $[0, 2\pi]$ to get DFT

continuous frequency signal in analysis domain. The term 'continuity' signifies a range ∞ in one way or other. The DTFT generated continuous spectrum holds infinite valid frequency values anywhere in the frequency range 2π. To let the spectrum usable by discrete (or digital) signal processors, the entire unique range of spectrum $[0, 2\pi]$ is sampled in digital angular frequency $\left(\omega = 2\pi \frac{F}{F_S}\right)$ domain. If the total number of frequency samples is N, distance between two adjacent samples obviously comes to $\frac{2\pi}{N}$. Therefore, we shall have $X(\omega)$ values with $\omega = 0, \frac{2\pi}{N}, \frac{2\pi}{N} \times 2, \frac{2\pi}{N} \times 3, \cdots \frac{2\pi}{N} \times (N-1)$, as shown in the Fig. 7.1. Therefore, the spectrum sampled can be derived mathematically from DTFT as follows.

$$X(\omega) = \sum_{n=0}^{\infty} x(n) e^{-j\omega n} \tag{7.1}$$

Putting $\omega = \frac{2\pi}{N} k$ for $k = 0, 1, 2, \cdots (N-1)$

$$X(k) = \sum_{n=0}^{N-1} x(n) e^{-j\frac{2\pi}{N}kn} \tag{7.2}$$

Similarly, the inverse DFT (IDFT) can be computed by the following equation:

$$x(n) = \frac{1}{N} \sum_{n=0}^{N-1} X(k) e^{j\frac{2\pi}{N}kn} \tag{7.3}$$

The normalization factor can be customized for the sake of simplicity of calculation. In the equation of IDFT computation (Eq. 7.3), we have used the $1/N$ factor whereas in the DFT expression as represented in Eq. 7.2, we have not used any normalization factor. This normalization factor can be customized keeping the relation between DFT and IDFT unchanged. We can add normalization factor of $1/\sqrt{N}$ to both DFT and IDFT equation, instead.

The forward Discrete Fourier Transform (DFT) is represented by the Eq. 7.2. DFT frequency sequences can be obtained from the time sequence by the said algorithm. Though the famous paper by Cooley-Tukey [1] claims the algorithm of DFT as a classical one (1965), long before that (1886) the eminent mathematician Carl Fredriche Gauss [2, 3] has proposed the concept of DFT not as a discrete form of Fourier transform, but as an altogether different and unique algorithm of transformation.

In the next section we shall discuss about the phasor $e^{-j\frac{2\pi}{N}kn}$ as a special type of unit circle generator, named as Twiddle factor. The Twiddle factor will help the computation of DFT and IDFT easy by introducing a matrix for multiplication by the time samples and frequency samples respectively.

7.3 Twiddle Factor

In Eq. 7.2, the term $e^{-j\frac{2\pi}{N}kn}$ is representing a unit phasor rotating in clockwise direction. If we notice the phasor closely, we will see some interesting properties of the phasor which are not present in any ordinary $e^{-j\theta}$ phasor. The special phasor multiplier in Eq. 7.2 is called as Twiddle factor represented by the notation W_N as defined in Eq. 7.4.

$$W_N = e^{-j\frac{2\pi}{N}} \tag{7.4}$$

The magnitude of the twiddle factor is always 1, but the phase varies with respect to the power of W_N. Here N is fixed for a particular operation. It means, each unit of W_N will make an angle of 360°/N with the reference in clockwise direction.[2] Therefore, by increasing power of the twiddle factor W_N, a unit circle will be completed with N equal divisions. Therefore getting the values of the multiplier of time sequences is made easier as shown in the Fig. 7.2. The basic algorithm of DFT and IDFT computation (as presented in Eqs. 7.2 and 7.3) now can be expressed in terms of twiddle factor as presented in Eqs. 7.5 and 7.6.

$$X(k) = \sum_{n=0}^{N-1} x(n) W^{kn} \tag{7.5}$$

$$x(n) = \frac{1}{N}\sum_{n=0}^{N-1} X(k) W^{-kn} \tag{7.6}$$

Considering Eq. 7.5 as the basic equation for deriving DFT, we can compute *X(k)s* for all values of $k \subset [0, N-1]$ as under.

[2] The direction of phasor is clockwise as it's defined as a negative exponential.

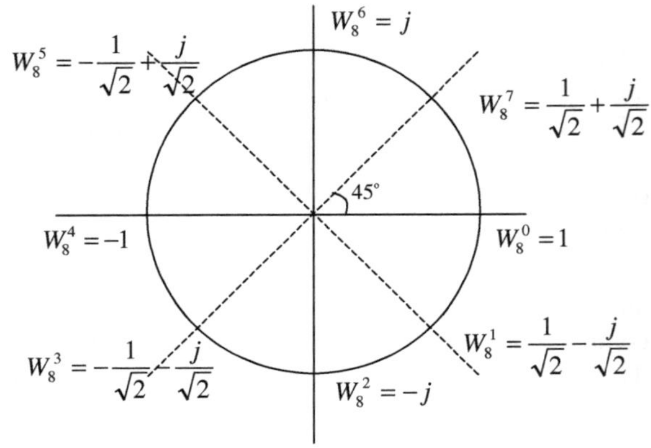

Fig. 7.2 Twiddle factors for Eight point DFT are calculated from a unit circle

$$X(0) = x(0) + x(1) + x(2) + \cdots + x(N-1) \tag{7.7a}$$

$$\begin{aligned} X(1) &= x(0)W_N^0 + x(1)W_N^1 + x(2)W_N^2 + \cdots + x(N-1)W_N^{N-1} \\ &= x(0) + x(1)W_N^1 + x(2)W_N^2 + \cdots + x(N-1)W_N^{N-1} \end{aligned} \tag{7.7b}$$

$$X(2) = x(0) + x(1)W_N^2 + x(2)W_N^4 + \cdots + x(N-1)W_N^{2(N-1)} \tag{7.7c}$$

$$X(3) = x(0) + x(1)W_N^3 + x(2)W_N^6 + \cdots + x(N-1)W_N^{3(N-1)} \tag{7.7d}$$

$$\vdots$$

$$\vdots$$

$$X(N-1) = x(0) + x(1)W_N^{(N-1)} + x(2)W_N^{2(N-1)} + \cdots + x(N-1)W_N^{(N-1)^2} \tag{7.7e}$$

From the above set of equations (Eq. 7.7), the DFT computation an be formalized by matrix multiplication as

$$\begin{bmatrix} X(0) \\ X(1) \\ X(2) \\ \vdots \\ X(N-1) \end{bmatrix} = \begin{bmatrix} 1 & 1 & 1 & \cdots & 1 \\ 1 & W_N^1 & W_N^2 & \cdots & W_N^{(N-1)} \\ 1 & W_N^2 & W_N^4 & \cdots & W_N^{2(N-1)} \\ \vdots & \vdots & \vdots & & \vdots \\ 1 & W_N^{(N-1)} & W_N^{2(N-1)} & \cdots & W_N^{(N-1)^2} \end{bmatrix} \begin{bmatrix} x(0) \\ x(1) \\ x(2) \\ \vdots \\ x(N-1) \end{bmatrix} \tag{7.8}$$

$$\therefore \vec{X} = \tilde{W} \times \vec{x} \tag{7.9}$$

By the above procedure, the frequency sample (DFT) vector $\vec{X}$ can by computed from given time sample vector $\vec{x}$ and twiddle matrix $\widetilde{W}$. The values of each

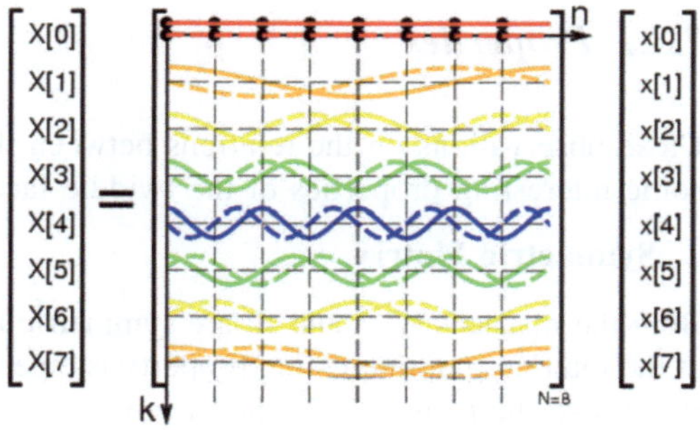

Fig. 7.3 Computation of 8-point DFT by sine and cosine wave distribution

element of the twiddle matrix comprises of a real (cosine) and an imaginary (sine) term. Taking an example of 8-point DFT, the values of each element are presented in the pictorial wave representation (Fig. 7.3). In the figure above, the real part (cosine wave) of W_8^{nk} is denoted by solid line and imaginary part (sine wave) of W_8^{nk} by dotted line. The cosine and sine waves are of same frequency for fixed k. For varying k, the initial phase for sine and cosine are changed along with frequency.

Employing similar methodology, we can express IDFT in terms of inverse twiddle matrix as shown in the following equations.

$$\begin{bmatrix} x(0) \\ x(1) \\ x(2) \\ \vdots \\ x(N-1) \end{bmatrix} = \frac{1}{N}\begin{bmatrix} 1 & 1 & 1 & \cdots & 1 \\ 1 & W_N^{-1} & W_N^{-2} & \cdots & W_N^{-(N-1)} \\ 1 & W_N^{-2} & W_N^{-4} & \cdots & W_N^{-2(N-1)} \\ \vdots & \vdots & \vdots & & \vdots \\ 1 & W_N^{-(N-1)} & W_N^{-2(N-1)} & \cdots & W_N^{-(N-1)^2} \end{bmatrix}\begin{bmatrix} X(0) \\ X(1) \\ X(2) \\ \vdots \\ X(N-1) \end{bmatrix} \tag{7.10}$$

$$\therefore \vec{x} = \tilde{W}^{-1} \times \vec{X} \tag{7.11}$$

From the Eqs. 7.10, or 7.11, we can compute IDFT from given frequency domain samples and inverse twiddle matrix. Now, if we can find some relationships between forward and inverse twiddle matrix, then the computation would be simpler. Let's look at any element of the inverse twiddle matrix.

$$\begin{aligned} W_N^{-1} &= e^{j\frac{2\pi}{N}} = \cos\frac{2\pi}{N} + j\sin\frac{2\pi}{N} = \left(\cos\frac{2\pi}{N} - j\sin\frac{2\pi}{N}\right)^* \\ &= \left(e^{-j\frac{2\pi}{N}}\right)^* = W_N^* \end{aligned} \tag{7.12}$$

The complex conjugate relationship is also reflected to the entire twiddle matrix as

$$\tilde{W}^{-1} = \frac{1}{N}\tilde{W}^* \tag{7.13}$$

7.3.1 Properties

Close observations on the relations between the twiddle matrix elements extracts some interesting properties of the twiddle factor.

1. **Symmetric Matrix**

Both the matrices $\tilde{W}^{-1}$ and $\tilde{W}$ are symmetric matrices. When we holistically look at the matrices, an interesting property is emerged. The matrix remains unchanged if we take the *transpose* of the matrix.

$$\tilde{W} = \tilde{W}^T, \text{ and} \tag{7.14a}$$

$$\tilde{W}^{-1} = \left[\tilde{W}^{-1}\right]^T \tag{7.14b}$$

2. **Periodicity**

As the elements of the twiddle matrix are obtained from a unit circle, if we go on increasing the power of the twiddle factor, after a complete rotation, the values will be repeated. Therefore, a two type periodicity is observed as presented in Eqs. 7.15 and 7.16.

$$\text{(a)} \quad W_N^{k+N} = e^{-j\frac{2\pi}{N}(k+N)} = e^{-j2\pi}e^{-j\frac{2\pi}{N}k} = W_N^k \tag{7.15}$$

$$\text{(b)} \quad W_N^{k+xN} = e^{-j\frac{2\pi}{N}(k+xN)} = e^{-j2\pi x}e^{-j\frac{2\pi}{N}k} = W_N^k \tag{7.16}$$

3. **Half-circle Anti-symmetry**

Let us consider any point on the unit circle shown in Fig. 7.2. Let the phasor move freely in any direction by an amount of π, i.e., 180°, i.e., half-circle, it would move towards exactly opposite end of the circle, and obviously the value would be just the negative to the reference phasor. Consider W_8^1 and move the phasor by angle 180°, we get W_8^5 from the Fig. 7.2. The figure says, $W_8^1 = \frac{1}{\sqrt{2}} - \frac{j}{\sqrt{2}}$ and $W_8^5 = -\frac{1}{\sqrt{2}} + \frac{j}{\sqrt{2}} = -W_N^1$. The rule is maintained for any twiddle factor element. The rule is proved in the following equations.

$$W_N^{k+\frac{N}{2}} = e^{-j\frac{2\pi}{N}\left(k+\frac{N}{2}\right)} = e^{-j\pi}e^{-j\frac{2\pi}{N}k} = -W_N^k \tag{7.17}$$

4. **Power-N relationship**

Square of a twiddle factor used for N-point DFT computation can also be expressed as a twiddle factor used for N/2-point DFT.

$$W_N^2 = e^{-j\frac{2\pi}{N}\times 2} = e^{-j\frac{2\pi}{N/2}} = W_{N/2} \tag{7.18}$$

The rule is followed not only for power 2, this is applicable for any power of W.

The above mentioned properties make the DFT computation much easier than the classical one (as presented in Eq. 7.2). To support the claim we have now worked out a problem for better understanding of the DFT computation methodology and moreover, to understand the importance of higher dimension resolution in frequency samples.

Example 7.1

Find DFT of a sequence $x(n) = \begin{cases} 1; 0 \le n \le 2 \\ 0; elsewhere \end{cases}$.

For (i) N = 4 and (ii) N = 8, plot $|X(k)|$ and $\angle X(k)$ and comment on the results.

$$x(n) = \{1, 1, 1\}$$

1. When we consider 4-point DFT (N = 4), by zero padding, $x(n) = \{1, 1, 1, 0\}$ (Fig. 7.4)

$$\begin{bmatrix} X(0) \\ X(1) \\ X(2) \\ X(3) \end{bmatrix} = \begin{bmatrix} 1 & 1 & 1 & 1 \\ 1 & W_4^1 & W_4^2 & W_4^3 \\ 1 & W_4^2 & W_4^4 & W_4^6 \\ 1 & W_4^3 & W_4^6 & W_4^9 \end{bmatrix} \begin{bmatrix} x(0) \\ x(1) \\ x(2) \\ x(3) \end{bmatrix} \tag{7.19}$$

$$\Rightarrow \begin{bmatrix} X(0) \\ X(1) \\ X(2) \\ X(3) \end{bmatrix} = \begin{bmatrix} 1 & 1 & 1 & 1 \\ 1 & W_4^1 & W_4^2 & W_4^3 \\ 1 & W_4^2 & W_4^0 & W_4^2 \\ 1 & W_4^3 & W_4^2 & W_4^1 \end{bmatrix} \begin{bmatrix} x(0) \\ x(1) \\ x(2) \\ x(3) \end{bmatrix}$$

$$\Rightarrow \begin{bmatrix} X(0) \\ X(1) \\ X(2) \\ X(3) \end{bmatrix} = \begin{bmatrix} 1 & 1 & 1 & 1 \\ 1 & -j & -1 & j \\ 1 & -1 & 1 & -1 \\ 1 & j & -1 & -j \end{bmatrix} \begin{bmatrix} x(0) \\ x(1) \\ x(2) \\ x(3) \end{bmatrix}$$

$$\Rightarrow \begin{bmatrix} X(0) \\ X(1) \\ X(2) \\ X(3) \end{bmatrix} = \begin{bmatrix} 1 & 1 & 1 & 1 \\ 1 & -j & -1 & j \\ 1 & -1 & 1 & -1 \\ 1 & j & -1 & -j \end{bmatrix} \begin{bmatrix} 1 \\ 1 \\ 1 \\ 0 \end{bmatrix}$$

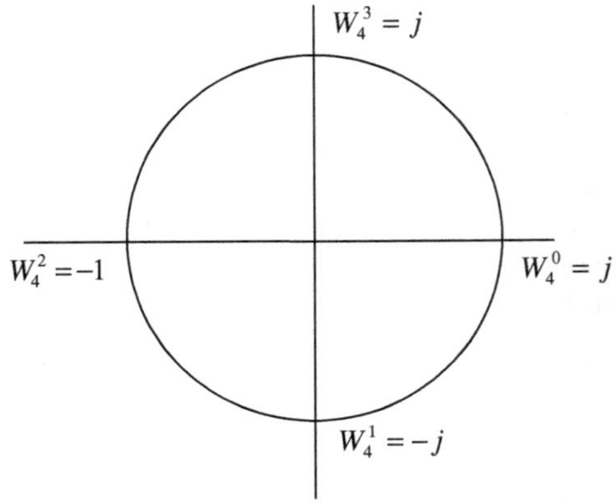

Fig. 7.4 Twiddle factors for 4 point DFT are calculated from a unit circle

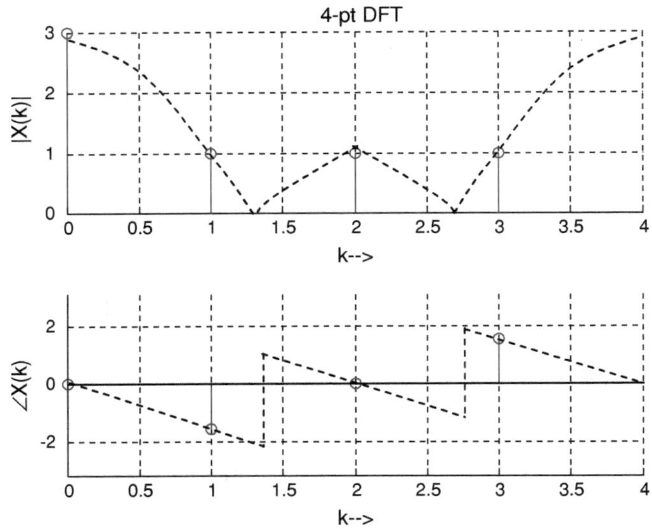

Fig. 7.5 Magnitude and phase spectrum of 4 pt DFT

$$
\begin{aligned}
\therefore X(0) &= 1 + 1 + 1 = 3 = 3\angle 0 \\
X(1) &= 1 - j - 1 = -j = 1\angle - \pi/2 \\
X(2) &= 1 - 1 + 1 = 1 = 1\angle 0 \\
X(3) &= 1 + j - 1 = j = 1\angle \pi/2
\end{aligned}
\tag{7.20}
$$

From the above equations, we can separately express the magnitude and phase of DFT samples as (Fig. 7.5)

$$\begin{aligned} |X(k)| &= \{3, 1, 1, 1\} \\ \angle X(k) &= \{0, \pi/2, 0, \pi/2\} \end{aligned} \tag{7.21}$$

2. When we consider 8-point DFT (N = 8), we have to make the length f the given sequence eight by zero padding, $x(n) = \{1, 1, 1, 0, 0, 0, 0, 0\}$

$$\begin{bmatrix} X(0) \\ X(1) \\ X(2) \\ X(3) \\ X(4) \\ X(5) \\ X(6) \\ X(7) \end{bmatrix} = \begin{bmatrix} 1 & 1 & 1 & 1 & 1 & 1 & 1 & 1 \\ 1 & W_8^1 & W_8^2 & W_8^3 & W_8^4 & W_8^5 & W_8^6 & W_8^7 \\ 1 & W_8^2 & W_8^4 & W_8^6 & W_8^8 & W_8^{10} & W_8^{12} & W_8^{14} \\ 1 & W_8^3 & W_8^6 & W_8^9 & W_8^{12} & W_8^{15} & W_8^{18} & W_8^{21} \\ 1 & W_8^4 & W_8^8 & W_8^{12} & W_8^{16} & W_8^{20} & W_8^{24} & W_8^{28} \\ 1 & W_8^5 & W_8^{10} & W_8^{15} & W_8^{20} & W_8^{25} & W_8^{30} & W_8^{35} \\ 1 & W_8^6 & W_8^{12} & W_8^{18} & W_8^{24} & W_8^{30} & W_8^{36} & W_8^{42} \\ 1 & W_8^7 & W_8^{14} & W_8^{21} & W_8^{28} & W_8^{35} & W_8^{42} & W_8^{49} \end{bmatrix} \begin{bmatrix} x(0) \\ x(1) \\ x(2) \\ x(3) \\ x(4) \\ x(5) \\ x(6) \\ x(7) \end{bmatrix} \tag{7.22}$$

To get the values of different powers of twiddle factor, W_8, one unit circle is divided in eight equal divisions as the objective is to compute 8-point DFT. Thus, phase difference between adjacent phasors is 360°/8 = 45°. In the Fig. 7.6, all the W_8 factors are calculated and thereafter the matrix obtained from Eq. 7.22 is filled up for complete calculation of DFT (Fig. 7.6). Before putting the values, the matrix is simplified using only the periodicity property of twiddle factor as presented in Eq. 7.16.

$$\Rightarrow \begin{bmatrix} X(0) \\ X(1) \\ X(2) \\ X(3) \\ X(4) \\ X(5) \\ X(6) \\ X(7) \end{bmatrix} = \begin{bmatrix} 1 & 1 & 1 & 1 & 1 & 1 & 1 & 1 \\ 1 & W_8^1 & W_8^2 & W_8^3 & W_8^4 & W_8^5 & W_8^6 & W_8^7 \\ 1 & W_8^2 & W_8^4 & W_8^6 & W_8^0 & W_8^2 & W_8^4 & W_8^6 \\ 1 & W_8^3 & W_8^6 & W_8^1 & W_8^4 & W_8^7 & W_8^2 & W_8^5 \\ 1 & W_8^4 & W_8^0 & W_8^4 & W_8^0 & W_8^4 & W_8^6 & W_8^4 \\ 1 & W_8^5 & W_8^2 & W_8^7 & W_8^4 & W_8^1 & W_8^6 & W_8^3 \\ 1 & W_8^6 & W_8^4 & W_8^2 & W_8^0 & W_8^6 & W_8^4 & W_8^2 \\ 1 & W_8^7 & W_8^6 & W_8^5 & W_8^4 & W_8^3 & W_8^2 & W_8^1 \end{bmatrix} \begin{bmatrix} x(0) \\ x(1) \\ x(2) \\ x(3) \\ x(4) \\ x(5) \\ x(6) \\ x(7) \end{bmatrix} \tag{7.23}$$

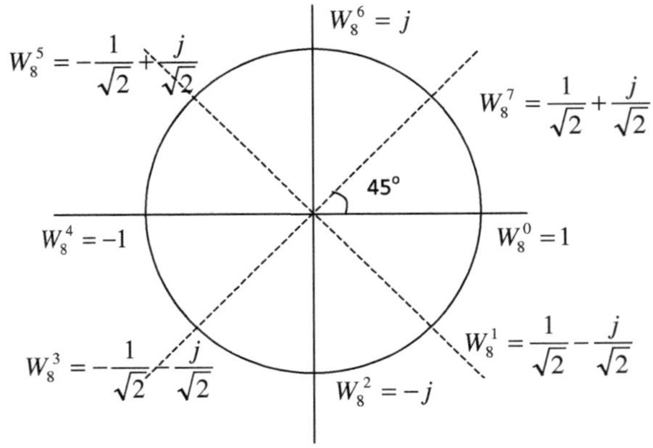

Fig. 7.6 Twiddle factors for 8 point DFT are calculated from a unit circle

$$
\Rightarrow \begin{bmatrix} X(0) \\ X(1) \\ X(2) \\ X(3) \\ X(4) \\ X(5) \\ X(6) \\ X(7) \end{bmatrix} = \begin{bmatrix}
1 & 1 & 1 & 1 & 1 & 1 & 1 & 1 \\
1 & \left(\frac{1}{\sqrt{2}}-\frac{j}{\sqrt{2}}\right) & -j & \left(-\frac{1}{\sqrt{2}}-\frac{j}{\sqrt{2}}\right) & -1 & \left(-\frac{1}{\sqrt{2}}+\frac{j}{\sqrt{2}}\right) & j & \left(\frac{1}{\sqrt{2}}+\frac{j}{\sqrt{2}}\right) \\
1 & -j & -1 & j & 1 & -j & -1 & j \\
1 & \left(-\frac{1}{\sqrt{2}}-\frac{j}{\sqrt{2}}\right) & j & \left(\frac{1}{\sqrt{2}}-\frac{j}{\sqrt{2}}\right) & -1 & \left(\frac{1}{\sqrt{2}}+\frac{j}{\sqrt{2}}\right) & -j & \left(-\frac{1}{\sqrt{2}}+\frac{j}{\sqrt{2}}\right) \\
1 & -1 & 1 & -1 & 1 & -1 & j & -1 \\
1 & \left(-\frac{1}{\sqrt{2}}+\frac{j}{\sqrt{2}}\right) & -j & \left(\frac{1}{\sqrt{2}}+\frac{j}{\sqrt{2}}\right) & -1 & \left(\frac{1}{\sqrt{2}}-\frac{j}{\sqrt{2}}\right) & j & \left(-\frac{1}{\sqrt{2}}-\frac{j}{\sqrt{2}}\right) \\
1 & j & -1 & -j & 1 & j & -1 & -j \\
1 & \left(\frac{1}{\sqrt{2}}+\frac{j}{\sqrt{2}}\right) & j & \left(-\frac{1}{\sqrt{2}}+\frac{j}{\sqrt{2}}\right) & -1 & \left(-\frac{1}{\sqrt{2}}-\frac{j}{\sqrt{2}}\right) & -j & \left(\frac{1}{\sqrt{2}}-\frac{j}{\sqrt{2}}\right)
\end{bmatrix} \begin{bmatrix} 1 \\ 1 \\ 1 \\ 0 \\ 0 \\ 0 \\ 0 \\ 0 \end{bmatrix} \tag{7.24}
$$

$$
\begin{aligned}
X(0) &= 3 = 3\angle 0 \\
X(1) &= 1 + \left(\frac{1}{\sqrt{2}} - \frac{j}{\sqrt{2}}\right) - j = 1.707 - j1.707 = 2.414\angle - \pi/4 \\
X(2) &= 1 - j - 1 = -j = 1\angle - \pi/2 \\
X(3) &= 1 + \left(-\frac{1}{\sqrt{2}} - \frac{j}{\sqrt{2}}\right) + j = (1 - 0.707) + j(1 - 0.707) = 0.293 + j0.293 \\
&= 0.4143\angle \pi/4 \\
X(4) &= 1 - 1 + 1 = 1 = 1\angle 0 \\
X(5) &= 1 + \left(-\frac{1}{\sqrt{2}} + \frac{j}{\sqrt{2}}\right) - j = 0.293 - j0.293 = 0.4143\angle - \pi/4 \\
X(6) &= 1 + j - 1 = j = 1\angle \pi/2 \\
X(7) &= 1 + \left(\frac{1}{\sqrt{2}} + \frac{j}{\sqrt{2}}\right) + j = 1.707 + j1.707 = 2.414\angle \pi/4
\end{aligned}
$$

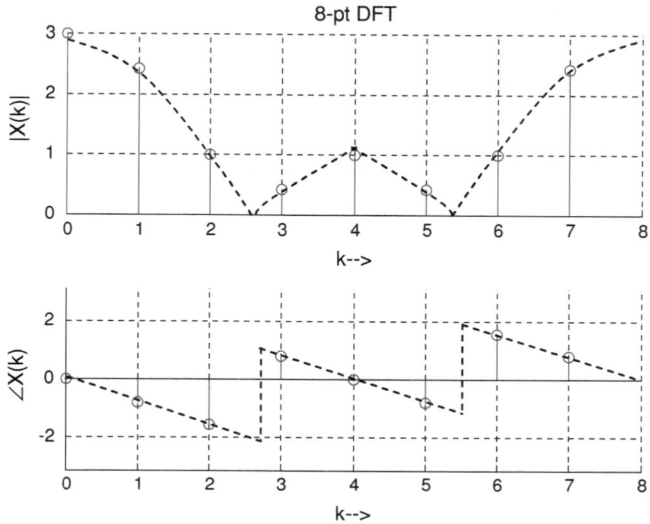

Fig. 7.7 Magnitude and phase spectrum of 8 pt DFT

From the above set of equations, we can separately express the magnitude and phase response of DFT as (Fig. 7.7)

$$\begin{aligned} |X(k)| &= \{3, 2.414, 1, 0.4143, 1, 0.4143, 1, 2.414\} \\ \angle X(k) &= \{0, -\pi/4, -\pi/2, \pi/4, 0, -\pi/4, \pi/2, \pi/4\} \end{aligned} \tag{7.25}$$

Comments:
Based on the two DFT spectra (4 point and 8 point) obtained from the same sequence, we can notice that it is difficult to extrapolate the entire frequency spectrum with N = 4. In that case the "spectrum resolution" is very poor. In order to increase the resolution, we must increase the value of N. In Fig. 7.7, it is possible to extrapolate the spectrum where N = 8. Therefore, we can conclude that, zero padding gives us high density spectrum and provides a better displayed version of plotting.

7.4 Properties of DFT

From the necessity point of view of discrete Fourier transform, we have already seen that, it is actually a truncated part of a continuous periodic spectrum. From the very concept itself, lots of properties are observed. Some of the interesting properties of DFT that are used to process a finite duration sequence, are discussed.

7.4.1 Periodicity

If $X(k)$ be the N-point DFT of a finite duration sequence $x(n)$, then

$$\begin{aligned} x(n+N) &= x(n); \forall n \\ X(k+N) &= X(k); \forall k \end{aligned} \tag{7.26}$$

7.4.2 Linearity

If two finite duration sequences $x_1(n)$ and $x_2(n)$ are linearly combined, i.e.,

$$x_3(n) = ax_1(n) + bx_2(n) \tag{7.27}$$

where, a and b are constants.

Then the DFT of $x_3(n)$ is

$$X_3(k) = aX_1(k) + bX_2(k) \tag{7.28}$$

If $x_1(n)$ and $x_2(n)$ are of length N_1 and N_2 respectively, then the length of the *weighted sequence* $x_3(n)$ will be $N_3 = \max(N_1, N_2)$.

7.4.3 Circular Shift of a Sequence

When we have approached towards the Discrete Fourier Transform from the Discrete Time Fourier Transform in the Sect. 7.2, we have understood that the Fourier transform (DTFT) obtained from discrete time sequence is a periodic one with periodicity 2π in ω. We know that, when continuous time signal is sampled with sampling interval T_s, the spectrum of the resulting discrete time sequence becomes a periodic function of frequency with period $2\pi/T_s$. Similarly, the basic time frequency duality property of Fourier transform suggests that, when $X(\omega)$ is sampled with period $2\pi/N$, the corresponding discrete time sequence $x_p(n)$ becomes periodic in time domain with periodicity N, as expressed in Eq. 7.29 (Fig. 7.8).

$$x_p(n) = \sum_{l=-\infty}^{\infty} x(n - lN) \tag{7.29}$$

To visualize any periodic sequence like $x_p(n)$ is wrapping the finite duration sequence $x(n)$ around a circle in counter clockwise direction. By traversing the periphery of the circle repeatedly, we see the sequence repeats on a circular (modulo N) time axis. Thus, $x_p(n)$ can be expressed as

$$x_P(n) = x[(n \bmod uloN)] \tag{7.30}$$

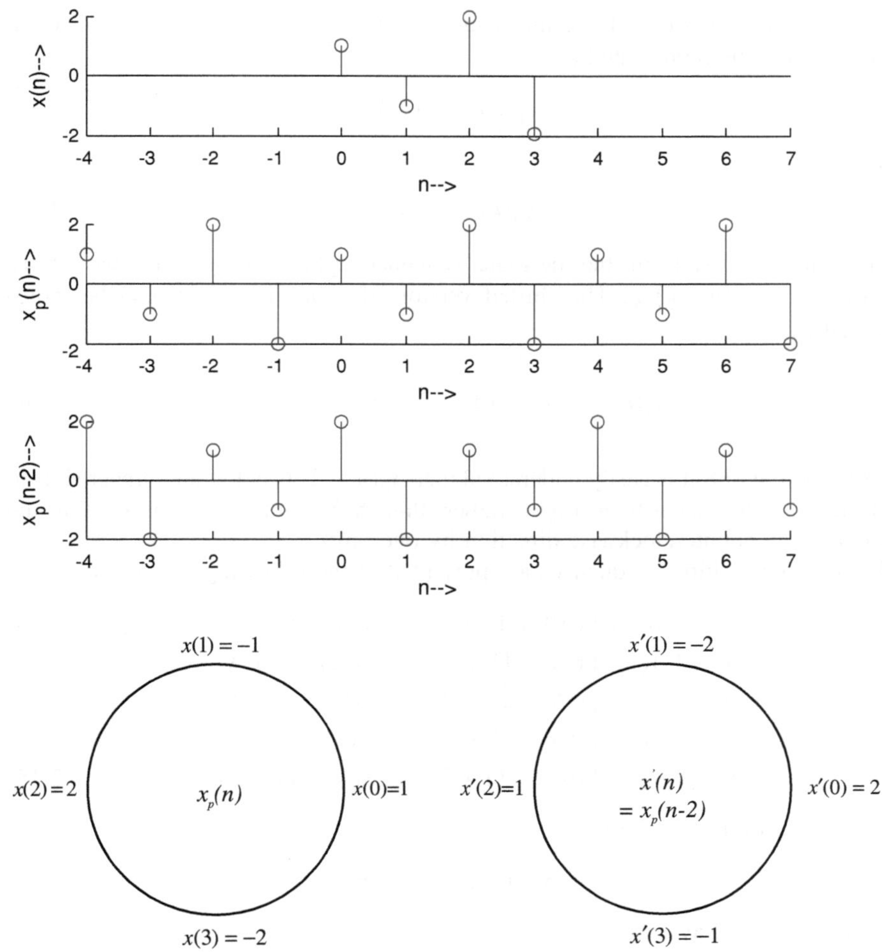

Fig. 7.8 Circular shift of a sequence represented by a unit circle

The above equation is valid if length of $x(n)$ is less than or equal (more likely) to N. If the argument n is more than or equal to N, we have to add or subtract suitable multiples of N from n until the result is between 0 and N − 1, as illustrated in the following examples.
Examples:

$$\begin{aligned} x(-3 \bmod ulo4) &= x(1) \\ x(10 \bmod ulo8) &= x(2) \\ x(-11 \bmod ulo4) &= x(1) \end{aligned} \tag{7.31}$$

For convenience, we will use the notation $((n))_N$ to denote n modulo N. Thus, Eq. 7.30 can be represented as

$$x_p(n) = x((n))_N \tag{7.32}$$

Similarly,

$$X_p(k) = X((k))_N \tag{7.33}$$

The shifted version of the periodic sequence $x_p(n)$ can now be written as $x_p(n-k) = x((n-k))_N$. The shifted version of $x_p(n)$, i.e., $x'_p(n)$ can be represented as

$$x'_p(n) = x_p(n-k) = \sum_{l=-\infty}^{\infty} x(n - lN - k) \tag{7.34}$$

From the Fig. 7.8 it is easily understood that, $x_p(n-2)$ is not at all a linear shift to the right with respect to n origin, rather, this shift is interpreted as shifting the sequence in counterclockwise direction by two units.
The circularly shifted sequences are presented in the following examples.

$$\begin{aligned} x(n) &= \{x(0), x(1), x(2), \ldots, x(N-1)\} \\ x((n-1))_N &= \{x(N-1), x(0), x(1), x(2), \ldots, x(N-2)\} \\ x((n-2))_N &= \{x(N-2), x(0), x(1), x(2), \ldots, x(N-3)\} \\ x((n-k))_N &= \{x(N-k), x(N-k+1), \ldots, x(N-k-1)\} \\ x((n-N))_N &= \{x(0), x(1), x(2), \ldots, x(N-1)\} = x(n) \end{aligned} \tag{7.35}$$

As seen in the Eq. 7.35,

$$x((n-N))_N = x(n)$$

In the same way,

$$x((n-m))_N = x(N-m+n) \tag{7.36}$$

Now, we can prove that,
IF

$$DFT[x(n)] = X(k)$$

Then,

$$DFT\left[x((n-m))_N\right] = e^{-j2\pi km/N} X(k) \tag{7.37}$$

Proof

$$DFT\left[x((n-m))_N\right] = \sum_{n=0}^{N-1} x((n-m))_N e^{\frac{-j2\pi kn}{N}}$$
$$= \sum_{n=0}^{m-1} x((n-m))_N e^{\frac{-j2\pi kn}{N}} + \sum_{n=m}^{N-1} x((n-m))_N e^{\frac{-j2\pi kn}{N}} \quad (7.38)$$

But,

$$x((n-m))_N = x(N-m+n)$$

Therefore,

$$\sum_{n=0}^{m-1} x((n-m))_N e^{\frac{-j2\pi kn}{N}} = \sum_{n=0}^{m-1} x(N-m+n) e^{\frac{-j2\pi kn}{N}}$$
$$= \sum_{l=N-m}^{N-1} x(l) e^{\frac{-j2\pi k(-N+m+l)}{N}} = \sum_{l=N-m}^{N-1} x(l) e^{\frac{-j2\pi k(m+l)}{N}} \quad (7.39)$$

Similarly,

$$\sum_{n=m}^{N-1} x((n-m))_N e^{\frac{-j2\pi kn}{N}} = \sum_{l=0}^{N-1-m} x((l))_N e^{\frac{-j2\pi k(m+l)}{N}} \quad (7.40)$$

Substituting Eq. 7.39 and Eq. 7.40 into Eq. 7.38,

$$DFT\left[x((n-m))_N\right] = \sum_{l=N-m}^{N-1} x(l) e^{\frac{-j2\pi k(m+l)}{N}} + \sum_{l=0}^{N-1-m} x((l))_N e^{\frac{-j2\pi k(m+l)}{N}}$$
$$= e^{\frac{-j2\pi km}{N}} \sum_{l=0}^{N-1} x(l) e^{\frac{-j2\pi kl}{N}} = e^{\frac{-j2\pi km}{N}} X(k) \quad (7.41)$$

7.4.4 Time Reversal of a Sequence

The time reversal of an N-point sequence $x(n)$ is attained by wrapping the sequence x(n) around the circle in clockwise direction. It is denoted as $x((-n))_N$.

$$x((-n))_N = x(N-n); 0 \le n \le N-1 \quad (7.42)$$

If

$$DFT[x(n)] = X(k)$$

Then

$$DFT\left[x((-n))_N\right] = DFT[x(N-n)] = X((-k))_N = X(N-k) \tag{7.43}$$

Proof

$$\begin{aligned} DFT[x(N-n)] &= \sum_{n=0}^{N-1} x(N-n)e^{\frac{-j2\pi kn}{N}} \\ &= \sum_{m=0}^{N-1} x(m)e^{\frac{-j2\pi k(N-m)}{N}} = \sum_{m=0}^{N-1} x(m)e^{\frac{j2\pi km}{N}} \\ &= \sum_{m=0}^{N-1} x(m)e^{\frac{-j2\pi m(N-k)}{N}} = X(N-k) \end{aligned} \tag{7.44}$$

7.4.5 Circular Frequency Shift

If

$$DFT[x(n)] = X(k)$$

Then

$$DFT\left[x(n)e^{\frac{2\pi \ln}{N}}\right] = X((k-l))_N \tag{7.45}$$

Proof

$$\begin{aligned} DFT\left[x(n)e^{\frac{2\pi \ln}{N}}\right] &= \sum_{n=0}^{N-1} x(n)e^{\frac{-j2\pi kn}{N}}e^{\frac{2\pi \ln}{N}} = \sum_{n=0}^{N-1} x(n)e^{\frac{-j2\pi n(k-l)}{N}} \\ &= X((k-l))_N \end{aligned}$$

7.4.6 Complex Conjugate Property

If

$$DFT[x(n)] = X(k)$$

Then

$$DFT[x^*(n)] = X^*(N-k) = X^*((-k))_N \tag{7.46}$$

Proof

$$DFT[x^*(n)] = \sum_{n=0}^{N-1} x^*(n)e^{\frac{-j2\pi kn}{N}} = \left[\sum_{n=0}^{N-1} x(n)e^{\frac{j2\pi kn}{N}}\right]^*$$

$$= \left[\sum_{n=0}^{N-1} x(n)e^{\frac{-j2\pi n(N-k)}{N}}\right]^* = X^*(N-k) = X^*((-k))_N$$

7.4.7 Circular Convolution

Let, $x_1(n)$ and $x_2(n)$ are finite duration sequences both of length N with DFTs $X_1(k)$ and $X_2(k)$. Now, we find a sequence $x_3(n)$ for which DFT is $X_3(k)$Where,

$$X_3(k) = X_1(k)X_2(k) \tag{7.47}$$

We know from the concept of periodic convolution,

$$x_{3p}(n) = \sum_{m=0}^{N-1} x_{1p}(m)x_{2p}(n-m)$$

$$\Rightarrow x_3((n))_N = \sum_{m=0}^{N-1} x_1((m))_N x_2((n-m))_N \tag{7.48}$$

For $0 \le m \le N-1$, $x_3((n))_N = x_3(n)$ and $x_1((m))_N = x_1(m)$
Therefore,

$$x_3(n) = \sum_{m=0}^{N-1} x_1(m)x_2((n-m))_N \tag{7.49}$$

The above equation represents the circular convolution represented as

$$x_3(n) = \sum_{m=0}^{N-1} x_1(m)x_2((n-m))_N = x_1(n)\langle N\rangle x_2(n) \tag{7.50}$$

From Eqs. 7.48 and 7.50, we can therefore write,

$$DFT[x_1(n)\langle N\rangle x_2(n)] = X_1(k)X_2(k) \tag{7.51}$$

7.4.8 Circular Correlation

Let, $x_1(n)$ and $x_2(n)$ are finite duration complex valued sequences both of length N with DFTs $X_1(k)$ and $X_2(k)$.

Then

$$DFT[\tilde{r}_{x_1x_2}(l)] = DFT\left[\sum_{m=0}^{N-1} x_1(m)x_2^*((m-l))_N\right] = X_1(k)X_2^*(k) \tag{7.52}$$

7.4.9 Multiplication Between Two Sequences

If

$$DFT[x_1(n)] = X_1(k)$$
$$DFT[x_2(n)] = X_2(k)$$

Then

$$DFT[x_1(n)x_2(n)] = \frac{X_1(k)\langle N\rangle X_2(k)}{N} \tag{7.53}$$

7.4.10 Perseval's Theorem

If

$$DFT[x_1(n)] = X_1(k)$$
$$DFT[x_2(n)] = X_2(k)$$

Then

$$DFT\left[\sum_{n=0}^{N-1} x_1(n)x_2^*(n)\right] = \frac{1}{N}\sum_{k=0}^{N-1} X_1(k)X_2^*(k) \tag{7.54}$$

7.5 Two Dimensional (2D) DFT

As seen from the previous discussion, the transformed domain representation of an electrical signal (obviously as a function of time) can be obtained by Fourier transform, in general. If the system, through which the transformed domain (frequency domain) signal is to be processed (like filtering), is a digital signal processor, then sampling in frequency domain is necessary and that process is called as Discrete Fourier transform (DFT). In the present section, we are going to discuss about frequency concept where the concept of time (like electrical signal) is absent. Here one digital image is considered which is essentially a 2D discrete signal represented not as a function of time, but as a function of space (x and y).

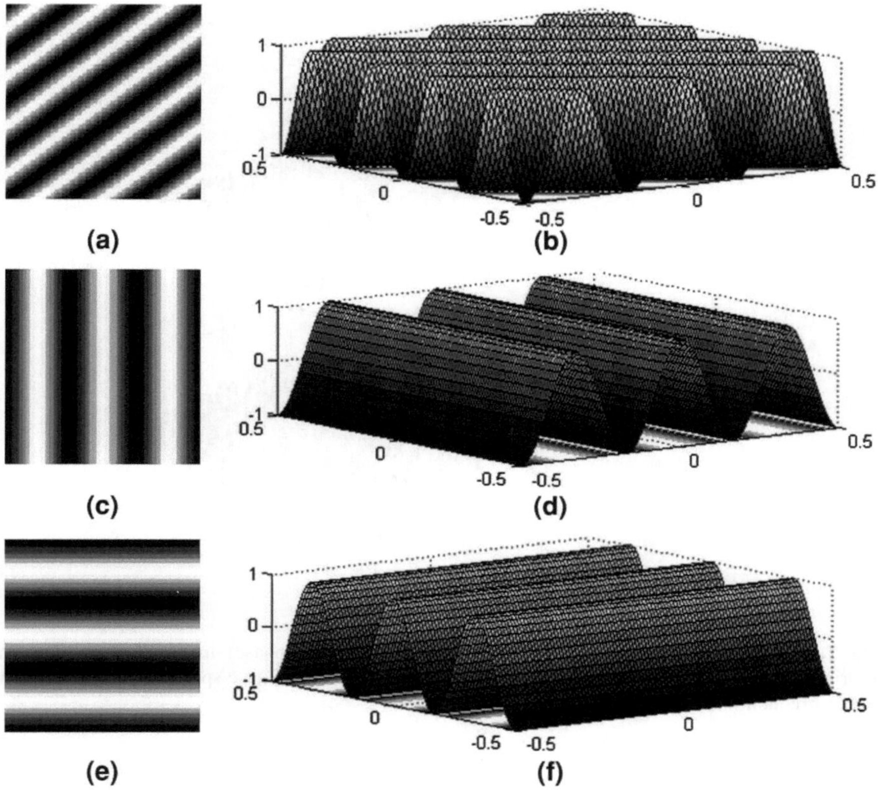

Fig. 7.9 Concept of 'spatial frequency' in images **a** Image showing frequency in Diagonal direction, **b** surface plot of (**a**), **c** Image showing frequency in X-direction, **d** surface plot of (**c**), **e** Image showing frequency in Y-direction

Some images and their surface plots are shown in the Fig. 7.9. The concept of single tone two dimensional frequency can be understood from the figures. The first image (Fig. 7.9a) is having sinusoid behavior both in row (x direction) and column (y direction), therefore the 3D[3] surface plot (Fig. 7.9b) sinusoid appears in the diagonal direction. In the next two images (Fig. 7.9c, e), frequency concept can be analyzed in either x-direction of y direction. The surface plots (Fig. 7.9d, f) are the evidences. Thus, we confirm the claim that, static image has also the concept of frequency, though the concept of time does not exist for static images. This particular frequency concept is known as '*spatial frequency*' [5] of images.

In general, we can represent a 2D spatial signal in frequency domain as given in the following equations.

[3] Dimension 1: X-direction of the image, Dimension 2: Y-direction of the image, Dimension 3: Intensity of the image.

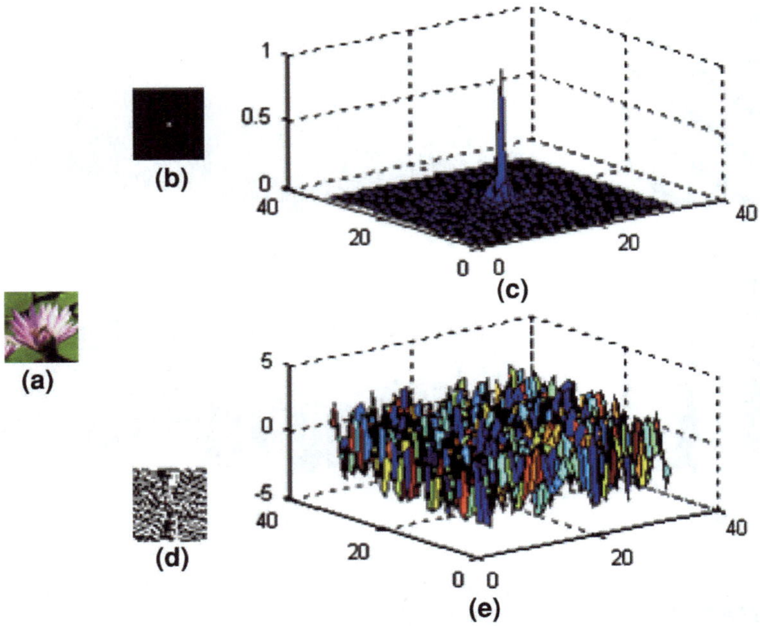

Fig. 7.10 An image is forming two images (magnitude and phase) in transformed domain **a** Original image, **b** Magnitude spectrum, **c** surface plot of (**b**), **d** Phase spectrum, **e** Surface plot of (**d**) Frequency expansion

$$F(u,v) = \sum_{x=0}^{R-1}\sum_{y=0}^{C-1} f(x,y)e^{-j2\pi(uxx_0+vyy_0)} \tag{7.55}$$

$$f(x,y) = \frac{1}{UV}\int_{0}^{U-1}\int_{0}^{V-1} F(u,v)e^{j2\pi(uxx_0+vyy_0)}dudv \tag{7.56}$$

where x_0 and y_0 are the spatial intervals between consecutive signal samples in the x and y directions, respectively, and $U = 1/x_0$ and $V = 1/y_0$ are sampling rates in the two directions, and they are also the periods of the spectrum $F(u, v)$. R and C are the number of rows and columns in the digital image.

From the basic discussion of DTFT, we are now convinced that, transformed spectrum of discrete samples in time or space domain always generates periodic signal with periodicity 2π. The concept is still true for two dimensional signals like image. Here, two dimensional DFT will also create two dimensional frequency samples. The 2D frequency samples (with the same size of the image, i.e., number of rows and columns of image will be similar to the number of vertical and horizontal frequency samples respectively. If we just plot the magnitude and phase part of the 2D DFT spectrum samples, that would generate two 2D signals, i.e., images, as shown in Fig. 7.10. From the figure above, it is observed that, the magnitude spectrum is more or less homogeneous, whereas the phase spectrum has a good number of variations. In other words, we can say the phase spectrum has higher amount of *entropy*. As we know, [6] entropy[4] is the measure of information, the phase spectrum must contain significant amount of information about the image. We'll support our claim in the section of case studies.

The image is represented as aperiodic discrete signal (function of space) $f(x,y)$ which generates the frequency samples $F(k,l)$ by two dimensional DFT, as represented in the following equations.

$$F(k,l) = \frac{1}{\sqrt{MN}} \sum_{y=0}^{N-1} \sum_{x=0}^{M-1} f(x,y) e^{-j2\pi\left(\frac{mk}{M}+\frac{nl}{N}\right)} \tag{7.57}$$

On the other hand, the original image can also be retrieved from the frequency samples (2D DFT, both magnitude and phase) by applying 2D IDFT as expressed in the following equation.

$$f(x,y) = \frac{1}{\sqrt{MN}} \sum_{l=0}^{N-1} \sum_{k=0}^{M-1} F(k,l) e^{j2\pi\left(\frac{mk}{M}+\frac{nl}{N}\right)} \tag{7.58}$$

where, M and N are the number of frequency samples in x and y directions, respectively. Both $F(k,l)$ and $f(x,y)$ are of same size $M \times N$, i.e., the size of the image.

7.5.1 Physical Interpretation: 2D-FT

To interpret physically, here we have considered the Fourier transform of continuous but non-periodic signal (like analog 2D signal) which can be easily generalized to digital image, too. The space domain inverse transform can be obtained from the following equation.

$$f(x,y) = \int_{-\infty}^{\infty} \int_{-\infty}^{\infty} F(u,v) e^{j2\pi(xu+yv)} du dv \tag{7.59}$$

where, u and v are the frequencies in the x and y directions, respectively. The double integration is a weighted[5] linear combination of the exponent

$$e^{j2\pi(xu+yv)} = \cos\{2\pi(xu+yv)\} + j\sin\{2\pi(xu+yv)\} \tag{7.60}$$

4 Entropy is measured as non-homogeneity or randomness in a signal/ message [6].

5 weight $F(u,v)$.

To understand the physical meaning of this 2D Fourier transform, we first consider the complex exponential $e^{j2\pi(xu+yv)}$ and its complex weight $F(u,v)$ separately:

7.5.1.1 The Complex Exponential

The complex exponential is composed of two sinusoidal functions:

$$e^{j2\pi(xu+yv)} = \cos\{2\pi(xu+yv)\} + j\sin\{2\pi(xu+yv)\} \tag{7.61}$$

First we define the following:

$$\begin{cases} \omega = \sqrt{u^2+v^2} \\ \theta = \tan^{-1}\left(\dfrac{v}{u}\right) \end{cases} \text{and} \begin{cases} u = \omega\cos\theta \\ v = \omega\sin\theta \end{cases} \tag{7.62}$$

We can now define two vectors:

(a) The vector associated with point (x,y) in the 2D spatial domain:

$$\vec{r} = (x,y) \tag{7.63}$$

(b) The unit vector along the direction of point (u,v) in the 2D spatial frequency domain:

$$\vec{n} = (\cos\theta, \sin\theta) = \left(\frac{u}{\omega}, \frac{v}{\omega}\right) \tag{7.64}$$

Therefore, the complex exponential $e^{j2\pi(xu+yv)}$ can be expressed in terms of these two vectors, as

$$\begin{aligned} e^{j2\pi(xu+yv)} &= \cos\{2\pi(xu+yv)\} + j\sin\{2\pi(xu+yv)\} \\ &= \cos\{2\pi\omega(\vec{r}.\vec{n})\} + j\sin\{2\pi\omega(\vec{r}.\vec{n})\} \end{aligned} \tag{7.65}$$

where, $(\vec{r}.\vec{n})$ is the inner product of the two vectors, representing the projection of point $\vec{r} = (x,y)$ onto the direction of $\vec{n}$. The function value $\cos\{2\pi(xu+yv)\} = \cos\{2\pi\omega(\vec{r}.\vec{n})\}$ at any spatial point $\vec{r} = (x,y)$ is the same as that of its projection $(\vec{r}.\vec{n})$ on the direction $\vec{n}$.) In other words, the function $\cos\{2\pi(xu+yv)\}$ represents a planar sinusoid in the x–y plane with (Fig. 7.11)

1. **Frequency** $\omega = \sqrt{u^2+v^2}$
2. **Direction** $\theta = \tan^{-1}\left(\frac{v}{u}\right)$ (i.e., $\vec{n}$.)

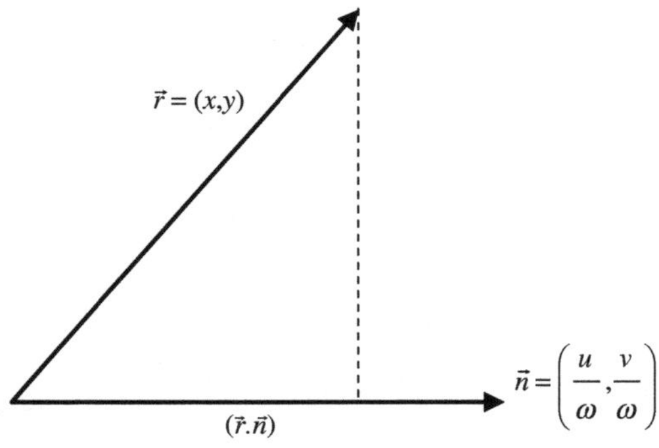

Fig. 7.11 Interpretation of 2D-FT as projection of point $\vec{r} = (x, y)$ onto the direction of $\vec{n}$

Example 7.2 Compute 2D-FT of the given function (Fig. 7.12)

$$f(x,y) = \begin{cases} 1; & -\frac{a}{2} \leq x \leq \frac{a}{2}, -\frac{b}{2} \leq y \leq \frac{b}{2} \\ 0; & elsewhere \end{cases}$$

$$F(u,v) = \int_{-\infty}^{\infty} \int_{-\infty}^{\infty} f(x,y) e^{-j2\pi(ux+vy)} dxdy$$

$$= \int_{-a/2}^{a/2} e^{-j2\pi ux} dx \int_{-b/2}^{b/2} e^{-j2\pi vy} dx = \frac{\sin(\pi ua)}{\pi u} \frac{\sin(\pi vb)}{\pi v}$$

7.5.2 Space-Frequency Expansion-Contraction in Image

The problem solved in the Example 7.2 is very important from the image processing point of view. As discussed in the continuous time Fourier transform (Chap. 3), we have shown that, as we go on reducing the width of a rectangular pulse (gate function), the width of the *sinc* spectrum is increased. From that very point of understanding, we have come to an interesting conclusion that a time restricted signal is frequency expanded.

In the previous sub-section, we have introduced the concept of spatial frequency in image. Therefore, here the gate function corresponds to a rectangular white patch placed at the center. In the Fig. 7.13, we have shown three different images and their corresponding two dimensional DFTs. In image of Fig. 7.13a, the white patch is of higher width in the X-direction. The corresponding DFT image as

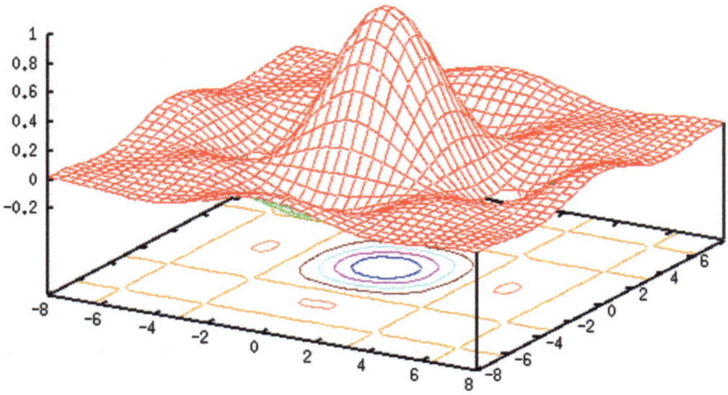

Fig. 7.12 2D FT of f(x,y) given in example 7.2

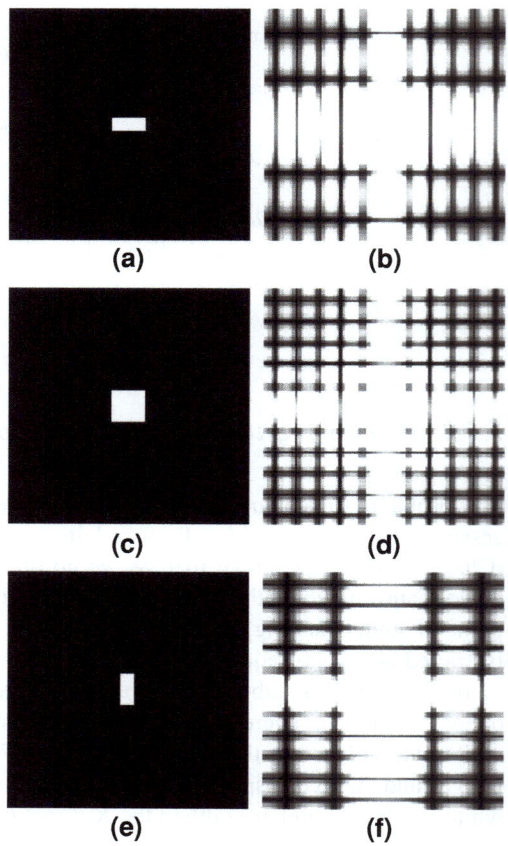

Fig. 7.13 Space contraction in image intensity (any direction) corresponds to frequency expansion: image table

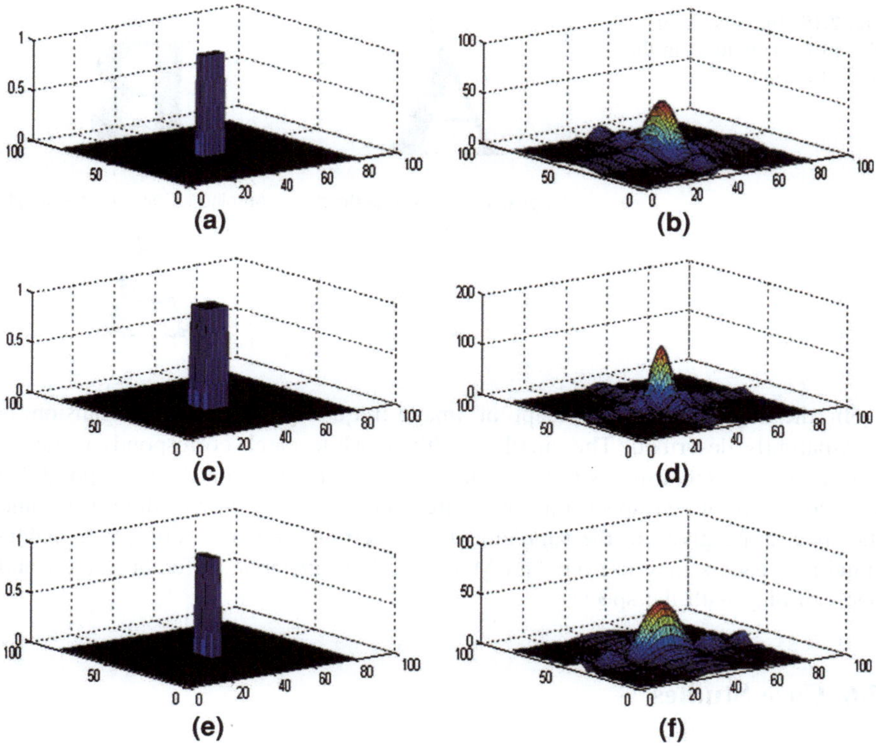

Fig. 7.14 Space contraction in image intensity (any direction) corresponds to frequency expansion: surface table

shown in Fig. 7.13b, shows higher width in the Y-direction. The image pair of Fig. 7.13e and f is the evidence of the contraction-expansion concept. It is just showing the opposite width contraction for opposite width rectangle. In the second image as shown in Fig. 7.13c, the central patch is a perfect square. So, the DFT image (Fig. 7.13d) is showing same width of sing type characteristics in both the directions.

We have seen some alternative black and white patches in the DFT spectrums. To understand the shape of the two dimensional (2D) DFT spectrum, the image view is not sufficient. The opaqueness is arrived for introduction of a three dimensional concept of signal representation. Dimension one represents X-direction (number of columns) of the image, dimension two represents Y-direction (number of rows) of the image, the third dimension is also there which is the image intensity represented as function of X and Y.

Thus, to understand the third dimension, we have taken the help of surface plot of the image as well as the DFT. The images of Fig. 7.14 are taken here for illustration.

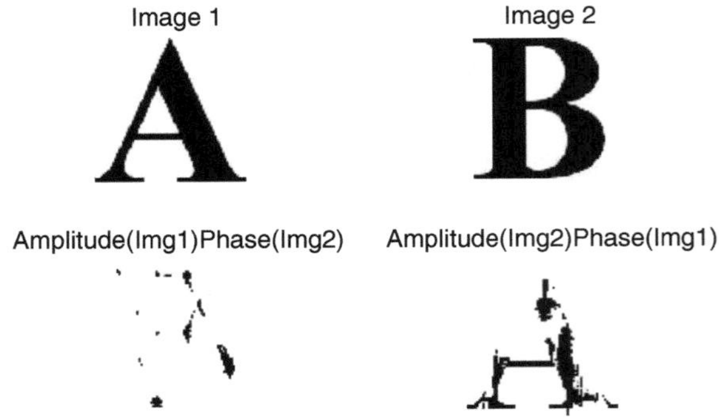

Fig. 7.15 Importance of phase over amplitude in the DFT spectrum

In the Fig. 7.14, the concept of image-frequency contraction-expansion is transparently described. The smaller width of white patch corresponds to larger width of *sinc* spectrum; similarly larger width of white patch corresponds to smalller width of *sinc* spectrum. The source code for generation of the images and the spectra are given in the supplementary electronic material. For better understanding, readers may use the said Matlab code to generate different images and one can play with the spectra.

7.6 Case Studies

7.6.1 Importance of Phase Over Amplitude in DFT Spectrum

As understood from the entire discussion of continuous time Fourier transform (CTFT), discrete time fourier transform (DTFT) and finally discrete Fourier transform (DFT) the spectra generated by the analysis are always complex. Therefore each of the spectrums must have one 2D amplitude and one 2D phase image correspondence (spectra). In this particular case, we have taken two images, computed the DFT over both of them and interchanged the amplitude and phase parts between the images. We can think of having two face images; one of a tiger and another of a human. Say, the phase spectrum of the tiger is combined with the amplitude spectrum of the human face. The newly developed image (frequency domain) is now passed through the inverse DFT (IDFT) operation. Guess, what happened. Astonishingly, we shall have the prominent tiger face and not the human face at the spatial domain, though the entire magnitude part of the human face is present in frequency domain (Fig. 7.15). It justifies the fact the importance of phase is higher than the importance of magnitude in any type of signal. In the previous chapter of continuous time Fourier series (CTFS) also we have shown how by the method of Kovesi, the phase crossover can detect significant features like edges from an image. The present experiment is the generalized validation of the claim.

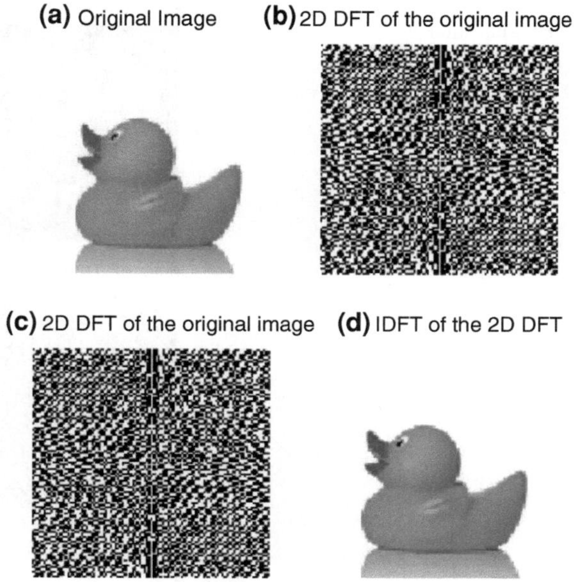

Fig. 7.16 Illustration of 2D IDFT over complete spectrum to recover undistorted image

The present experiment is given at the rear end of the present chapter. Readers can play with the Matlab source code customizing the input images, for better understanding.

7.6.2 Image Filtering

To understand the power of DFT we have presented the concept of image filtering in a non-conventional way. In the Fig. 7.16, we have taken an image of a duck. The 2D DFT spectrum is presented in the figure. When we apply inverse DFT (IDFT) over the spectrum, the original duck image is recovered. Now, the central shifted DFT spectrum is taken (Fig. 7.17a) and the central part is taken by multiplying a centre on circular mask of same size. The central frequency represents low frequency components of the image. Next, IDFT is done over the center on truncated 2D DFT. We have got the duck image, but the edges, i.e., the intensity transition parts are blurred (Fig. 7.17b). This is easily understandable that the edged are the high frequency regions, as there certain change in intensity occurs with respect to small change in space. But in the inverse transformed image, the constant illumination parts of the image, like body of the duck, shadows, are prominent (Fig. 7.17b).

In the next experiment we have done just the opposite. We have removed the centre circular part of the DFT spectrum (Fig. 7.17c) and IDFT is operated. The resultant image consists of the edges (sharp intensity transitions) only. The border

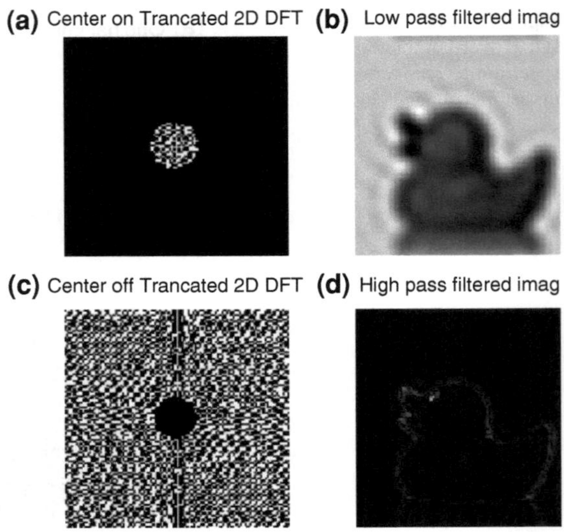

Fig. 7.17 Image filtering by DFT truncation

line of duck, eye and border of shadow is detected. But the constant illumination parts are absent as desired (Fig. 7.17d).

In the supplementary electronic content, the source code is given in Matlab. The radius of the circle can be changed for the same image or any other type of images for allowing a desired amount of low frequency to high frequency ratio of an image.

7.7 Computational Complexity

Looking at the twiddle matrix as presented in the Eqs. 7.8 and 7.10, in the operation of DFT and IDFT both, the multiplicand matrix size is $N \times N$. The DFT operation using twiddle matrix is shown below.

$$\begin{bmatrix} X(0) \\ X(1) \\ X(2) \\ \vdots \\ X(N-1) \end{bmatrix} = \begin{bmatrix} 1 & 1 & 1 & \cdots & 1 \\ 1 & W_N^1 & W_N^2 & \cdots & W_N^{(N-1)} \\ 1 & W_N^2 & W_N^4 & \cdots & W_N^{2(N-1)} \\ \vdots & \vdots & \vdots & & \vdots \\ 1 & W_N^{(N-1)} & W_N^{2(N-1)} & \cdots & W_N^{(N-1)^2} \end{bmatrix} \begin{bmatrix} x(0) \\ x(1) \\ x(2) \\ \vdots \\ x(N-1) \end{bmatrix}$$

The pattern of inverse twiddle matrix used for IDFT is very similar to that of the twiddle matrix used for DFT. As seen from the above equation, all the elements of the first row and first column of the twiddle matrix (of size $N \times N$) is '1'. Therefore, that particular row will be responsible for real multiplication not complex multiplication. From the processor or system (like mobile phones, PC and many other electronic handheld devices) point of view the computational cost for complex multiplication and addition is much higher than that of real multiplication and addition.

7.7.1 Considering Real and Complex Operations

7.7.1.1 Total number of multiplications

From the matrix shown above (Eq. 7.8) for each row, the algorithm does N number of multiplications. Therefore, for the total matrix (N number of rows) total number of multiplications is N^2.

7.7.1.2 Total number of additions

From the matrix shown above (Eq. 7.8) for each row, the algorithm does N number of additions between the multiplied elements $W_N^i \times x(i)$. Therefore, for the total matrix (N number of rows) total number of additions is N^2.

Therefore the total computational cost is = $N^2 + N^2 = 2N^2$.

7.7.2 Considering Only Complex Operations

7.7.2.1 Total number of multiplications

From the matrix shown above (Eq. 7.8) for each row, the algorithm does $(N - 1)$ number of multiplications [4], except the first element of each row. Therefore, for the total matrix ($N - 1$ number of rows except the first row) total number of multiplications is $(N - 1)^2$.

7.7.2.2 Total number of additions

From the matrix shown above (Eq. 7.8) for each row, the algorithm does N number of additions between the multiplied elements $W_N^i \times x(i)$. Therefore, for the total matrix (N number of rows) total number of additions is N^2.

Therefore the total computational cost is = $(N - 1)^2 + N^2$.

Generally the number of samples taken for practical DFT is at least $N = 2^{20}$ (1 Mega Samples). Therefore, neglecting the value 1, we can generalize the computational cost as $2N^2$. It's really huge for high value of N. For $N = 2^{20}$, the total computational cost is $2^{41} = 2{,}199{,}023{,}255{,}552$.

The principle of DFT operation should now be reviewed to increase the speed of the operation. In the next chapter, a fast method of doing DFT, i.e., Fast Fourier Transform (FFT) is discussed. At the conclusion of the next chapter, the computation complexity will be compared with the classical DFT.

7.8 MATLAB Codes

7.8.1 Concept of Frequency in Two Dimensional Signal (Image)

```
% Concept of frequency in 2D signals (images)
% Apurba Das (22/05/2011)
close all;
clear all;
clc;
[x y]=meshgrid(-.5:.01:.5,-.5:.01:.5);
%%
% diagonal
z=cos(2*pi*(3*x+4*y));
figure(1)
subplot(211)
imshow(z,[min(min(z)) max(max(z))])
subplot(212)
surfc(x,y,z)
xlabel('row-->');
ylabel('column-->');
figure(4)
subplot(321)
imshow(z,[min(min(z)) max(max(z))])
subplot(322)
surfc(x,y,z)

%%
% freq.: x direction
figure(2),
z=cos(2*pi*(3*x));
subplot(211)
imshow(z,[min(min(z)) max(max(z))])
subplot(212)
surfc(x,y,z)
xlabel('row-->');
```

```
ylabel('column-->');
figure(4)
subplot(323)
imshow(z,[min(min(z)) max(max(z))])
subplot(324)
surfc(x,y,z)
%%

% freq.: y direction
figure(3),
z=cos(2*pi*(3*y));
subplot(211)
imshow(z,[min(min(z)) max(max(z))])
subplot(212)
surfc(x,y,z)
xlabel('row-->');
ylabel('column-->');
figure(4)
subplot(325)
imshow(z,[min(min(z)) max(max(z))])
subplot(326)
% colormap jet;
surfc(x,y,z)
```

7.8.2 Importance of Phase Over Amplitude in DFT Spectrum

```
% DFT phase and amp test
% Apurba Das
% 31/05/2011

close all;
clc;
clear all;
a=rgb2gray(imread('a.bmp'));
subplot(221)
imshow(a);
```

```
title('Image 1');
a1=fft2(a);
amp_A=abs(a1);
ph_A=angle(a1);
% figure,
% imshow(ph_A);
ph_A_ff=ifft2(ph_A);
% figure,
% imshow(ph_A_ff);

b=rgb2gray(imread('b.bmp'));
subplot(222)
imshow(b);
title('Image 2');

% imshow(b);
b1=fft2(b);
amp_B=abs(b1);
ph_B=angle(b1);
new_ff=(amp_A).*exp(i*ph_B);
new=ifft2(new_ff);
subplot(223)
imshow(new);
title('Amplitude(Img1)Phase(Img2)');

new_ff=(amp_B).*exp(i*ph_A);
new=ifft2(new_ff);
subplot(224)
imshow(new);
title('Amplitude(Img2)Phase(Img1)');
```

7.8.3 Image Filtering

```
% fft --> LPF and HPF
% Apurba Das
% 31/05/2011

[filename, pathname] = uigetfile({'*.bmp';}, ...
'Brows Image');
```

```
file=cat(2,pathname,filename);
a=imread(file);
if size(a,3)==3
    a=rgb2gray(a);
end
[ri ci]=size(a);
figure(1)
subplot(221)
imshow(a);
title('Original Image');
f1=fft2(a);
f2=fftshift(f1);
subplot(222)
f22=abs(f2);
% imshow(abs(f2),[ min(min(f22)) max(max(f22))]);
imshow(f2)
title('2D DFT of the original image');
% surf(f2)
f11=abs(ifft2(f1));
% f11=abs(ifft2(f2));
subplot(223)
imshow(f2)
title('2D DFT of the original image');
subplot(224)
imshow(f11,[ min(min(f11)) max(max(f11))]);
title('IDFT of the 2D DFT');
%%
% pause,
figure;
mask=fspecial('disk',10);
[rm cm]=size(mask);
rm=rm-1;cm=cm-1;
mask1=mask./max(max(mask));
mm=zeros(ri,ci);
mm(((ri/2)-(rm/2)):((ri/2)+(rm/2)),((ci/2)-...
(cm/2)):((ci/2)+(cm/2)))=mask1;
lp=f2.*mm;
% lp=imfilter(f2,mask);
subplot(221)
imshow(lp);
```

```
title('Center on Trancated 2D DFT');
lp1=abs(ifft2(lp));

subplot(222)
imshow(lp1,[ min(min(lp1)) max(max(lp1))]);
title('Low pass filtered image');
mm1=ones(ri,ci)-mm;
hp=f2.*mm1;
% lp=imfilter(f2,mask);
subplot(223)
imshow(hp);
title('Center off Trancated 2D DFT');
hp1=abs(ifft2(hp));
% f11=abs(ifft2(f2));
subplot(224)
imshow(hp1,[ min(min(hp1)) max(max(hp1))]);
title('High pass filtered image');
```

References

1. Cooley, J.W., Tukey, J.W.: An algorithm for the machine calculation of complex Fourier series. Math. Comput. **19**(90), 297–301 (1965)
2. Carl, F.G.: Nachlass: Theoria interpolationis methodo nova tractata, Werke band 3, pp. 265–327. Königliche Gesellschaft der Wissenschaften, Göttingen (1866)
3. Heideman, M.T., Johnson, D.H., Burrus, C.S.: Gauss and the history of the fast Fourier transform. IEEE ASSP Mag. **1**(4), 14–21 (1984)
4. Heideman, M.T., Burrus, C.S.: On the number of multiplications necessary to compute a length-2^n DFT. IEEE Trans. Acoust. Speech. Sig. Proc. **34**(1), 91–95 (1986)
5. Gonzalez, R.C., Woods, R.E.: Digital Image Processing. Pearsen Education, Inc, Upper Saddle River (2002)
6. Das, A.: Digital Communication: Principles and System Modelling. Springer, Berlin (2010)

Chapter 8
Fast Fourier Transform

8.1 Introduction

In the computation of classical Discrete Fourier Transform (DFT), the computational complexity is huge especially for complex multiplication and complex addition the processor takes large amount of time cycles. In the present chapter faster methods of doing DFT [1] are discussed. As we know, to expedite any algorithm, parallel processing is required. Either the entire process or the data can be decomposed in non-overlapping parallel parts and processor can handle them simultaneously. Fast Fourier Transform (FFT) uses data parallelization mostly. Cooley and Tukey [2] have popularized the algorithms which are based on decomposing the transformation into smaller transforms hierarchically and combining them to get the total transform. FFT reduces the time taken drastically by logarithmic factor of number of frequency samples.

8.2 The FFT Algorithm: Radix 2—Decimation is Time

Radix-2 Decimation in Time (DIT) Fast Fourier Transform (FFT) algorithm is operated to compute N-point DFT where N can be expressed as a power of 2, i.e., $N = 2^M$ where M is an integer.

Here $x(n)$ is considered as an N-point sequence[1] where N is expressed as a power of 2. Next, $x(n)$ is decimated in terms of even and odd samples each of length $N/2$ as presented in Eq. 8.1. Suffix 'e' and 'o' indicates even and odd sequence indices respectively.

$$\begin{aligned} x_e(m) &= x(2m); m = 0, 1, 2, \cdots \left({}^N\!/_2 - 1\right) \\ x_o(m) &= x(2m+1); m = 0, 1, 2, \cdots \left({}^N\!/_2 - 1\right) \end{aligned} \tag{8.1}$$

[1] If number of time samples is less than number of frequency samples (N), zero-padding is used to make the lengths of both the sequences equal.

A. Das, *Signal Conditioning*, Signals and Communication Technology,
DOI: 10.1007/978-3-642-28818-0_8, © Springer-Verlag Berlin Heidelberg 2012

The classical N-point DFT is represented by

$$X(k) = \sum_{n=0}^{N-1} x(n)W_N^{nk}; n = 0, 1, 2, \cdots (N-1) \tag{8.2}$$

Separating x(n) into even and odd indices, we obtain

$$\begin{aligned}
X(k) &= \sum_{m=0}^{N/2-1} x(2m)W_N^{2mk} + \sum_{m=0}^{N/2-1} x(2m+1)W_N^{(2m+1)k} \\
\Rightarrow X(k) &= \sum_{m=0}^{N/2-1} x_e(m)W_{N/2}^{mk} + W_N^k \sum_{m=0}^{N/2-1} x_o(m)W_{N/2}^{mk} \\
\Rightarrow X(k) &= X_e(K) + W_N^k X_o(k)
\end{aligned} \tag{8.3}$$

Each of the sums in the Eq. 8.3 is an N/2 point DFT. Therefore, the index of frequency sample, k must have the range from 0 to (N/2) − 1. But, finally our objective is to compute N point DFT. Therefore Eq. 8.3 is not at all sufficient for complete DFT computation. Only for the computation of first N/2 points, Eq. 8.3 is usable. Our objective now would be to compute the other N/2 point DFT simultaneously without wasting extra time cycles. With this objective, we put $(k + N/2)$ replacing k in Eq. 8.3. The range of k should be same as in Eq. 8.3. If that is done, we are actually getting two values from two sections simultaneously (Eqs. 8.3 and 8.4). For $k = 0$ we'll have two different DFT points from two different equations as $k = 0$, $N/2$. By this way, varying k from 0 to N/2, we can get the complete DFT.

$$\begin{aligned}
X\left(k+\frac{N}{2}\right) &= \sum_{m=0}^{N/2-1} x_e(m)W_{N/2}^{m\left(k+\frac{N}{2}\right)} + W_N^{\left(k+\frac{N}{2}\right)} \sum_{m=0}^{N/2-1} x_o(m)W_{N/2}^{m\left(k+\frac{N}{2}\right)} \\
\Rightarrow X\left(k+\frac{N}{2}\right) &= \sum_{m=0}^{N/2-1} x_e(m)W_{N/2}^{mk} - W_N^k \sum_{m=0}^{N/2-1} x_o(m)W_{N/2}^{mk} \\
\Rightarrow X(k) &= X_e(K) - W_N^k X_o(k)
\end{aligned} \tag{8.4}$$

By considering N = 8, we can observe from Eqs. 8.3 and 8.4 that, X(0) & X(4), X(1) & X(5), X(2) & X(6), X(3) & X(7) have same inputs. The operation can be represented using a butterfly diagram as shown in the following figure. In the Fig. 8.1, two point DFT computation using Radix 2 DIF-FFT algorithm is shown. Taking N = 2 from the above two equations, we can write

$$X(0) = x(0) + W_2^0 x(1) = x(0) + x(1) \tag{8.5}$$

$$X(1) = x(0) - W_2^0 x(1) = x(0) - x(1) \tag{8.6}$$

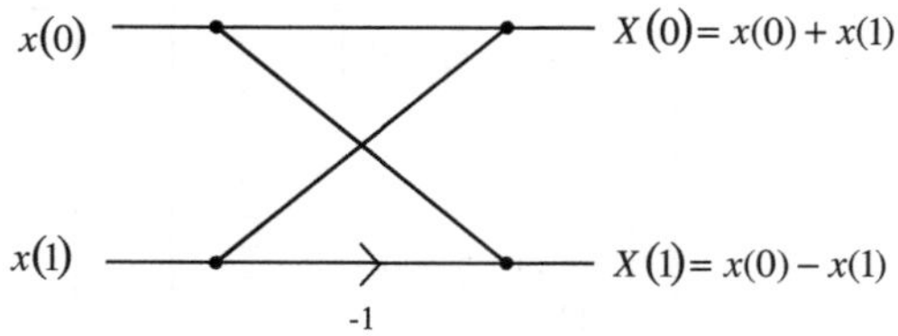

Fig. 8.1 Unit of butterfly representing 2-point DFT using DIT-FFT algorithm

As understood from the methodology of doing Fast Fourier Transform (FFT), it can be considered as a data parallelization algorithm. To compute N point DFT, we can divide the sequences in odd and even clusters in the first layer. Next each of the groups can be further divided in their respective even and odd sub groups and the same process can be realized throughout the computation up to 2-point DFT. The reason of choosing the termination of algorithm at 2-point is obvious as seen from Eqs. 8.5, 8.6 and Fig. 8.1. The aforementioned equations show there is no complex multiplication and addition in the computation of 2-points DFT.The method of the said hierarchical grouping of DFT odd–even sequences is shown in Fig. 8.2.

Four point DFT Computation by Radix 2 DIT-FFT Algorithm
To compute 4-point DFT, the hierarchy of groups depending on the even and odd indices of the sequences is to be done according to a very simple tree. We can directly divide the sequence $\{x(0), x(1), x(2), x(3)\}$ by two parts as $\{x(0), x(2)\}$ and $\{x(1), x(3)\}$. Directly we have reached two-point DFT. So no further divisions are required. Now, by applying Eqs. 8.3 and 8.4 we have realized the butterfly structure of 4-point DFT computation (Fig. 8.3). From the Figs. 8.2 and 8.3, it is understood that, the new algorithm takes care of the speed of computation by employing the concept of data parallelization.

Eight point DFT Computation by Radix 2 DIT-FFT Algorithm
To compute 8-point DFT, the hierarchy of groups depending on the even and odd indices of the sequences is to be done according to the tree shown in Fig. 8.4. The sequences are divided in three layers until we get four 2-point DFT sections as shown in Fig. 8.5.

8.2.1 Bit Reversal

In DIT-FFT algorithm we find that the frequency samples are in order but the time samples are hierarchically decimated by odd and even index of each stage of sample sequences. For an 8-point DFT, the input sequence is $x(0), x(4), x(2), x(6), x(1), x(5), x(3), x(7)$. We can see that $N = 2^{\Omega}; \Omega = 3$ here. The input sequence is formed by a rule of bit reversal which essentially signifies reading each of the indexes from MSB to LSB instead of the proper order, as shown in Table 8.1.

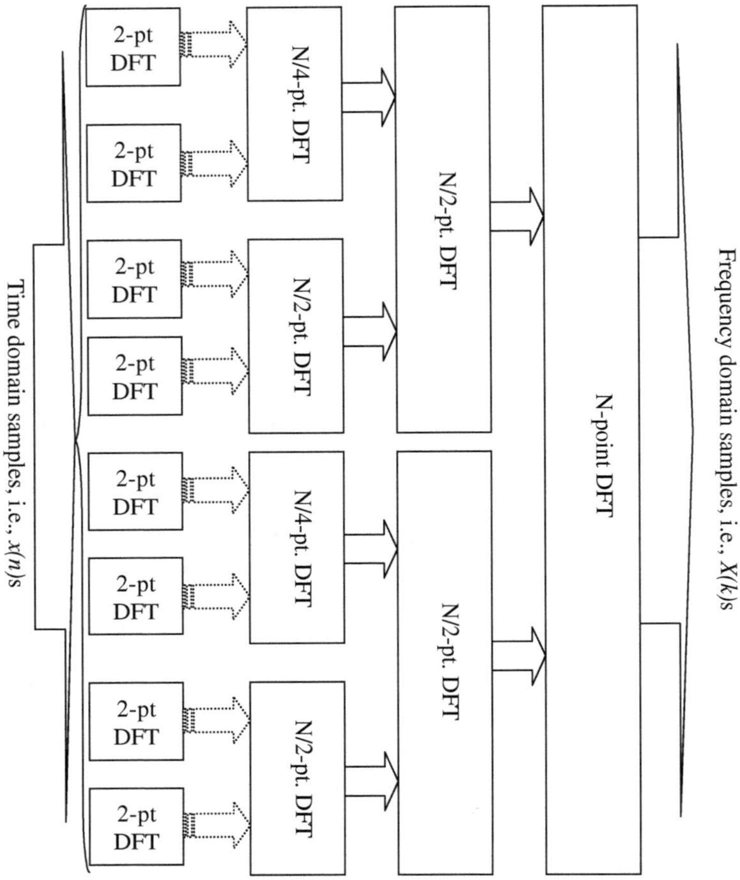

Fig. 8.2 Fast fourier transform as data parallelization

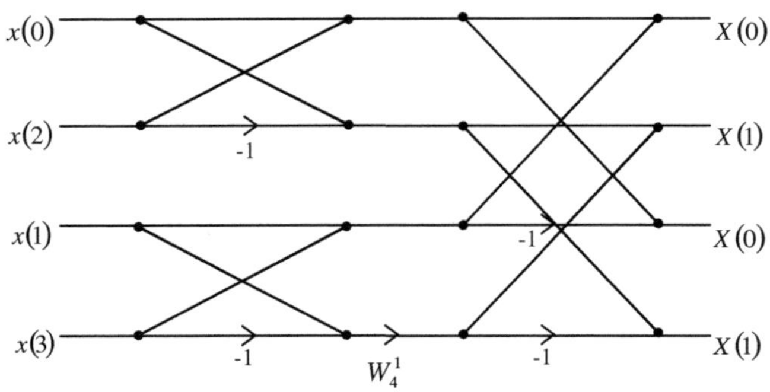

Fig. 8.3 DIT-FFT method to compute 4-point

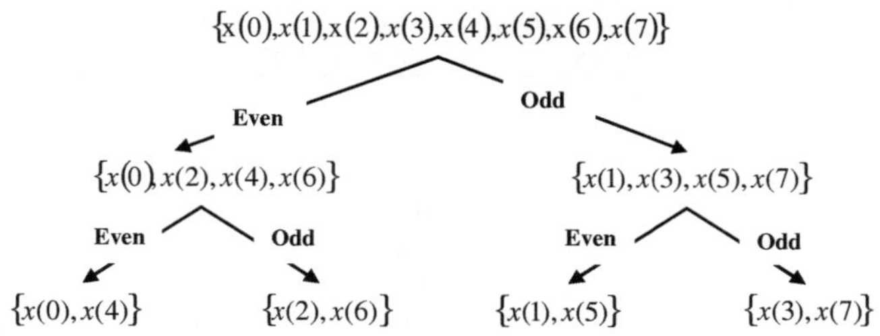

Fig. 8.4 Hierarchical grouping by even and odd sequence index for 8 point Radix 2 DIF-FFT

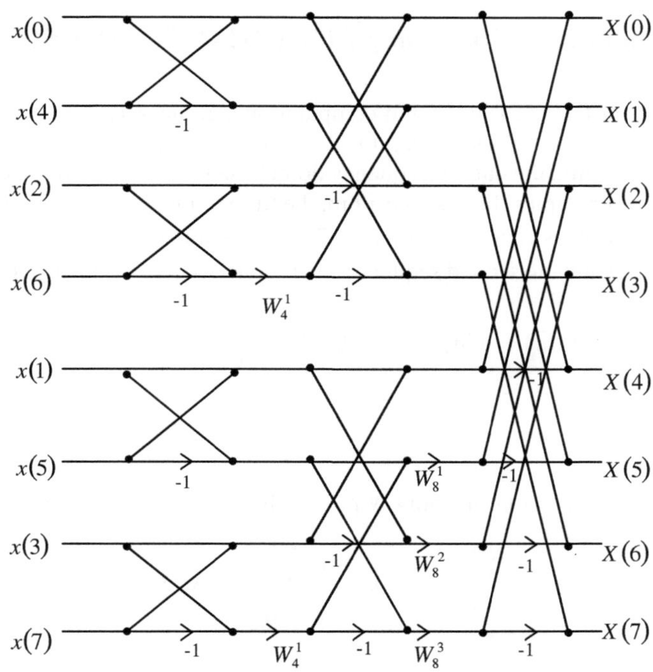

Fig. 8.5 DIT-FFT method to compute 8-point

8.2.2 Steps of Doing Radix-2 DIT-FFT

1. The number of input samples $N = 2^M$, where M is the number of stages.
2. The input sequence is decimated and shuffled by bit reversal.
3. The number of stages in the flow graph is given by $M = \log_2 N$.
4. Each stage consists of $\frac{N}{2}$ butterflies.
5. Inputs/outputs for each butterfly are separated by 2^{m-1} samples, where m represents the stage index.

Table 8.1 Bit reversal process for 8-point DIT-FFT

Input sample index	Binary representation	Bit reversed binary	Bit reversed sample index
0	000	000	0
1	001	100	4
2	010	010	2
3	011	110	6
4	100	001	1
5	101	101	5
6	110	011	3
7	111	111	7

8.3 Decimation in Frequency FFT (DIF-FFT) Algorithm

The principle of DIT-FFT algorithm computation significantly includes division of the sequence $x(n)$ into subsequences. On the other hand, in DIF-FFT (Decimation in time FFT) algorithm, the output frequency domain samples $X(k)$ is divided in smaller and smaller sub-samples. In this algorithm, the time sequence $x(n)$ is partitioned into smaller subsequences each of length N/2. The first sequence $x_1(n)$ consists of first $\frac{N}{2}$ samples of $x(n)$ and the second sequence $x_2(n)$ consists of last $\frac{N}{2}$ samples of $x(n)$ i.e.,

$$x_1(n) = x(n); \quad n = 0, 1, 2, \ldots \left(\frac{N}{2} - 1\right) \tag{8.7}$$

$$x_2(n) = x\left(n + \frac{N}{2}\right); \quad n = 0, 1, 2, \ldots \left(\frac{N}{2} - 1\right) \tag{8.8}$$

Therefore, the expression of doing N point DFT of $x(n)$ now can be written as

$$\begin{aligned} X(k) &= \sum_{m=0}^{N/2-1} x(m) W_N^{mk} + \sum_{m=N/2}^{N-1} x(m) W_N^{mk} \\ &= \sum_{n=0}^{N/2-1} x_1(n) W_N^{nk} + \sum_{n=0}^{N/2-1} x_2(n) W_N^{(n+N/2)k} \\ &= \sum_{n=0}^{N/2-1} x_1(n) W_N^{nk} + W_N^{Nk/2} \sum_{n=0}^{N/2-1} x_2(n) W_N^{nk} \\ &= \sum_{n=0}^{N/2-1} x_1(n) W_N^{nk} + e^{-j\frac{2\pi Nk}{2N}} \sum_{n=0}^{N/2-1} x_2(n) W_N^{nk} \\ &= \sum_{n=0}^{N/2-1} x_1(n) W_N^{nk} + e^{-j\pi k} \sum_{n=0}^{N/2-1} x_2(n) W_N^{nk} \end{aligned} \tag{8.9}$$

when k is even, $e^{-j\pi k} = 1$

$$X(2k) = \sum_{n=0}^{N/2-1} \{x_1(n) + x_2(n)\} W_N^{2nk}$$

$$= \sum_{n=0}^{N/2-1} \{x_1(n) + x_2(n)\} W_{\frac{N}{2}}^{nk} \tag{8.10}$$

$$\left[\because W_N^2 = W_{N/2}\right]$$

The above equation is the $\frac{N}{2}$ point DFT of the $\frac{N}{2}$ point sequence obtained by adding the first and last half of the input sequence $x(n)$.
when k is odd, $e^{-j\pi k} = -1$

$$X(2k+1) = \sum_{n=0}^{N/2-1} \{x_1(n) - x_2(n)\} W_N^{(2k+1)n}$$

$$= \sum_{n=0}^{N/2-1} \{x_1(n) - x_2(n)\} W_N^{2kn} W_N^n$$

$$= \sum_{n=0}^{N/2-1} \{x_1(n) - x_2(n)\} W_{N/2}^{kn} W_N^n \tag{8.11}$$

The above equation is the $\frac{N}{2}$ point DFT of the sequence obtained by subtracting the second half of the input sequence from the first half and multiplying the resulting sequence by W_N^n. The Eqs. 8.10 and 8.11 show that the even and off frequency samples (DFT) can be obtained from the $\frac{N}{2}$ point DFTs of $f(n)$ and $g(n)$ respectively, where

$$f(n) = \{x_1(n) + x_2(n)\} \quad n = 0, 1, 2, \ldots \frac{N}{2} - 1 \tag{8.12a}$$

$$g(n) = \{x_1(n) - x_2(n)\} W_N^n \quad n = 0, 1, 2, \ldots \frac{N}{2} - 1 \tag{8.12b}$$

Equations 8.12a and b now can be represented by a butterfly as shown in the Fig.... This is the unit of the basic DIF-FFT algorithm (Fig. 8.6).

8.3.1 Steps of Doing Radix-2 DIF-FFT

6. The number of input samples $N = 2^M$, where M is the number of stages.
7. The input sequence is in natural chronological order.
8. The number of stages in the flow graph is given by $M = \log_2 N$.

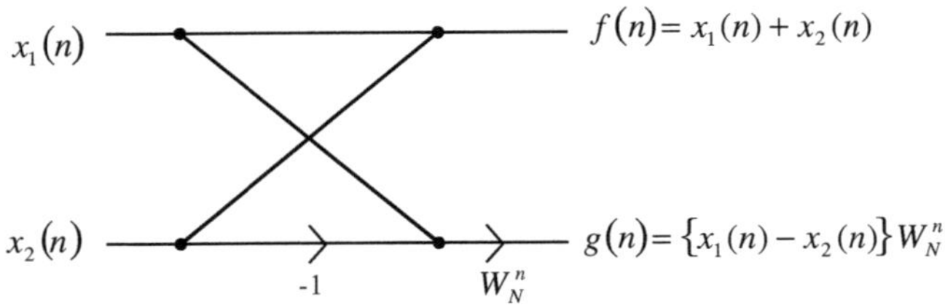

Fig. 8.6 Unit of butterfly representing for DIF-FFT

9. Each stage consists of $\frac{N}{2}$ butterflies.
10. Inputs/outputs for each butterfly are separated by 2^{M-m} samples, where m represents the stage index.

Example 8.1 Compute the DFT of the sequence $x(n) = \{1, 2, 3, 4, 4, 3, 2, 1\}$ using Radix-2 DIT-FFT algorithm.
As the given problem is to find out 8 point DFT, to compute associated twiddle factors, one unit circle needs to be divided in 8 equal divisions, as indicated in Chap. 7.

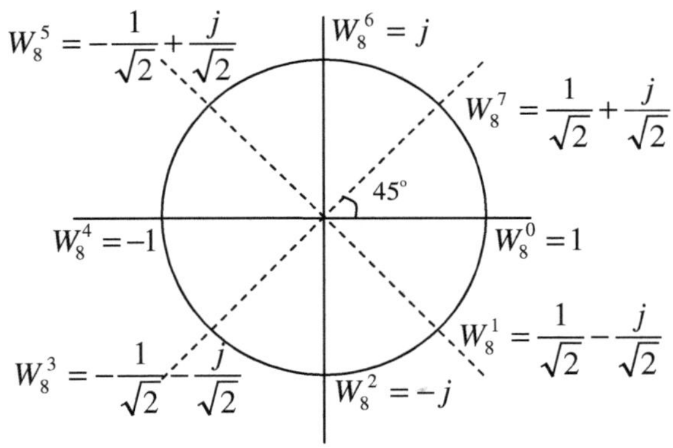

The basic operation of DIT-FFT is

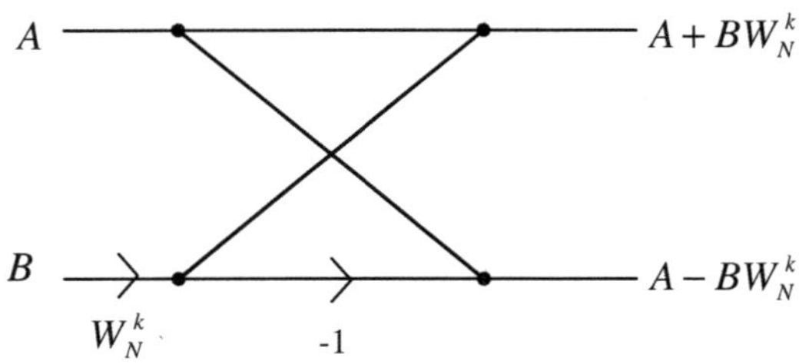

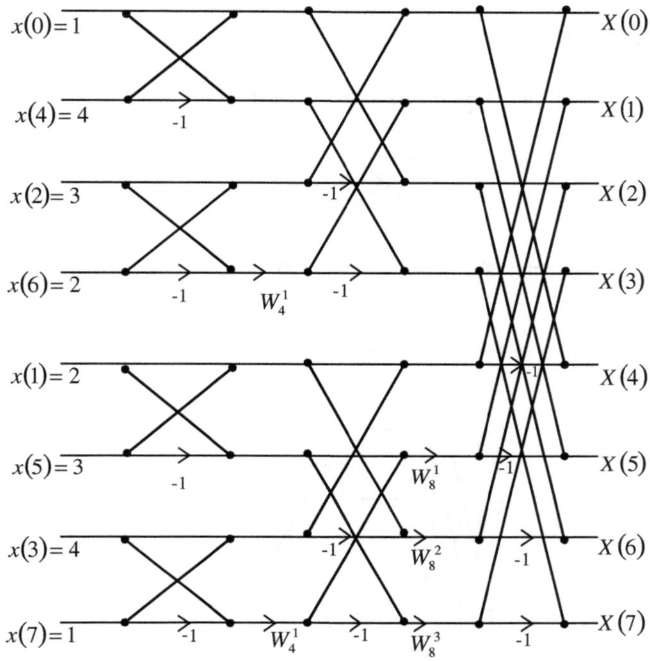

Input sequence	Output of stage-1	Output of stage-2	Final output
1	1 + 4 = 5	5 + 5 = 10	10 + 10 **= 20**
4	1 − 4 = −3	−3 + (−j)1 = −3 − j	−3 − j−(0.707 − j0.707) (−1−3j) = **−5.828 − j2.414**
3	3 + 2 = 5	5 − 5 = 0	**0**
2	3 − 2 = 1	−3 − (−j)1 = −3 + j	(−3 + j) + (−0.707 − j0.707) (−1 + 3j) **= −0.172 − j0.414**
2	2 + 3 = 5	5 + 5 = 10	10 − 10 **= 0**
3	2 − 3 = −1	−1 + (−j)3 = −1 − 3j	(−3−j) − (0.707 − j0.707) (−1−3j) **= −0.172 + j0.414**
4	4 + 1 = 5	5 − 5 = 0	**0**
1	4 − 1 = 3	−1 − (−j)3 = −1 + 3j	(−3 + j) − (−0.707 − j0.707) (−1 + 3j) **= −5.838 + j2.414**

Therefore, the 8-point DFT of the sequence $x(n)$ is

$$X(k) = \left\{ \begin{array}{l} 20, -5.828 - j2.414, 0, -0.172 - j0.414, 0, \\ -0.172 + j0.414, 0, -5.828 + j2.414 \end{array} \right\}$$

Example 8.2 Compute the DFT of the sequence $x(n) = \{1, 2, 3, 4, 4, 3, 2, 1\}$ using Radix-2 DIF-FFT algorithm.
The basic operation of DIF-FFT is

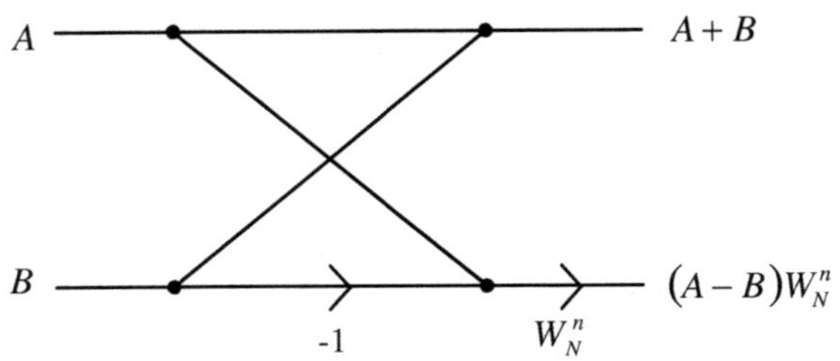

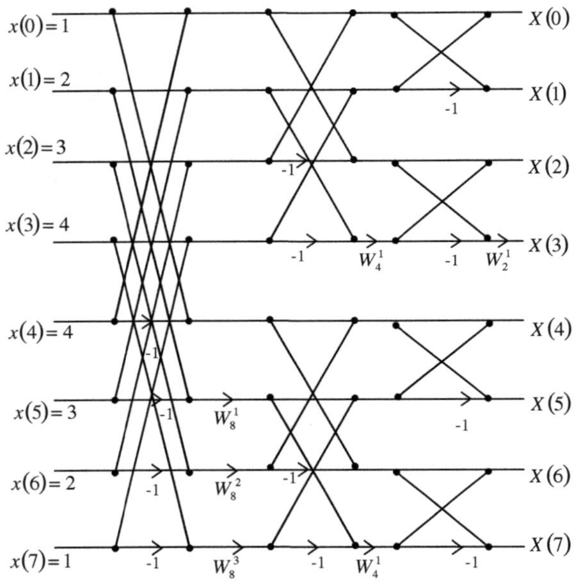

Input sequence	Output of stage-1	Output of stage-2	Final output
1	$1 + 4 = 5$	$5 + 5 = 10$	$10 + 10 =$ **20**
2	$2 + 3 = 5$	$5 + 5 = 10$	$10 - 10 =$ **0**
3	$3 + 2 = 5$	$5 - 5 = 0$	**0**
4	$4 + 1 = 5$	$(5 - 5)W_4^1 = 0$	**0**
4	$(1 - 4)W_8^0 = -3$	$-3 + (-j) = -3 - j$	$-3 - j + 2.828 + j1.414 =$ **−5.828 − j2.414**
3	$(2 - 3)(0.707 - j0.707) = -0.707 + j.707$	$(-0.707 + j.707 - 2.121 - j2.121)(-j) = 2.828 + j1.414$	$-3 - j - 2.828 - j1.414 = -$ **0.172 + j0.414**

(continued)

(continued)

Input sequence	Output of stage-1	Output of stage-2	Final output
2	(3 − 2)(−j) = −j	−3 − (−j) = −3 + j	−3 + j + 2.828 − j1.414 = **−0.172 − j0.414**
1	(4 − 1)(−0.707 − j0.707) = −2.121 − j2.121	(0.707 + j0.707 +2.121 + j2.121) (−j) = 2.828 − j1.414	−3 + j − 2.828 + j1.414 = − **5.828 + j2.414**

Therefore, the 8-point DFT of the sequence $x(n)$ is

$$X(k) = \left\{ \begin{matrix} 20, -5.828 - j2.414, 0, -0.172 - j0.414, 0 \\ , -0.172 + j0.414, 0, -5.828 + j2.414 \end{matrix} \right\}$$

8.4 Computational Complexity

The basic object of introducing FFT was to reduce the computational cost of the DFT operation significantly. It was also discussed in Chap. 7 that, the computational cost is really large for complex multiplication and addition with respect to real multiplication and addition. Therefore, in FFT also, the computational cost would be calculated in terms of number of complex multiplications and additions required to complete the entire operation of N-point DFT.

8.4.1 Number of Complex Multiplication

In each unit of butterfly of each stage of the FFT flow diagram, complex multiplication is essential. As the structure of DIT-FFT and DIF-FFT are anti-symmetric, i.e., in DIT-FFT 2-point DFT to N-point DFT flow through N/2 and other lower order DFTs is designed, whereas in DIF-FFT N-point DFT to 2-point DFT flow through N/2 and other lower order DFTs is designed. Ultimately, average number of butterfly units and total number of stages are indifferent.

Total number of complex multiplication $=$
Number of complex multiplication per butterfly $\times$ Number of butterfly units per stage $\times$ Number of stages

$$= 1 \times \frac{N}{2} \times \log_2 N = \frac{N}{2} \log_2 N \quad (8.13)$$

In classical DFT, number of complex multiplication is N^2. For $N = 1{,}024 = 2^{10}$, number of complex multiplication for classical DFT is $2^{20} = 1{,}048{,}576$. But for FFT, the number of complex multiplication is $\frac{1{,}024}{2} \times \log_2 2^{10} = 5,120$.

8.4.2 Number of Complex Addition

In each unit of butterfly of each stage of the FFT flow diagram, complex addition is also of same importance is for complex multiplication. For the reason stated in the previous sub-section, the average number of butterfly units and total number of stages are identical for Radix-2 DIT-FFT and Radix-2 DIF-FFT algoriithm.

$$\begin{aligned}&\text{Total number of complex addition } = \\ &\text{Number of complex addition per butterfly } \times \text{ Number of butterfly} \\ &\text{units per stage } \times \text{ Number of stages} \\ &= 2\times\frac{N}{2}\times\log_2 N = N\log_2 N\end{aligned} \tag{8.14}$$

In classical DFT, number of complex multiplication is N^2. For $N = 1{,}024 = 2^{10}$, number of complex multiplication for classical DFT is $2^{20} = 1{,}048{,}576$. But for FFT, the number of complex multiplication is $1,024 \times \log_2 2^{10} = 10,240$.

8.5 Circular Convolution

Let, $x_1(n)$ and $x_2(n)$ be two sequences each of finite duration N with FFTs (DFTs) $X_1(k)$ and $X_2(k)$ respectively. Now, to find the sequence $x_3(n)$ from its DFT $X_3(k)$, IDFT needs to be operated where

$$X_3(k) = X_1(k)X_2(k) \tag{8.15}$$

From the previous chapter (Sect. 7.4.7), it's clear that,

$$\begin{aligned}x_{3p}(n) &= \sum_{m=0}^{N-1} x_{1p}(m)x_{2p}(n-m)\\ \Rightarrow x_3((n))_N &= \sum_{m=0}^{N-1} x_{1p}((m))_N x_{2p}((n-m))_N\end{aligned} \tag{8.16}$$

As, $x_i((n))_N = x_i(n); 0 \le n \le N-1; i = 1, 2, 3$, Eq. 8.16 can be written as

$$x_3(n) = \sum_{m=0}^{N-1} x_1(m)x_2((n-m))_N \tag{8.17}$$

The above equation represents the ***circular convolution*** between $x_1(n)$ and $x_2(n)$ notified as

$$x_3(n) = \sum_{m=0}^{N-1} x_1(m)x_2((n-m))_N = x_1(n)\langle N\rangle x_2(n) \tag{8.18}$$

From Eqs. 8.17 and 8.18, we can therefore write,

$$DFT[x_1(n)\langle N\rangle x_2(n)] = X_1(k)X_2(k) \tag{8.19}$$

The methods used to find out circular convolution of two sequences are

1. Concentric circle method
2. Matrix multiplication method

8.5.1 Concentric Circle Method

Given two sequences $x_1(n)$ and $x_2(n)$ the circular convolution $x_3(n) = x_1(n)\langle N\rangle x_2(n)$ can be found by the following steps:

1. Graph all the N samples of $x_1(n)$ as equally spaced points around an outer circle on counter-clockwise direction.
2. Starting from the same point as $x_1(n)$, graph all the N samples of $x_2(n)$ as equally spaced points around an inner circle in clockwise direction.
3. Multiply corresponding samples on the two circles and sum the products to produce output at the starting position.
4. Rotate the inner circle by one sample at a time in either direction (say counterclockwise) to follow the process of multiply & summation as indicated in step 3 to get the next output sample.
5. Repeat step 4 until the inner circle manages to generate all the N output samples.

8.5.2 Matrix Multiplication Method

In this method, the circular convolution of two sequences $x_1(n)$ and $x_2(n)$ can be obtained by representing the sequences in matrix form as presented in the following equation.

$$\begin{bmatrix} x_2(0) & x_2(N-1) & x_2(N-2) & \cdots & x_2(2) & x_2(1) \\ x_2(1) & x_2(0) & x_2(N-1) & \cdots & x_2(3) & x_2(2) \\ x_2(2) & x_2(1) & x_2(0) & \cdots & x_2(4) & x_2(3) \\ \vdots & \vdots & \vdots & \vdots & \vdots & \vdots \\ x_2(N-2) & x_2(N-3) & x_2(N-4) & \cdots & x_2(0) & x_2(N-1) \\ x_2(N-1) & x_2(N-2) & x_2(N-3) & \cdots & x_2(1) & x_2(0) \end{bmatrix} \begin{bmatrix} x_1(0) \\ x_1(1) \\ x_1(2) \\ \vdots \\ x_1(N-2) \\ x_1(N-1) \end{bmatrix} = \begin{bmatrix} x_3(0) \\ x_3(1) \\ x_3(2) \\ \vdots \\ x_3(N-2) \\ x_3(N-1) \end{bmatrix} \tag{8.20}$$

The sequence $x_2(n)$ is repeated via circular shift of samples and represented in $N \times N$ matrix form. The sequence $x_1(n)$ is represented as column matrix. The multiplication of these two matrices gives the sequence $x_3(n)$.

Example 8.3 Compute circular convolution of the two sequences $x_1(n) = \{1, 2, 2, 1\}$ and $x_2(n) = \{1, 2, 3, 1\}$ using (a) concentric circle method, (b) matrix method.

(a) Concentric Circle Method

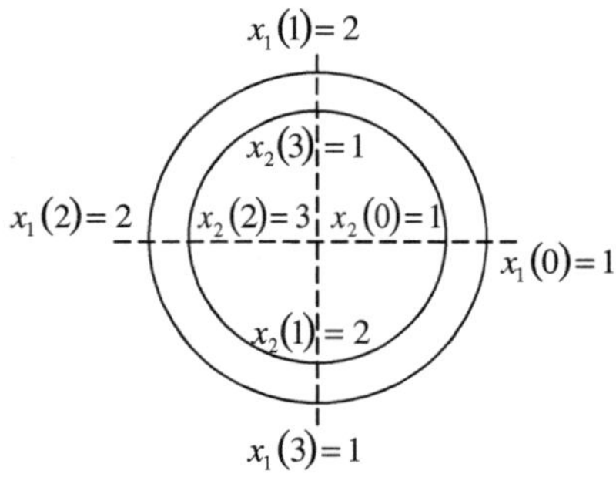

$$y(0) = 1 \times 1 + 2 \times 1 + 2 \times 3 + 1 \times 2 = 11$$

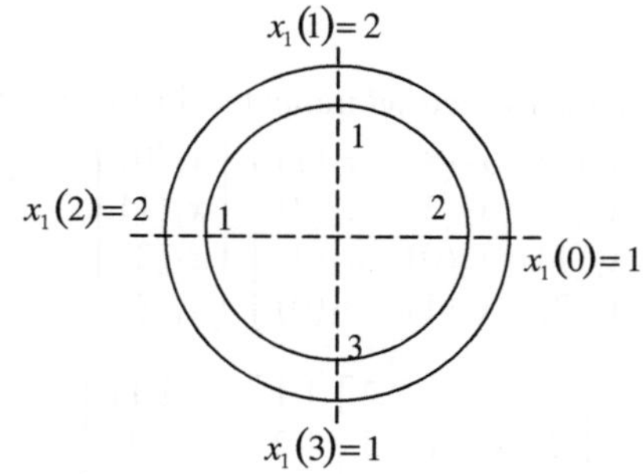

$$y(1) = 1 \times 2 + 2 \times 1 + 2 \times 1 + 1 \times 3 = 9$$

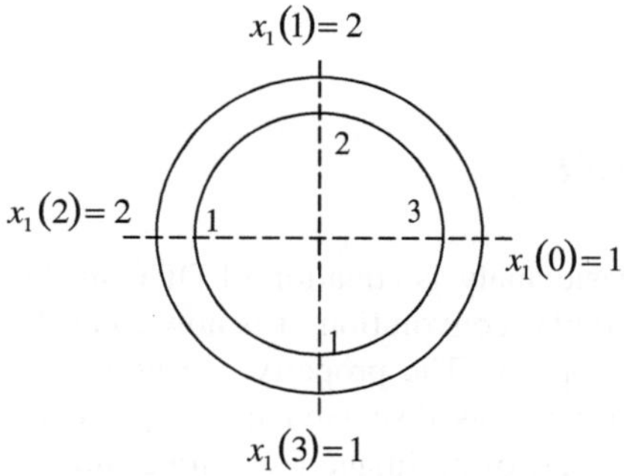

$$y(2) = 1 \times 3 + 2 \times 2 + 2 \times 1 + 1 \times 1 = 10$$

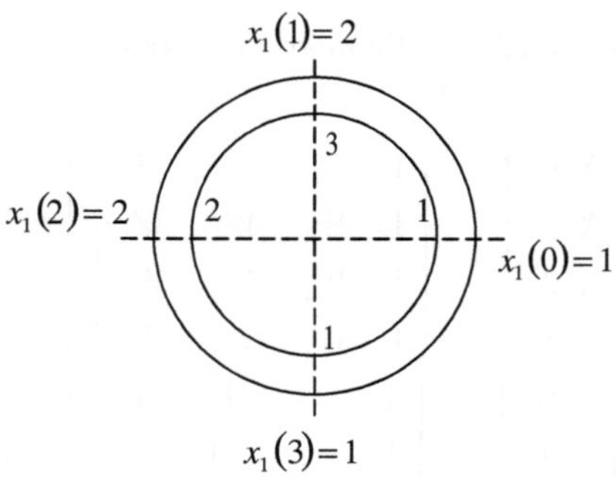

$$\begin{aligned} y(3) &= 1 \times 1 + 2 \times 3 + 2 \times 2 + 1 \times 1 = 12 \\ y(n) &= \{11, 9, 10, 12\} \end{aligned} \tag{8.21}$$

(b) Matrix Method

For $N = 4$, the matrices are formed as indicated in Eq. 8.20

$$\begin{bmatrix} x_2(0) & x_2(3) & x_2(2) & x_2(1) \\ x_2(1) & x_2(0) & x_2(3) & x_2(2) \\ x_2(2) & x_2(1) & x_2(0) & x_2(3) \\ x_2(3) & x_2(2) & x_2(1) & x_2(0) \end{bmatrix} \begin{bmatrix} x_1(0) \\ x_1(1) \\ x_1(2) \\ x_1(3) \end{bmatrix} = \begin{bmatrix} y(0) \\ y(1) \\ y(2) \\ y(3) \end{bmatrix}$$

$$\begin{bmatrix} 1 & 1 & 3 & 2 \\ 2 & 1 & 1 & 3 \\ 3 & 2 & 1 & 1 \\ 1 & 3 & 2 & 1 \end{bmatrix} \begin{bmatrix} 1 \\ 2 \\ 2 \\ 1 \end{bmatrix} = \begin{bmatrix} 11 \\ 9 \\ 10 \\ 12 \end{bmatrix} \tag{8.22}$$

$$\therefore y(n) = \{11, 9, 10, 12\}$$

8.6 Case Studies

8.6.1 FFT Over FFT

From the concept of basic analysis equation of DFT, as discussed in Chap. 7 and the discussion about circular convolution, it is now clear that DFT and hence FFT has some periodicity property. The property can be observed in a different way when we do some operation as discussed in the present case over a particular sequence (one dimensional) of an image (two dimensional).

Here, to discuss the concept of double FFT,[2] let's take an one dimensional sequence $x(n)$ of length 4, defined as

$$x(n) = \{x(0), x(1), x(2), x(3)\} \tag{8.23}$$

Now, applying the forward twiddle matrix of size 4×4 over the sequence $x(n)$ we obviously get 4-point DFT as

$$\begin{bmatrix} X(0) \\ X(1) \\ X(2) \\ X(3) \end{bmatrix} = \begin{bmatrix} 1 & 1 & 1 & 1 \\ 1 & W_4^1 & W_4^2 & W_4^3 \\ 1 & W_4^2 & W_4^4 & W_4^6 \\ 1 & W_4^3 & W_4^6 & W_4^9 \end{bmatrix} \begin{bmatrix} x(0) \\ x(1) \\ x(2) \\ x(3) \end{bmatrix}$$
$$\Rightarrow \begin{bmatrix} X(0) \\ X(1) \\ X(2) \\ X(3) \end{bmatrix} = \begin{bmatrix} 1 & 1 & 1 & 1 \\ 1 & -j & -1 & j \\ 1 & -1 & 1 & -1 \\ 1 & j & -1 & -j \end{bmatrix} \begin{bmatrix} x(0) \\ x(1) \\ x(2) \\ x(3) \end{bmatrix} \tag{8.24}$$

[2] This is not at all similar to 2D FFT.

$$\begin{aligned}
\therefore X(0) &= x(0) + x(1) + x(2) + x(3) \\
X(1) &= x(0) - j \times x(1) - x(2) + jx(3) \\
X(2) &= x(0) - x(1) + x(2) - x(3) \\
X(3) &= x(0) + jx(1) - x(2) - jx(3)
\end{aligned} \tag{8.25}$$

Now, we can define the frequency samples as another valid sequence of complex numbers. Here, we have defined a function $Y(l)$ as

$$Y(l) = \frac{1}{N} DFT(X(k)) = \frac{1}{N} DFT(DFT(x(n))) \tag{8.26}$$

Therefore,

$$\begin{bmatrix} Y(0) \\ Y(1) \\ Y(2) \\ Y(3) \end{bmatrix} = \frac{1}{4} \begin{bmatrix} 1 & 1 & 1 & 1 \\ 1 & -j & -1 & j \\ 1 & -1 & 1 & -1 \\ 1 & j & -1 & -j \end{bmatrix} \begin{bmatrix} X(0) \\ X(1) \\ X(2) \\ X(3) \end{bmatrix}$$

$$\begin{aligned}
\therefore Y(0) &= \frac{1}{4}[X(0) + X(1) + X(2) + X(3)] \\
&= \frac{1}{4}[x(0)\{1+1+1+1\} + x(1)\{1 - j - 1 + j\} + x(2)\{1 - 1 + 1 - 1\} \\
&\quad + x(3)\{1 + j - 1 - j\}] \\
&= \frac{1}{4} x(0) \times 4 = x(0)
\end{aligned} \tag{8.27}$$

Similarly,

$$\begin{aligned}
Y(1) &= \frac{1}{4}[X(0) - jX(1) - X(2) + jX(3)] \\
&= \frac{1}{4}[\{x(0) + x(1) + x(2) + x(3)\} - j\{x(0) - j \times x(1) - x(2) + jx(3)\} \\
&\quad - \{x(0) - x(1) + x(2) - x(3)\} + j\{x(0) + jx(1) - x(2) - jx(3)\}]
\end{aligned}$$

$$\begin{aligned}
Y(1) &= \frac{1}{4}[x(0)\{1 - j - 1 + j\} + x(1)\{1 - 1 + 1 - 1\} \\
&\quad + x(2)\{1 + j - 1 - j\} + x(3)\{1 + 1 + 1 + 1\}]
\end{aligned}$$

$$\therefore Y(1) = \frac{1}{4} x(3) \times 4 = x(3) \tag{8.28}$$

Similarly,

$$
\begin{aligned}
Y(2) &= \frac{1}{4}[X(0) - X(1) + X(2) - X(3)] \\
&= \frac{1}{4}[\{x(0) + x(1) + x(2) + x(3)\} - \{x(0) - j \times x(1) - x(2) + jx(3)\} \\
&\quad + \{x(0) - x(1) + x(2) - x(3)\} - \{x(0) + jx(1) - x(2) - jx(3)\}]
\end{aligned}
$$

$$
\begin{aligned}
\Rightarrow Y(2) = \frac{1}{4}[&x(0)\{1 - 1 - 1 + 1\} + x(1)\{1 + j - 1 - j\} + \\
&x(2)\{1 + 1 + 1 + 1\} + x(3)\{1 - j - 1 + j\}]
\end{aligned}
$$

$$
\therefore Y(2) = \frac{1}{4}x(2) \times 4 = x(2) \tag{8.29}
$$

And, finally

$$
\begin{aligned}
Y(3) &= \frac{1}{4}[X(0) + jX(1) - X(2) - jX(3)] \\
&= \frac{1}{4}[\{x(0) + x(1) + x(2) + x(3)\} + j\{x(0) - j \times x(1) - x(2) + jx(3)\} \\
&\quad - \{x(0) - x(1) + x(2) - x(3)\} - j\{x(0) + jx(1) - x(2) - jx(3)\}] \\
\Rightarrow Y(3) &= \frac{1}{4}[x(0)\{1 + j - 1 - j\} + x(1)\{1 + 1 + 1 + 1\} \\
&\quad + x(2)\{1 + j - 1 - j\} + x(3)\{1 - 1 + 1 - 1\}]
\end{aligned}
$$

$$
\therefore Y(3) = \frac{1}{4}x(1) \times 4 = x(1) \tag{8.30}
$$

From Eqs. 8.27–8.30, we can say,

$$
Y(l) = \frac{1}{N}DFT(DFT(x(n))) = \{x(0), x(3), x(2), x(1)\} \tag{8.31}
$$

The expression of double FFT described in Eq. 8.20 is really interesting. This is like counting the circular sequence in opposite direction, as shown in the Fig. 8.7.

Example 8.4 Compute normalized FFT over FFT of the sequence $x(n) = \{1, 2, 3, 4\}$

$$
\begin{aligned}
Y(l) &= \frac{FFT[FFT[x(n)]]}{N} \\
&= \frac{1}{4}\begin{bmatrix} 1 & 1 & 1 & 1 \\ 1 & -j & -1 & j \\ 1 & -1 & 1 & -1 \\ 1 & j & -1 & -j \end{bmatrix}\begin{bmatrix} 1 & 1 & 1 & 1 \\ 1 & -j & -1 & j \\ 1 & -1 & 1 & -1 \\ 1 & j & -1 & -j \end{bmatrix}\begin{bmatrix} x(0) \\ x(1) \\ x(2) \\ x(3) \end{bmatrix} \\
\Rightarrow Y(l) &= \frac{1}{4}\begin{bmatrix} 1 & 1 & 1 & 1 \\ 1 & -j & -1 & j \\ 1 & -1 & 1 & -1 \\ 1 & j & -1 & -j \end{bmatrix}\begin{bmatrix} 1 & 1 & 1 & 1 \\ 1 & -j & -1 & j \\ 1 & -1 & 1 & -1 \\ 1 & j & -1 & -j \end{bmatrix}\begin{bmatrix} 1 \\ 2 \\ 3 \\ 4 \end{bmatrix} \\
\Rightarrow Y(l) &= \frac{1}{4}\begin{bmatrix} 1 & 1 & 1 & 1 \\ 1 & -j & -1 & j \\ 1 & -1 & 1 & -1 \\ 1 & j & -1 & -j \end{bmatrix}\begin{bmatrix} 1+2+3+4 \\ 1-2j-3+4j \\ 1-2+3-4 \\ 1+2j-3-4j \end{bmatrix} \\
Y(l) &\Rightarrow \frac{1}{4}\begin{bmatrix} 1 & 1 & 1 & 1 \\ 1 & -j & -1 & j \\ 1 & -1 & 1 & -1 \\ 1 & j & -1 & -j \end{bmatrix}\begin{bmatrix} 10 \\ -2+2j \\ -2 \\ -2-2j \end{bmatrix} \\
\Rightarrow Y(l) &= \frac{1}{4}\begin{bmatrix} 10-2+2j-2-2-2j \\ 10-j(-2+2j)+2+j(-2-2j) \\ 10+2-2j-2+2+2j \\ 10+j(-2+2j)+2-j(-2-2j) \end{bmatrix} \\
\Rightarrow Y(l) &= \frac{1}{4}\begin{bmatrix} 4 \\ 16 \\ 12 \\ 8 \end{bmatrix} = \begin{bmatrix} 1 \\ 4 \\ 3 \\ 2 \end{bmatrix}
\end{aligned}
\tag{8.32}
$$

From Eqs. 8.31 and 8.32, we can interpret the transformation as a three step circular anti-symmetry. First of all, the sequence is arranged circularly, next the start point and end point is selected according to Fig. 8.8 and consider the direction of the sequence as clockwise. The resultant sequence will be the double FFT i.e., FFT over FFT of the given sequence, as shown in Fig. 8.8.

The source code of the case study is given in the next section and the supplementary electronic content. Readers can test the concept by their own inputs of any size.

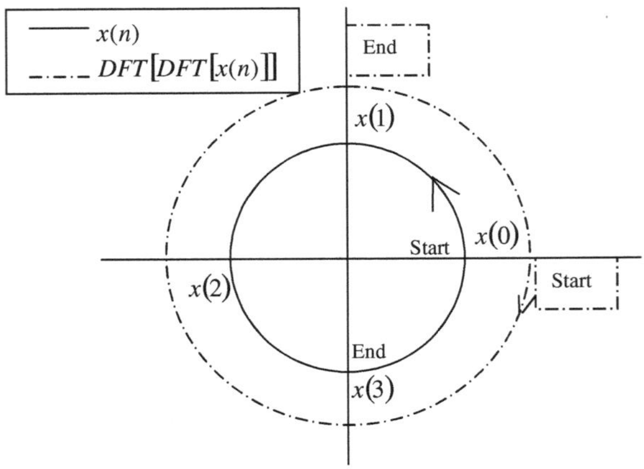

Fig. 8.7 FFT over FFT of a sequence shows circular anti-symmetry

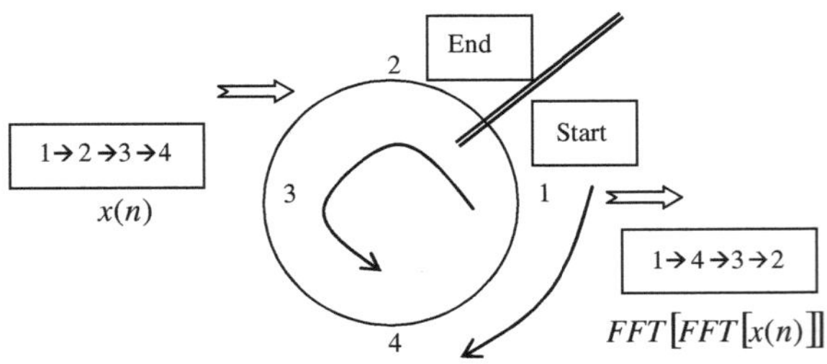

Fig. 8.8 Method to find double FFT of a sequence

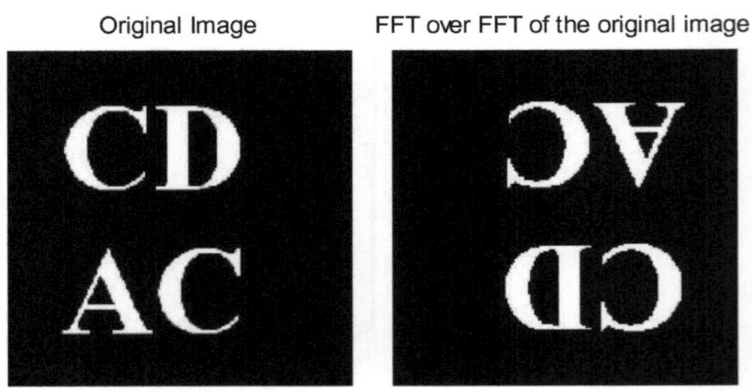

Fig. 8.9 FFT over FFT over an image is showing double mirror anti-symmetry

The anti-symmetry is also observed in two dimensional FFT. If we consider an image as a 2D sequence representation, by applying normalized double FFT[3] we get two way space reversals. We can get the double FFT of the 2D signal (image) by taking mirror image both in X-direction and Y-direction as shown in the Fig. 8.9.

The readers can use any 2D matrix or gray image to test the concept. The Matlab source code is given in the supplementary electronic material.

[3] It signifies FFT over FFT over 2D spatial samples.

Fig. 8.10 Multiplication using FFT

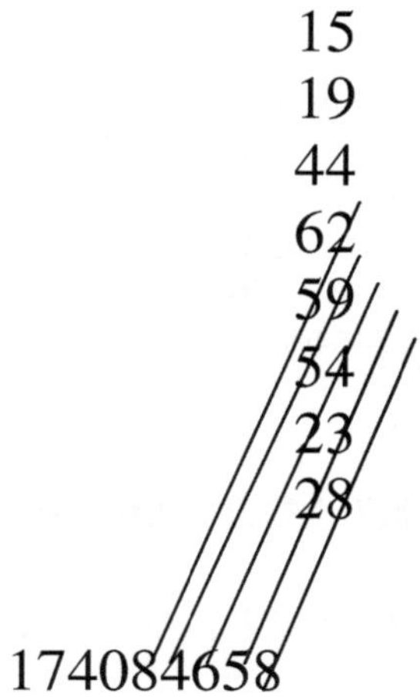

8.6.2 Multiplication Using FFT

Multiplication between two large numbers can be realized by FFT. Only an adjustment in their respective lengths is essential in terms of pre-processing. Suppose x and y are any two arrays of numbers represented based on specified base. Here the lengths of the numbers are Lx and Ly respectively.

1. Firstly we have to modify x, y by zero padding to meet both of the lengths to be a power of 2.
2. Compute FFT of x and y to get X and Y.
3. Multiply the FFT's of x and y, i.e., $X \times Y$.
4. Compute IDFT of $X \times Y$.
5. Find real value of $X \times Y$. Name it R_{XY}.
6. From a specific arrangement of the resulting array of sequence as shown in figure below, obtain the result of multiplication.

Example 8.5 Multiply the numbers 3,247 and 53,614 using FFT.

The lengths of numbers given are 4 and 5 respectively.

1. Let x = [3, 2, 4, 7], y = [5, 3, 6, 1, 4] are two sequences, derived from the two numbers given.
2. To fit both of them in a particular base, i.e., to a specific power of 2, we add 4 and 3 zeros to x and y respectively. So, the modified x and y are [3, 2, 4, 7, 0, 0, 0, 0] and [5, 3, 6, 1, 4, 0, 0] respectively.
3. XY = fft(x) × fft(y); Product of fft of x and y. This product is not using the rule of matrix multiplication rather it uses sample by sample multiplication rule.
4. RXY = real(ifft(XY));
5. RXY = [15, 19, 44, 62, 59, 54, 23, 28]
6. Add the numbers diagonally as shown in the figure below to get the final result of multiplication (Fig. 8.10).

From this we can get the result of multiplication of 3,247 × 53,614 = 174,084,658.

Multiplication of Large Numbers can also be obtained by the same procedure. There is no limitation in the length of the array as shown in the following example, as long as there is sufficient memory.

1234335543298867543233421334563333321 × 2344563453456345623333 3332342567897550435678987 6787 = 8939780041207075210621874622492647 2746408558851633869072951861210512190568240796915 19627.

8.7 MATLAB Codes

8.7.1 FFT Over FFT

```
% Double FFT
% Apurba Das
% 19/05/2011

%%
close all;
clear all;
clc;
a=input('Enter a sequence : ');
% a=[1 -2  5 7 2 11];
[h w]=size(a);
c=fft(a);
e=fft(fft(a))/w
%%

% disp('Enter to get an image');
% pause
[filename, pathname] = uigetfile({'*.bmp';}, ...
    'Brows Image');
file=cat(2,pathname,filename);
a=imread(file);
subplot(121)
imshow(a);
a=im2bw(a);
[h w]=size(a);
e=abs(fft2(fft2(a)))/(h*w);
subplot(122)
imshow(e,[min(min(e)) max(max(e))])
```

8.7.2 *Multiplication Using FFT*

```
function s=multifft(x,y,n)
% x=[1,2,8,4,5,7,2,4];
y=[2,3,5,6,7,8,5,1,2,4];multifft(x,y,1)
% x=[12,84,57,24]; y=[23,56,78,51,24];multifft(x,y,2)
t = cputime;
mx=length(x);
x=[x zeros(1,length(y)-1)];
y=[y zeros(1,mx-1)];
yx=fft(x).*fft(y);
yr=(real(ifft(yx)));
s='';
for i=length(yr):-1:2
s=[num2str(rem(yr(i),10^n)) s];
c=floor(yr(i)/(10^n));
yr(i-1)=yr(i-1)+c;
end
s=[num2str(yr(1)) s];
e = cputime-t
```

References

1. Gauss C.F.: Nachlass: theoria interpolationis methodo nova tractata, werke band, vol. 3, pp. 265–327. Göttingen: Königliche Gesellschaft der Wissenschaften (1866)
2. Cooley, J.W., Tukey, J.W.: An algorithm for the machine calculation of complex fourier series. Math. Comput. **19**(90), 297–301 (1965)

Chapter 9
Z-Transform

9.1 Introduction

Fourier and Laplace transform were proposed only for analyzing the continuous signal in frequency domain. But there was no transformation tool for analyzing signals in discrete time domain whereas there were necessity to analyze the discrete signals and their responses with different input signals. To meet the requirement, the Z-transform was introduced by the famous scientists Ragazzini and Zadeh in the sampled data control group at Columbia University in 1952 [1].

The Z-transform converts a discrete time domain signal into a complex frequency domain representation. It can be considered as discrete time equivalent of Laplace transform. By using this transformation, we can formulate a discrete LTI system in the frequency domain, determine the stability, the transient response, response to different input signal and the steady state characteristics of the system.

The name Z-transform may be derived from the idea of the letter "Z" which is sampled/digitalized version of the flipped letter 'S' (As Z-transform is discrete times equivalent of Laplace transform, which is denoted by S) or the another possible source is the presence of the letter "Z" in the names of both Ragazzini and Zadeh who published the seminal paper [1, 2].

9.2 Laplace Transform and S-Plane

Let's consider a time continuous signal $x(t)$.

A. Das, *Signal Conditioning*, Signals and Communication Technology,
DOI: 10.1007/978-3-642-28818-0_9, © Springer-Verlag Berlin Heidelberg 2012

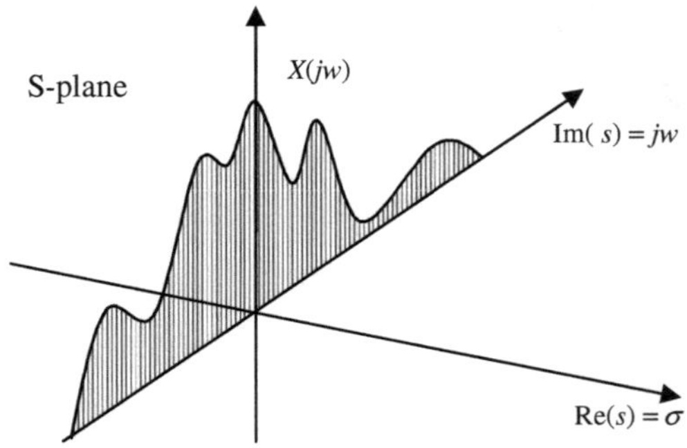

Fig. 9.1 Illustration of S-plane

The Fourier transform of the signal $x(t)$ is defined as:

$$F[x(t)] = \int_{-\infty}^{+\infty} x(t)e^{-j\omega\tau}d\tau$$

$$\Rightarrow X(f) = \int_{-\infty}^{+\infty} x(t)e^{-j2\pi f\tau}d\tau \tag{9.1}$$

$$[\text{Provided} \int_{-\infty}^{+\infty} |x(t)|dt < \infty]$$

Obviously many functions do not satisfy this condition and their Fourier transform do not exist. Such as $x(t) = t$, $x(t) = t^2$ and $x(t) = e^t$. In fact signals such as, $x(t) = 1$, $x(t) = u(t)$ and $x(t) = \cos(2\pi f_0 t)$ are not strictly integrable and their Fourier transforms all contain some non-conventional function such as $\delta(f)$.

To overcome this difficulty, we can multiply the given $x(t)$ by an exponential decaying factor $e^{-\sigma t}$ so that $x(t)$ may be forced to be integrable for certain values of the real parameter σ. Now the Fourier transform becomes:

$$\int_{-\infty}^{+\infty} x(t)e^{-\sigma t}e^{-j\omega t}dt = \int_{-\infty}^{+\infty} x(t)e^{-st}dt \tag{9.2}$$

The result of this integral is a function of a complex variable $s = \sigma + j\omega$, and is defined as the *Laplace transform* of the given signal, $x(t)$ denoted as:

$$X(s) = L[x(t)] = \int_{-\infty}^{+\infty} x(t)e^{-st}dt \tag{9.3}$$

provided the value of σ is such that the integral converges, i.e., the function $X(s)$ exists. $X(s)$ is a function defined in a 2-D complex plane, called the S-plane, spanned by σ for the real axis and ω for the imaginary axis as shown in Fig. 9.1.

9.2.1 Stability Criteria S-Plane

Let's consider a transfer function in s-domain to be expressed as $H(S) = A(S)/B(S)$. Now the roots of the equation $A(S) = 0$ are zeros and the roots of the equation $B(S) = 0$are the poles of the system defined by $H(S)$. Now, if the poles lie at left half of the S-plane then the system is stable otherwise unstable. The claim is verified by one simple example as follows.

Let us take a function

$$h(t) = e^{-at};\ a > 0 \tag{9.4}$$

If we plot this function with respect to running positive time t it is obvious that the function would be bounded (maximum value of the function must be finite for any value of t). The Laplace transform will then be

$$H(S) = \frac{1}{(s + a)} \tag{9.5}$$

So, the pole in the S-plane is at $= -a$ which lies in the negative real axis in the S-plane.

Next, an exponentially rising function is considered as

$$h(t) = e^{at} \tag{9.6}$$

Here the impulse response would not be bounded, i.e., it ensures system instability. The Laplace transform of this function would be

$$H(S) = \frac{1}{(s - a)} \tag{9.7}$$

So pole in the S-plane is $= a$

As, we see here that the function is diverging and hamper the stability of the system, so it is proved that the poles on the right half be unstable.

Similarly, by considering another function

$$h(t) = e^{-at}\sin(\omega t) \tag{9.8}$$

and its Laptace transform

$$H(S) = \frac{\omega}{(s + a)^2 + (\omega)^2} \tag{9.9}$$

We see that the poles at S-plane are complex pole at located left half of the plane and as $h(t)$ is a function of damped oscillation so, we can easily said that the system is stable, when it situated at left half of the plane.

9.3 Algorithm of Z-Transform

As understood from the discussion of the previous section, Z-transform can be characterized as the discrete version of Laplace transform. A common advantage of going for digital system is its design simplicity. Digital systems generally deal with fixed point numbers, not the floating point numbers [3].

To start with Laplace transform (Eq. 9.3), we get

$$X(s) = \int_{-\infty}^{\infty} x(t)e^{-st}dt$$

$$X(e^{sT_s}) = \sum_{n=-\infty}^{\infty} x(n)\left(e^{sT_s}\right)^{-n}$$

$$X(z) = \sum_{n=-\infty}^{\infty} x(n)z^{-n} \tag{9.10}$$

It is to be noted here that, the complex term e^{sT_s} has been considered as a single term z which makes the computation simple. Some numerical examples are given in the following.

Example 9.1 Compute Z-transform of the following sequence $x(n) = a^n u(n)$; where $u(n)$ is unit step sequence.

$$X(z) = \sum_{n=-\infty}^{\infty} x(n)z^{-n}$$

$$X(z) = \sum_{n=-\infty}^{\infty} a^n u(n)z^{-n}$$

$$X(z) = \sum_{n=0}^{\infty} a^n z^{-n} = \sum_{n=0}^{\infty} \left(az^{-1}\right)^n = \sum_{n=0}^{\infty} r^n; \text{ Taking } r = az^{-1}$$

$$X(z) = 1 + r + r^2 + r^3 + \cdots = \frac{1}{1-r} = \frac{1}{1-az^{-1}} = \frac{z}{z-a} \tag{9.11}$$

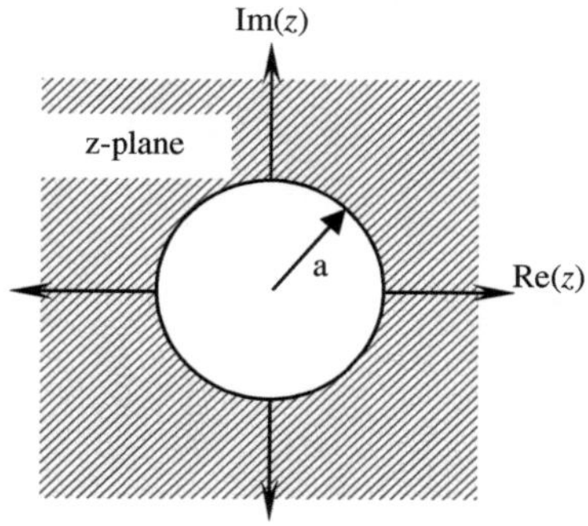

Fig. 9.2 RoC of a positive sided sequence

Example 9.2 Compute Z-transform of the following sequence $x(n) = -a^n u(-n-1)$; where $u(n)$ is unit step sequence.

$$X(z) = \sum_{n=-\infty}^{\infty} x(n) z^{-n}$$

$$X(z) = \sum_{n=-\infty}^{\infty} a^n u(-n-1) z^{-n}$$

$$X(z) = -\sum_{n=-\infty}^{-1} a^n z^{-n} = -\sum_{-n=1}^{\infty} \left(a^{-1}z\right)^{-n}$$

$$= -\sum_{m=1}^{\infty} \left(a^{-1}z\right)^{m} = -\sum_{m=1}^{\infty} p^m; \text{ Taking } p = a^{-1}z$$

$$X(z) = -\left(p + p^2 + p^3 + p^4 + \cdots\right)$$

$$= -p\left(1 + p + p^2 + p^3 + p^4 + \cdots\right) = \frac{-p}{1-p}$$

$$= \frac{p}{p-1} = \frac{a^{-1}z}{a^{-1}z - 1} = \frac{z}{z-a} \tag{9.12}$$

From Example 9.1 and 9.2 we see that the Z-transforms are same though the original functions are not same. So, when we compute the inverse Z-transform it is difficult to get the proper parent in discrete time domain. It seems therefore *many-to-one* mapping which is *incomplete* and *irreversible*. To supplement the calculated sum, another parameter is needed to make the definition of the transformation complete. The parameter is **Region of Convergence** (**RoC**).

In the Example 9.1 the infinite G.P series is converging and the result is valid if $|r| < 1$ i.e., $|z| > a$. As we know z is a complex quantity, it can be represented by $x + iy$.

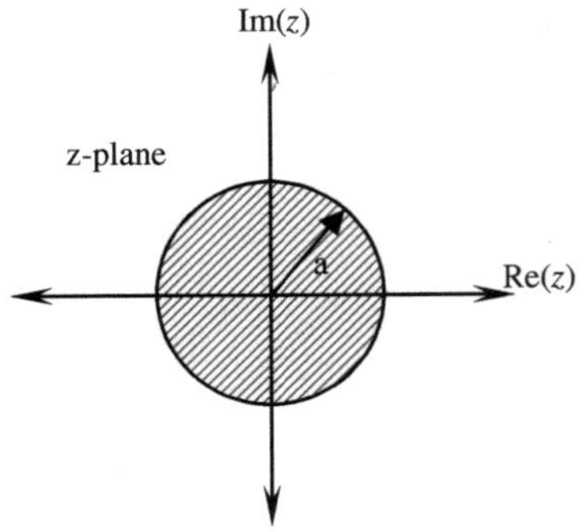

Fig. 9.3 RoC of a negative sided sequence

So

$$
\begin{aligned}
|z| &= \sqrt{x^2 + y^2} \\
&\Rightarrow x^2 + y^2 > a^2.
\end{aligned}
\tag{9.13}
$$

The RoC of Example 9.1 is shown in Fig. 9.2.

In the Example 9.2 the infinite G.P series would only converge if $|p| < 1$ i.e., $|z| < a$. In this case, the region where the Z-transform converges would differ to that of Example 9.1 as

$$x^2 + y^2 < a^2 \tag{9.14}$$

The RoC of Example 9.2 is shown in Fig. 9.3.

9.3.1 Physical Significance of Z-Transform

Z-transform is actually a summation of a sequence which may either be converging or diverging with certain conditions and constraints. The region of Z-plane where Z-transform is a valid finite quantity is called as the *Region of Convergence (RoC)*. As we know convergence series always approaches into stability. Therefore, by Z-transform we can find the stability zone of a discrete time systems while analyzing with them in frequency domain and we can perform our operation with the signal within that region only for favorable result.

9.3.2 Utility of Z-Transform

1. Convolution of discrete-time signals simply becomes multiplication of their Z-transforms.
2. Systematic method for finding the impulse response of LTI systems described by difference equations.
3. Characterize LTI discrete-time systems in the z-domain.
4. Characterize LTI discrete-time systems and their response to various input signals.

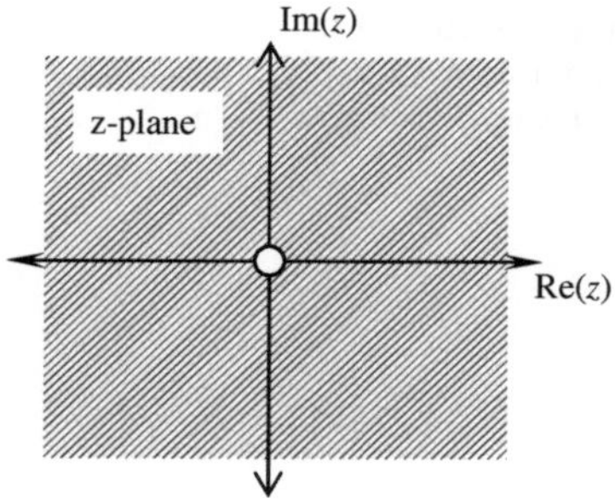

Fig. 9.4 RoC of the Z-transform of a perfect causal sequence

As understood from Eq. 9.13 and 9.14, the Region of Convergence (RoC) is always represented by a circle.

9.4 Region of Convergence (RoC) and Its Properties

The Region of Convergence (RoC) of $X(z)$ is the set of all values of z for which $X(z)$ attains a finite value.

1. The RoC is ring or disk in the Z-plane centered at the origin.
2. The RoC cannot contain any pole.
3. If $x(n)$ is a causal sequence then the RoC is the entire Z-plane except at $z = 0$.
4. If $x(n)$ is a non-causal sequence then the RoC is the entire Z-plane except at $z = \infty$.
5. If $x(n)$ is a finite duration, two sided sequence the RoC is entire Z-plane except at $z = 0$ and $z = \infty$.
6. If $x(n)$ is an infinite duration, two sided sequence the RoC will consist of a ring in the Z-plane, bounded on the interior and exterior by a pole, not containing any pole.
7. The RoC of a LTI stable system contains the unit circle.
8. The RoC must be a connected region.

9.5 RoC of Finite Duration Sequence

9.5.1 Causal Sequence

A right hand sequence is one for which $x(n) = 0; \forall n < n_0$ where n_0 is positive or negative but finite. If n_0 is greater than or equal to zero, the resulting sequence is causal or a positive time sequence. For such type of sequence the RoC is entire Z-plane except '0' radius circle or z = 0 (Fig. 9.4).

Fig. 9.5 RoC of the Z-transform of a non-causal sequence

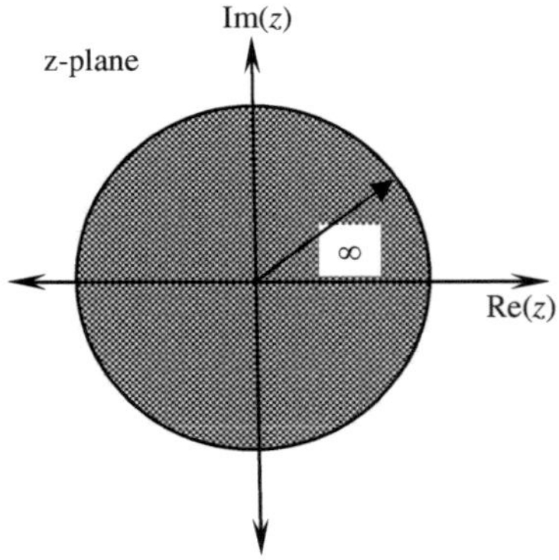

Example 9.3 Compute Z-transform of the following sequence

$$x(n) = \{1, 2, 3\}$$
$$\uparrow$$

Expanding

$$X(z) = \sum_{n=0}^{\infty} x(n)z^{-n}$$
$$= x(0) + x(1)z^{-1} + x(2)x^{-2} \tag{9.15}$$

The sequence given in the problem $x(0) = 1$; $x(1) = 2$; $x(2) = 3$

Substituting these values in the Eq. 9.15 we get

$$X(z) = 1 + 2z^{-1} + 3z^{-2} \tag{9.16}$$

$X(Z)$ converges for all values of z except at $z = 0$.

9.5.2 Anti-Causal Sequence

A left hand sequence (or anti-causal) is one for which $x(n) = 0$; $\forall n \geq n_0$ where n_0 is positive and $x(n) \neq 0$; $\forall n < 0$ if $n_0 < 0$. For such type of sequence the RoC is entire Z-plane except ∞ radius circle, i.e., except $z = \infty$ (Fig. 9.5).

Example 9.4 Compute Z-transform of the following non-causal (may have causal part and must have anti-causal part) sequence

$$x(n) = \{1, 2, 3\}$$
$$\uparrow$$

Expanding

$$\begin{aligned} X(z) &= \sum_{n=-\infty}^{0} x(n)z^{-n} \\ &= x(0) + x(-1)z^1 + x(-2)x^2 \end{aligned} \tag{9.17}$$

The sequence values are $x(0) = 3$; $x(-1) = 2$; $x(-2) = 1$.
Substituting these values in the Eq. 9.17 we get

$$X(z) = 3 + 2z^1 + z^2 \tag{9.18}$$

$X(z)$ converges for all values of z except at $z = \infty$.

9.5.3 Double Sided Sequence

A signal that has finite duration on both the left and right is known as double sided sequence. For such type of sequence the RoC is entire Z-plane except at z=0 and $z = \infty$.

Example 9.5 Compute Z-transform of the following sequence

$$x(n) = \{-3, -2, 1, 2, 3\}$$
$$\uparrow$$

Expanding

$$\begin{aligned} X(z) &= \sum_{n=-\infty}^{\infty} x(n)z^{-n} \\ &= x(-1)z^1 + x(-2)z^2 + x(0) + x(1)z^{-1} + x(2)z^{-2} \end{aligned} \tag{9.19}$$

From the given problem, $x(0) = 1$; $x(-1) = -2$; $x(-2) = -3$; $x(1) = 2$; $x(2) = 3$
Substituting these values in the Eq. 9.19 we get

$$X(z) = -3z^2 - 2z + 1 + 2z^{-1} + 3z^{-2}$$

The $X(z)$ converges for all values of z except at z $= 0$ and $z = \infty$.

9.6 Properties of Z-Transform

Some important properties of the Z-transform would help us in analysis and synthesis of discrete time signals and systems. In the following discussion $X(z)$ denotes the Z-transform of $x(n)$ and the RoC of $X(z)$ is indicated by R_X.i.e.,

$$x(n) \overset{Z}{\leftrightarrow} X(Z) \quad RoC = R_x$$

9.6.1 *Intersection of RoC*

Let's consider a sequence $x(n)$ which can be expressed as a summation of two sequences as, $x(n) = x_1(n) + x_1(n)$. The Z-transform pairs of the two sequences are given be

$$x_1(n) \overset{Z}{\leftrightarrow} X_1(z) \quad RoC = R_{X1}$$
$$x_2(n) \overset{Z}{\leftrightarrow} X_2(z) \quad RoC = R_{X2}$$

Then,

$$x(n) \overset{Z}{\leftrightarrow} X(z) = X_1(z) + X_2(z) \tag{9.20}$$

The RoC of $X(z)$ would then be the intersection of the two RoCs as

$$R_X = R_{X1} \cap R_{X2} \tag{9.21}$$

Example 9.6 Compute Z-transform of the following sequence $x(n) = 3^n u(n) - 5^n u(-n-1)$; where $u(n)$ is unit step sequence.

$$\begin{aligned} x(n) &= 3^n u(n) - 5^n u(-n-1) \\ &\Rightarrow x(n) = x_1(n) + x_2(n) \end{aligned}$$

Considering $x_1(n) = 3^n u(n)$ and $x_2(n) = -5^n u(-n-1)$

$$\therefore X_1(z) = \frac{z}{z-3}$$
$$\text{and } X_2(z) = \frac{z}{z-5} \tag{9.22}$$

Therefore from the intersection RoC property, we can express the Z-transform of the given sequence as

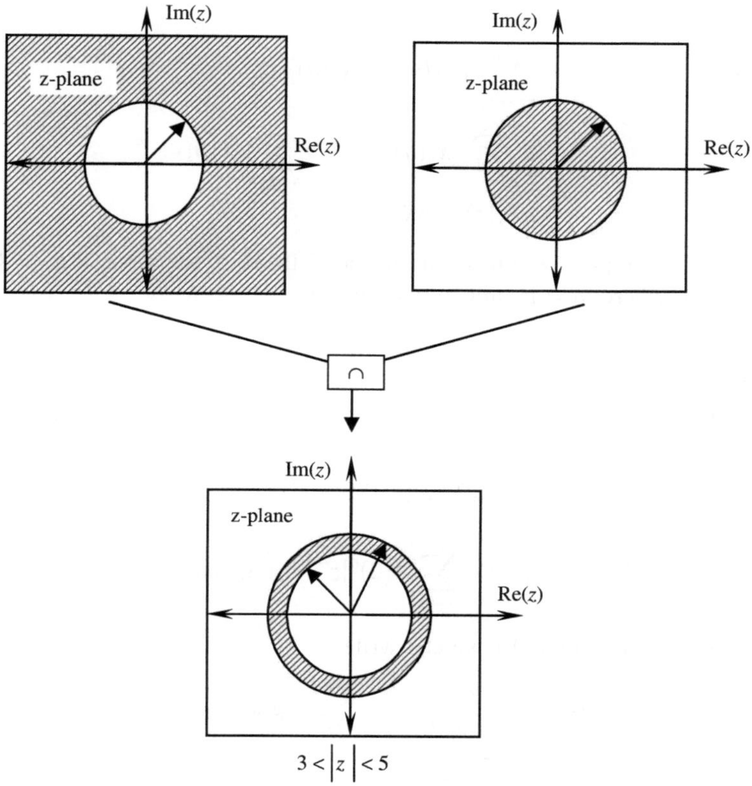

Fig. 9.6 Illustration of intersection between *RoCs*

$$X(z) = X_1(z) + X_2(z) = \frac{z}{z-3} + \frac{z}{z-5} \tag{9.23}$$

From the Examples 9.1 and 9.2 we can find the RoC of a casual signal is exterior of a circle and the RoC of an anti-casual signal is interior of a circle. The RoC of $X(z)$ is $3<|z|<5$ which is the intersection of the two RoCs R_{X1} and R_{X2} as illustrated in Fig. 9.6.

9.6.2 Linearity

If $X_1(z) = Z\{x_1(n)\}$ and $X_2(z) = Z\{x_2(n)\}$ then

$$Z\{ax_1(n) + bx_2(n)\} = aX_1(z) + bX_2(z) \quad \text{RoC} = R_{x1} \cap R_{x2}$$

Proof

$$\begin{aligned} Z\{ax_1(n)+bx_2(n)\} &= \sum_{n=-\infty}^{\infty}[ax_1(n)+bx_2(n)]z^{-n} \\ &= a\sum_{n=-\infty}^{\infty}[x_1(n)]z^{-n}+b\sum_{n=-\infty}^{\infty}[x_2(n)]z^{-n} \\ &= aX_1(z)+bX_2(z) \end{aligned}$$

Example 9.7 Compute Z-transform of the following sequence $x(n)=\{3(3)^n -4(2)^n\}u(n)$; where $u(n)$ is unit step sequence.

We know

$$\begin{aligned} X(z) &= \sum_{n=-\infty}^{\infty}x(n)z^{-n} \\ &= \sum_{n=-\infty}^{\infty}[3(3)^n-4(2)^n]u(n)z^{-n} \\ &= \sum_{n=0}^{\infty}[3(3)^n]z^{-n}-\sum_{n=0}^{\infty}[4(2)^n]z^{-n} \text{ using Linearity Property} \end{aligned}$$

With the help of Example 9.1 we can write

$$\begin{aligned} X(z) &= \frac{3}{1-3z^{-1}}-\frac{4}{1-2z^{-1}} \quad \text{RoC: } |z|>3 \\ &= \frac{3z}{z-3}-\frac{4z}{z-2} \end{aligned} \tag{9.24}$$

9.6.3 Time Shift or Translation

If $X(z)=Z\{x(n)\}$ and the initial condition for $x(n)$ are zeros, then

$$Z\{x(n-m)\}=z^{-m}X(z)$$

where m is a positive or negative integer.

Proof

$$\begin{aligned} Z\{x(n-m)\} &= \sum_{n=-\infty}^{\infty}x(n-m)z^{-n} \\ &= z^{-m}\sum_{n=-\infty}^{\infty}x(n-m)z^{-(n-m)} \end{aligned}$$

Let $(n-m)=l$, then we have

$$\begin{aligned} Z\{x(n-m)\} &= z^{-m} \sum_{l=-\infty}^{\infty} x(l) z^{-l} \\ &= z^{-m} X(z) \end{aligned} \tag{9.25}$$

Example 9.8 Compute Z-transform of the following sequence $x(n) = u(n-2)$; where $u(n)$ is unit step sequence.

We know $Z[u(n)] = \frac{z}{z-1}$ for detail solve see Example 9.1.
So, $Z[u(n-2)] = z^{-2}\frac{z}{z-1} = \frac{z^{-1}}{z-1}$ Using Time Shift Property

9.6.4 Multiplication by an Exponential Sequence

If $X(z) = Z\{x(n)\}$, then

$$Z\{a^n x(n)\} = X\left(a^{-1}z\right)$$

Proof

$$\begin{aligned} Z\{a^n x(n)\} &= \sum_{n=-\infty}^{\infty} a^n x(n) z^{-n} \\ &= \sum_{n=-\infty}^{\infty} x(n)\left(a^{-1}z\right)^{-n} \\ &= X\left(a^{-1}z\right) \end{aligned}$$

Example 9.9 Compute Z-transform of the following sequence $x(n) = r^n \cos(n\theta)$ $u(n)$; where $u(n)$ is unit step sequence.

Let $y(n) = \cos(n\theta)$
We know

$$\begin{aligned} Y(z) &= \sum_{n=-\infty}^{\infty} y(n) z^{-n} \\ &= \sum_{n=-\infty}^{\infty} \cos(n\theta) u(n) z^{-n} \\ &= \sum_{n=-\infty}^{\infty} \cos(n\theta) u(n) z^{-n} \\ &= \sum_{n=0}^{\infty} \left[\frac{e^{in\theta} + e^{-in\theta}}{2}\right] z^{-n} \\ &= \frac{1}{2}\left[\sum_{n=0}^{\infty} \left(e^{i\theta} z^{-1}\right)^n + \sum_{n=0}^{\infty} \left(e^{-i\theta} z^{-1}\right)^n\right] \end{aligned}$$

But both the series converges if $\left|e^{i\theta}z^{-1}\right| < 1$ and $\left|e^{-i\theta}z^{-1}\right| < 1$ which implies $|z| > 1$.

Now,

$$\begin{aligned}
Y(z) &= \frac{1}{2}\left[\frac{1}{1-e^{i\theta}z^{-1}} + \frac{1}{1-e^{-i\theta}z^{-1}}\right] \\
&= \frac{1}{2}\left[\frac{1-e^{-i\theta}z^{-1}+1-e^{i\theta}z^{-1}}{1-e^{i\theta}z^{-1}-e^{-i\theta}z^{-1}+z^{-2}}\right] \quad \text{RoC} : |z| > 1 \\
&= \frac{1}{2}\left[\frac{2-2z^{-1}\cos\theta}{1-2z^{-1}\cos\theta+z^{-2}}\right] \quad \because e^{i\theta}+e^{-i\theta} = 2\cos\theta \\
&= \frac{1-z^{-1}\cos\theta}{1-2z^{-1}\cos\theta+z^{-2}}
\end{aligned}$$

So, $Z[\cos(n\theta)u(n)] = \dfrac{1-z^{-1}\cos\theta}{1-2z^{-1}\cos\theta+z^{-2}}$

Now using Multiplication property we get

$$\begin{aligned}
Z[r^n\cos(n\theta)u(n)] &= \frac{1-(r^{-1}z)^{-1}\cos\theta}{1-2(r^{-1}z)^{-1}\cos\theta+(r^{-1}z)^{-2}} \\
&= \frac{1-rz^{-1}\cos\theta}{1-2rz^{-1}\cos\theta+r^2z^{-2}}
\end{aligned} \tag{9.26}$$

9.6.5 Time Reversal

If $X(z) = Z\{x(n)\}$, then

$$Z\{x(-n)\} = X\left(z^{-1}\right) \tag{9.27}$$

Proof

$$\begin{aligned}
Z\{x(-n)\} &= \sum_{n=-\infty}^{\infty} x(-n)z^{-n} \\
&= \sum_{l=-\infty}^{\infty} x(l)z^{l} \\
&= \sum_{l=-\infty}^{\infty} x(l)\left(z^{-1}\right)^{-l} \\
&= X\left(z^{-1}\right)
\end{aligned}$$

9.6.6 Differentiation of X(z)

If $X(z) = Z\{x(n)\}$, then

$$Z\{nx(n)\} = -z\frac{d}{dz}X(z) \tag{9.28}$$

Proof

$$X(Z) = \sum_{n=-\infty}^{+\infty} x(n)Z^{-n}$$

Differentiating the Z-transform we have

$$\begin{aligned}\frac{d}{dz}X(z) &= \sum_{n=-\infty}^{\infty} x(n)\frac{d}{dz}(z^{-n}) \\ &= \sum_{n=-\infty}^{\infty} x(n)(-n)z^{-n-1} \\ &= -\frac{1}{z}\sum_{n=-\infty}^{\infty} nx(n)(-n)z^{-n}\end{aligned}$$

Multiplying on both side with $-z$ we get

$$\begin{aligned}-z\frac{d}{dz}X(z) &= \sum_{n=-\infty}^{\infty} nx(n)(-n)z^{-n} \\ &= Z\{nx(n)\} \end{aligned} \tag{9.29}$$

Example 9.10 Compute Z-transform of the following sequence $x(n) = na^n u(n)$; where $u(n)$ is unit step sequence.

We know $z[a^n u(n)] = \frac{1}{1-az^{-1}}$
Using the differentiation property, we have

$$Z\{nx(n)\} = -z\frac{d}{dz}X(z)$$

Z-transform of the given sequence

$$\begin{aligned}Z\{na^n u(n)\} &= -z\frac{d}{dz}\left(\frac{1}{1-az^{-1}}\right) \\ &= \frac{az^{-1}}{(1-az^{-1})^2} \quad \text{RoC}: |z| > |a| \end{aligned} \tag{9.30}$$

9.7 System Representation by Z-Transform

The discrete time systems are generally represented by difference equation as given under. Here the excitation (input) and the response (output) of the system are $x(n)$ and $y(n)$ respectively [4].
or,

$$y(n) + a_1 y(n-1) + a_2 y(n-2) \cdots + a_N y(n-N) = b_0 x(n) + b_1 x(n-1) + b_2 x(n-2) \cdots + b_M x(n-M)$$

$$y(n) = -\sum_{k=1}^{N} a_k y(n-k) + \sum_{l=0}^{M} b_l x(n-l). \tag{9.31}$$

Taking Z-transform both side and applying the time shifting property,

$$Y(z) = -\sum_{k=1}^{N} a_k Y(z) z^{-k} + \sum_{l=0}^{M} b_l X(z) z^{-l}$$

or,

$$Y(z)\left\{1 + \sum_{k=1}^{N} a_k z^{-k}\right\} = \sum_{l=0}^{M} b_l X(z) z^{-l}$$

or,

$$H(z) = \frac{Y(z)}{X(z)} = \frac{\sum_{l=0}^{M} b_l z^{-l}}{1 + \sum_{k=1}^{N} a_k z^{-k}} \tag{9.32}$$

This $H(z)$ is known as transfer function of the system. We can also obtain the impulse response $h(t)$ of the system simply by inverse Z-transform (IZT) of the transfer function $H(z)$.

Example 9.11 Find the transfer function and impulse response of a system defined by the difference equation

$$y(n) = 0.8y(n-1) + x(n)$$

Taking Z-transform both side,

$$Y(z) = 0.8Y(z)z^{-1} + X(z)$$

or,

$$Y(z)\{1 - 0.8z^{-1}\} = X(z)$$

or,

$$H(z) = \frac{Y(z)}{X(z)} = \frac{1}{1 - 0.8z^{-1}}$$

By taking the IZT, we get the impulse response as

$$h(t) = (0.8)^n u(n) \tag{9.33}$$

9.7.1 Solution of Difference Equations Using Z-Transform

To solve linear constant coefficient difference equations, finding out the particular and homogeneous solutions are necessary steps. Instead, we use the time shift property of one sided Z-transform to the difference equation so that we can solve for the transform of the output. Finally the output sequence can be obtained by inverse Z-transform (IZT).

Example 9.12 Determine $y(n), n \geq 0$ by one sided Z-transform for the given expressions

1.

$$y(n) = \frac{1}{2}y(n-1) + x(n)$$

2.

$$x(n) = \left(\frac{1}{3}\right)^n u(n)$$

3.

$$y(-1) = 1$$

Given, $y(n) = \frac{1}{2}y(n-1) + x(n)$
Taking Z-transform both sides,

$$Y(z) = \frac{1}{2}\left(Y(z)z^{-1} + y(-1)\right) + X(z) \tag{9.34}$$

also,

$$y(-1) = 1 \tag{9.35}$$

and

$$X(z) = \frac{z}{z - \frac{1}{3}} \tag{9.36}$$

From the above three equations,

$$Y(z) = \frac{1}{2}\left(Y(z)z^{-1} + 1\right) + \frac{z}{z - \frac{1}{3}}$$

or,

$$\left(1 - 0.5z^{-1}\right)Y(z) = 0.5 + \frac{z}{z - \frac{1}{3}}$$

or,

$$Y(z) = \frac{0.5}{(1 - 0.5z^{-1})} + \frac{z}{\left(z - \frac{1}{3}\right)(1 - 0.5z^{-1})}$$

or,

$$Y(z) = \frac{0.5z}{(z - 0.5)} + \frac{z^2}{\left(z - \frac{1}{3}\right)(z - 0.5)}$$

or,

$$\frac{Y(z)}{z} = \frac{0.5}{(z - 0.5)} + \frac{z}{\left(z - \frac{1}{3}\right)(z - 0.5)}$$

or,

$$\frac{Y(z)}{z} = \frac{0.5}{(z - 0.5)} + \frac{3}{(z - 0.5)} - \frac{2}{\left(z - \frac{1}{3}\right)}$$

or,

$$Y(z) = \frac{0.5z}{(z - 0.5)} + \frac{3z}{(z - 0.5)} - \frac{2z}{\left(z - \frac{1}{3}\right)}$$

Taking the inverse Z-transform we get

$$\begin{aligned} y(n) &= \left[(0.5)^{n+1} + 3(0.5)^n - 2(1/3)^n\right]u(n) \\ y(n) &= [3.5(0.5)^n - 2(1/3)^n]u(n) \end{aligned} \tag{9.37}$$

9.8 Poles and Zeros

We have expressed (Eq. 9.32) the transfer function of the of a system in Z-domain as

$$H(z) = \frac{Y(z)}{X(z)} = \frac{\sum_{l=0}^{M} b_l z^{-l}}{1 + \sum_{k=1}^{N} a_k z^{-k}}$$

where a and b are the polynomial in Z.

The zeros of a transfer function $H(z)$ are the values of z for which the transfer function (or equivalently its numerator) is zero. Therefore the zeros are the roots of numerator polynomial. The poles of $H(z)$ are the values of the variable z for which $H(z)$ is infinite. This happens when the denominator of $H(z)$ is zero. Therefore the poles of $H(z)$ are the roots of the denominator polynomial. The system function described in Eq. 9.32 contains both poles and zeros and hence the corresponding system is called a pole–zero system, with M zeros and N poles.
If in Eq. 9.32 $a_k = 0$ for $1 \leq k \leq N$ then

$$H(z) = \sum_{l=0}^{M} b_l z^{-l} \tag{9.38}$$

In this case $H(z)$ contains M zeros only. Such a system is called an all zero system. It has a finite duration impulse response (FIR) and it is called an FIR system.
If in Eq. 9.32 $b_k = 0$ for $1 \leq k \leq M$ for then

$$H(z) = \frac{b_0}{1 + \sum_{k=1}^{N} a_k z^{-k}} \tag{9.39}$$

In this case $H(z)$ contains N poles only. Such a system is called an all pole system. Due to the presence of poles it has an infinite duration impulse response (IIR) and it is called an IIR system.

9.9 Stability Criteria

9.9.1 Stability Theorem

A linear time-invariant system with the system function $H(z)$ is said to be Bounded Input Bounded Output (*BIBO*) stable if the pole of $H(z)$ falls into the unit circle in Z-plane. Therefore, the expression in discrete tie domain expressing the criteria of BIBO stability [5]

$$\sum_{n=0}^{\infty} |h(n)| < \infty$$

would be reflected in z-domain as

$$Poles|_{H(z)} \leq 1 \tag{9.40}$$

9.10 Bounded Input Bounded Output Stability

If we get bounded output for exciting a system by bounded input then the system is BIBO stable. A sequence is said to be 'bounded' if $x(n)$ is finite for any values of n.

We have described the concept taking an example sequence $h(n) = a^n u(n)$, where $u(n)$ is a unit step function and a is an arbitrary scalar. Here we have considered four different cases as follows.

Case 1: $a < -1$
Let $a = -2$

$$h(0) = (-2)^0 = 1$$
$$h(1) = (-2)^1 = -2$$
$$h(2) = (-2)^2 = 4$$

So we see that with increasing the value of n the value of $h(n) = a^n u(n)$ is increasing monotonically alternatively creating impulses at positive and negative polarity. So, this system is unstable.

Case 2: $-1 < a < 0$
Let $a = -0.5$

$$h(0) = (-0.5)^0 = 1$$
$$h(1) = (-0.5)^1 = -0.5$$
$$h(2) = (-0.5)^2 = 0.25$$

Thus, here we see that with increasing the value of n the value of $h(n) = a^n u(n)$ is decreasing monotonically creating alternative impulses at positive and negative polarity. Therefore it satisfies the condition of BIBO stability.

Case 3: $a < 1$
Let $a = 0.5$

$$h(0) = (0.5)^0 = 1$$
$$h(1) = (0.5)^1 = 0.5$$
$$h(2) = (0.5)^2 = 0.25$$

Thus, here we see that with increasing the value of n the value of $h(n) = a^n u(n)$ is decreasing monotonically. Therefore it satisfies the condition of BIBO stability.

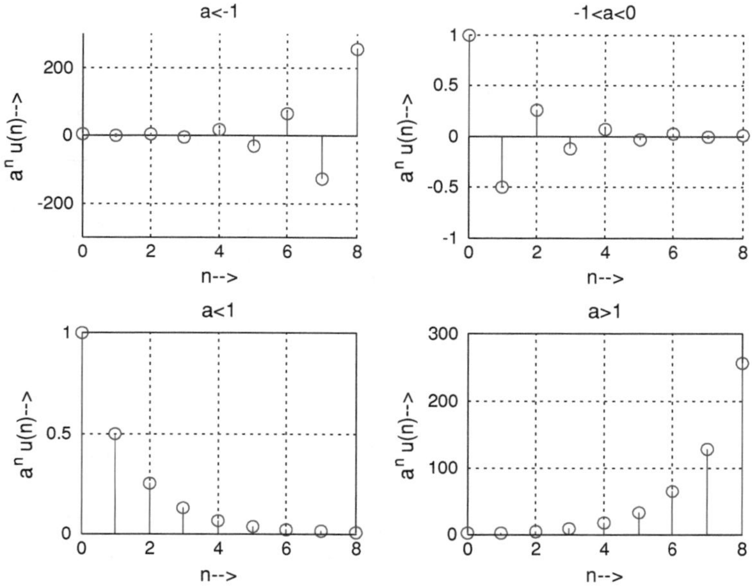

Fig. 9.7 Stability: case study of the function $a^n u(n)$ by varying a

Case 4: $a > 1$
Let $a = 2$

$$h(0) = (2)^0 = 1$$
$$h(1) = (2)^1 = 2$$
$$h(2) = (2)^2 = 4$$

Thus, here we see that with increasing the value of n the value of $h(n) = a^n u(n)$ is increasing monotonically (diverging). It is obvious that, at higher values of n (close to infinite) the function value would reach the value infinite. It *does not* satisfy the condition of BIBO stability.

Thus from above four cases as seen in Fig. 9.7, the function is converging and thus stable if $-1 < a < +1$. From the Example 9.1, we know that the Z-transform of the given function $h(n) = a^n u(n)$ is $H(z) = \frac{z}{z-a}$. From the transfer function we can see that the position of pole would be $z = a$. In the present section, we have shown the allowable range of 'a' for stability is $|a| < 1$. So we can conclude that a linear time-invariant system is BIBO stable if and only if all the poles of the system falls inside the unit circle as shown in Fig. 9.8.

Fig. 9.8 Location of pole in Z-plane for satisfying BIBO stability

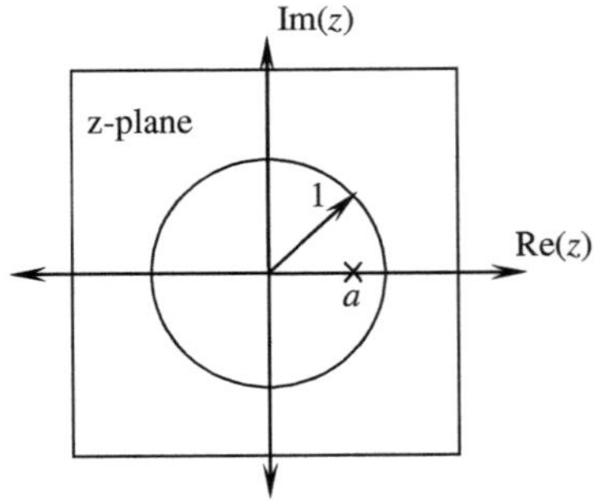

Fig. 9.9 Relationship between Z and S-plane

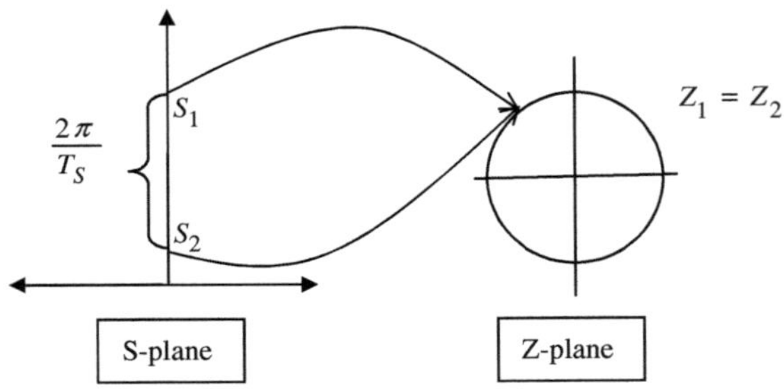

9.11 Relationship Between S and Z-Plane

Let S_1 and S_2 are two poles of a system in S-plane. As, we know the relationship $Z = e^{sT}$, the poles can be mapped to Z-plane by

$$Z_1 = e^{S_1 T_S} \tag{9.41}$$

$$Z_2 = e^{S_2 T_S} \tag{9.42}$$

Now, if these two poles are separated by an amount $\frac{2\pi}{T_S}$ in the imaginary axis, the mapped pole Z_2 would be

$$\begin{aligned} Z_2 &= e^{\left(S_1 + j\frac{2\pi}{T_S}\right)T_S} \\ &= e^{S_1 T_S} \times e^{j\frac{2\pi}{T_S}T_S} \\ &= e^{S_1 T_S} \times e^{j2\pi} \\ &= e^{S_1 T_S} \\ &= Z_1 \end{aligned} \tag{9.43}$$

From the Eq. 9.43, we can observe that the two distinct poles in S-plane is mapped to one pole in Z-plane which is not desired. This many-to-one mapping (Fig. 9.9) needs to be prevented to avoid irreversibility and for the sake of completion of the mapping. This problem has arisen for the significant distance

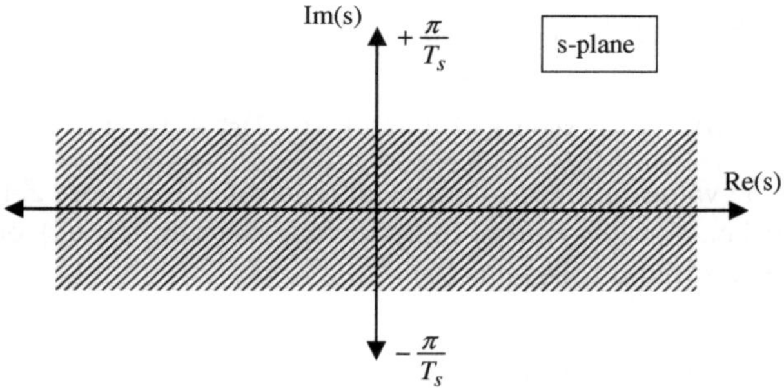

Fig. 9.10 Primary strip in S-plane to avoid aliasing in S–Z mapping

between the S-plane poles $\left(\frac{2\pi}{T_S}\right)$. The same problem would arise if the distance between S_1 and S_2 becomes any integral multiple of the factor $\left(\frac{2\pi}{T_S}\right)$. This problem of *aliasing* can only be solved by ensuring the distance between S_1 and S_2 to be less than $\frac{2\pi}{T_S}$. To ensure that a region is marked between $\pm\frac{\pi}{T_s}$ on the imaginary axis of the S-plane (shown as shaded region in Fig. 9.10. To have unique mapping between S and Z-plane, all the poles in S-plane must reside into the shaded strip known as *the primary strip* (Fig. 9.10).

9.12 Inverse Z-Transform

9.12.1 Long Division Method

From the expression of Z-transform as given in Eq. 9.10, one sided Z-transform can be expanded as

$$X_+(z) = \sum_{n=0}^{\infty} x(n)z^{-n} = x(0) + x(1)z^{-1} + x(2)z^{-2} + x(3)z^{-3} + x(4)z^{-4} + \cdots \tag{9.44}$$

Thus the value of x(n) at any instant of time are the coefficients of z^{-k}. If X(z) is given as a ratio of two polynomials, the coefficients x(0), x(1), ...x(n) can be obtained by synthetic division of the numerator by the denominator as follows

$$\begin{aligned} X(z) &= \frac{N(z)}{D(z)} = \frac{b_0 + b_1z^{-1} + b_2z^{-2} + b_3z^{-3} + \cdots + b_Nz^{-N}}{1 + a_1z^{-1} + a_2z^{-2} + a_3z^{-3} + \cdots + a_Nz^{-N}} \\ &= x(0) + x(1)z^{-1} + x(2)z^{-2} + x(3)z^{-3} + x(4)z^{-4} + \cdots \end{aligned} \tag{9.45}$$

Therefore, the values of the sequence the represent the inverse Z-transform are x(n) for $n \geq 0$ i.e., for *causal sequence*

The same Z-transform X(z) can also be presented as depending on the RoC specified as

$$X(z) = \frac{N(z)}{D(z)} = \frac{b_0 + b_1 z^{-1} + b_2 z^{-2} + b_3 z^{-3} + \cdots + b_N z^{-N}}{1 + a_1 z^{-1} + a_2 z^{-2} + a_3 z^{-3} + \cdots + a_N z^{-N}}$$
$$= \cdots + x(-n)z^n + \cdots x(-k)z^{-k} + \cdots + x(-2)z^2 + x(-1)z + x(0) \quad (9.46)$$

Therefore, the values of the sequence that represent the inverse Z-transform are x(n) for $n \leq 0$ i.e., for *non-causal* (may include the zero index) or *anti-causal* (excluding the zero index) *sequence*.

Example 9.13
Find the inverse Z-transform of

$$X(z) = \frac{z + 0.3}{z^2 - 0.5z - 0.5}; \ |z| > 1$$

From the RoC we can ensure the sequence $x(n)$ is causal. Therefore, from the basic expression of Z-transform, we can infer that the highest power of z would be -1 in transformed domain.

$$X(z) = \frac{z + 0.3}{z^2 - 0.5z - 0.5}$$

Dividing the numerator by denominator

$$\begin{array}{r l}
 & z^{-1} + 0.8z^{-2} + 0.9z^{-3} + 0.85z^{-4} \\
z^2 - 0.5z - 0.5 \,\big) & z + 0.3 \\
 & \underline{z - 0.5 - 0.5z^{-1}} \\
 & 0 + 0.8 + 0.5z^{-1} \\
 & \underline{0.8 - 0.4z^{-1} - 0.4z^{-2}} \\
 & +0.9z^{-1} + 0.4z^{-2} \\
 & +0.9z^{-1} \underline{- 0.45z^{-2} - 0.45z^{-3}} \\
 & 0.85z^{-2} + 0.45z^{-3} \\
 & 0.85z^{-2} \underline{- 0.425z^{-3} - 0.425z^{-3}} \\
 & 0.875z^{-3} + 0.425z^{-3}
\end{array}$$

From the result of the division

$$\therefore X(z) = z^{-1} + 0.8z^{-2} + 0.9z^{-3} + 0.85z^{-4} + \cdots = \sum_{n=1}^{\infty} x(n)z^{-n}$$

$$\therefore x(n) = \{0, 1, 0.8, 0.9, 0.85, \cdots\} \quad (9.47)$$

Example 9.14
Find the inverse Z-transform of

$$X(z) = \frac{z}{(z-3)(z-4)};\ |z| < 3$$

From the RoC we can ensure the sequence $x(n)$ is anti-causal. Therefore, from the basic expression of Z-transform, we can infer that the lowest power of z would be +1 in transformed domain.

$$X(z) = \frac{z}{(z-3)(z-4)} = \frac{z}{z^2 - 7z + 12}$$

$$\begin{array}{r l}
 & \frac{z}{12} + \frac{7z^2}{144} + \frac{37z^3}{1728} + .. \\
12 - 7z + z^2 \,\big) & z \\
 & z - \frac{7z^2}{12} + \frac{z^3}{12} \\ \hline
 & 0 + \frac{7z^2}{12} - \frac{z^3}{12} \\
 & \frac{7z^2}{12} - \frac{49z^3}{144} + \frac{7z^4}{144} \\ \hline
 & + \frac{37z^3}{144} - \frac{7z^4}{144} \\
 & + \frac{37z^3}{144} - \frac{259z^4}{1728} + \frac{37z^{-3}}{1728} \\ \hline
\end{array}$$

From the result of the division

$$\therefore X(z) = \frac{z}{12} + \frac{7z^2}{144} + \frac{37z^3}{1728} + \cdots = \sum_{n=-\infty}^{-1} x(n) z^{-n}$$

$$\therefore x(n) = \left\{ \cdots, \frac{37}{1728}, \frac{7}{144}, \frac{z}{12}, \underset{\uparrow}{0} \right\} \tag{9.48}$$

9.12.2 Convolution Method

Example 9.15

Find the inverse Z-transform of

$$X(z) = \frac{1}{1 - 3z^{-1} + 2z^{-2}}$$

We can express the transformed domain representation of the sequence as

$$X(z) = \frac{1}{1 - 3z^{-1} + 2z^{-2}} = \frac{1}{(1 - z^{-1})(1 - 2z^{-1})} = X_1(z) X_2(z)$$

where

$$X_1(z) = \frac{1}{(1-z^{-1})} \tag{9.49}$$

$$X_2(z) = \frac{1}{(1-2z^{-1})} \tag{9.50}$$

From Eqs. 9.49 and 9.50,

$$x_1(n) = u(n),$$
$$x_2(n) = 2^n u(n)$$

We know

$$\begin{aligned} x(n) &= x_1(n) \times x_2(n) \\ &= \sum_{k=-\infty}^{\infty} x_1(k)x_2(n-k) \\ &= \sum_{k=-\infty}^{\infty} u(k)2^{n-k}u(n-k) \\ &= \sum_{k=0}^{n} 2^{n-k} \\ &= 2^n \sum_{k=0}^{n} 2^{-k} \\ &= 2^n \sum_{k=0}^{n} \left(\frac{1}{2}\right)^k \\ &= 2^n \left\{ \frac{1-\left(\frac{1}{2}\right)^{n+1}}{1-\frac{1}{2}} \right\} u(n) \\ &= \left\{2^{n+1}-1\right\}u(n) \end{aligned} \tag{9.51}$$

References

1. Jury, E.I.: Theory and Application of the Z-Transform Method. Wiley, New York (1964)
2. Ragazzini, J.R., Franklin, G.F.: Sampled Data Control System. McGraw-Hill Book Company, New York (1958)
3. Oppenheim, A.V., Schafer, R.W.: Digital Signal Processing. Prentice Hall, New Jersey (1975)
4. Wang, R.: Initial and final value theorems. (2010-02-17). Retrieved 21 Oct 2011
5. Shanks, J.L., Treitel, S., Justice, J.H.: Stability and synthesis of two dimensional recursive filters. IEEE Trans. Audio Electroacoust. **AU-20**(3), 115–128 (1972)

Chapter 10
Wavelets: Multi-Resolution Signal Processing

10.1 Introduction

Fourier Transform has been considered to be a well accepted transformation both for time domain and spatial signal analysis since late 1950s. A relatively new transformation technique named as *wavelet transform* has been utilized even in a better way for 1D and 2D signal decomposition, compression, encoding and different methods of analysis and synthesis. Unlike the Fourier Transform, whose basis functions and sinusoids (Fig. 10.1), wavelet transforms are based on small waves (i.e., *wavelets*) of varying frequency and limited duration (Fig. 10.2). Conventional Fourier Transform lags of the localized analysis of signal in terms of frequency content. Discrete Wavelet Transform (DWT) refers to a class of transformations that differ not only in the transformation kernel employed but also in the fundamental nature of the basis functions and in the way in which they are applied. A set of scaled and shifted wavelet basis (Fig. 10.2) can be used to to represent any kind of signal in a time-frequency or space-frequency localized manner [1, 2].

In the present chapter, we have started with the concept of windowed Fourier transform or Short Time Fourier Transform (STFT). Indicating its drawbacks, we have proposed the suitable transformation technique for 1D and 2D signal processing as wavelet transform. Both continuous and discrete wavelet series and transforms are interpreted physically with elegant mathematical supports. Dyadic decomposition for image processing is discussed for sub-band coding. The last section of this chapter deals with an application of the discrete wavelet transform (DWT) to image compression. We have formally introduced and described the technique of Embedded Zero-tree Wavelet (EZW) coding for image compression based on levels of significance according to bit-budget.

A. Das, *Signal Conditioning*, Signals and Communication Technology,
DOI: 10.1007/978-3-642-28818-0_10, © Springer-Verlag Berlin Heidelberg 2012

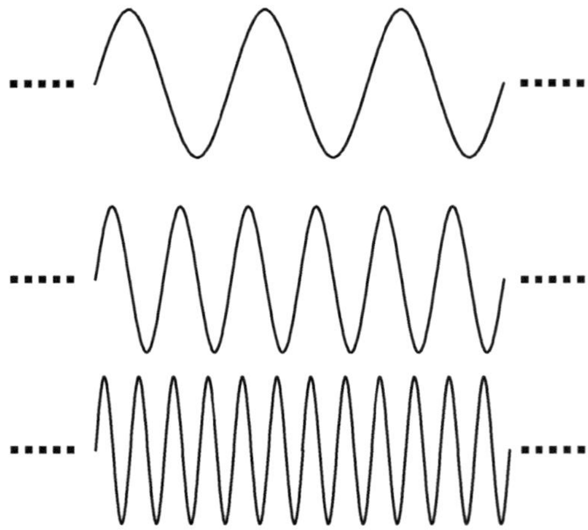

Fig. 10.1 Fourier basis functions: sinusoids of different frequencies and infinite durations

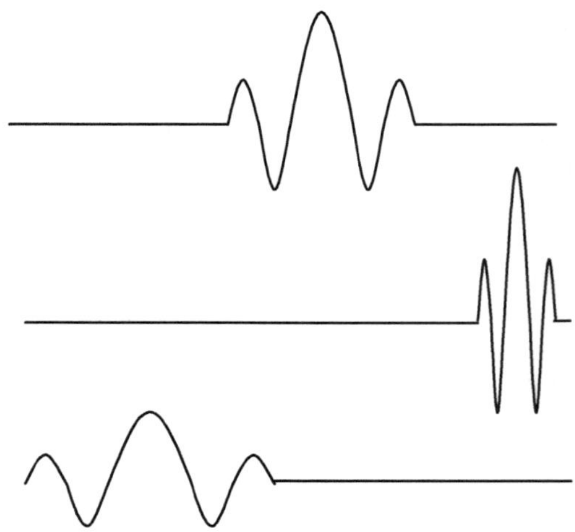

Fig. 10.2 Wavelet basis functions: "small waves" of different frequencies and finite durations

10.2 Short Time Fourier Transform

Generalized Fourier Transform is undoubtedly a wonderful tool for frequency domain analysis of any kind of single and double dimensional (1D and 2D) signals. Essentially, the transformation returns the amplitude and phase spectrum from a given time domain or spatial signal. To understand the necessity of another new and specialized kind of Fourier Transform we have formed an example as follows.

In the Fig. 10.3, we have shown two time domain signals (Fig. 10.3c, d) whose spectral representations (Fig. 10.3e, f) are identical in the sense of component of single tone and their relative amplitude. But if we just take a close look at the time domain representation of the said signals, they are significantly dissimilar.

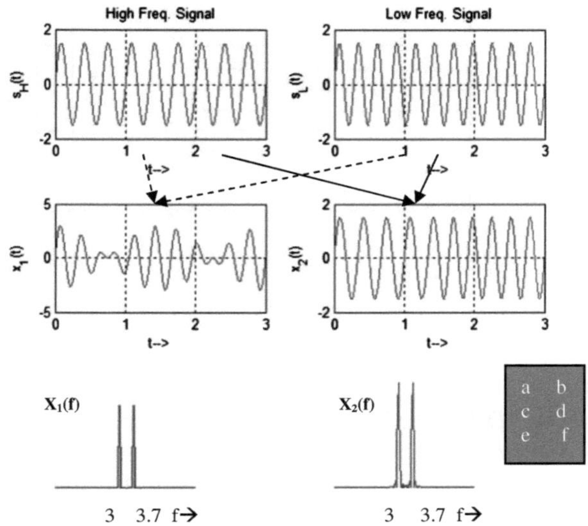

Fig. 10.3 Identical conventional Fourier spectra (**e** and **f**) from different signals (**c** and **d**) derived from **a** and **b**

Example 10.1
Let's look at the synthesis phenomena of the two signals $x_1(t)$ and $x_2(t)$ (Fig. 10.3). Both of them have been generated from same parent high frequency single tone signal $s_H(t)$ and low frequency single tone signal $s_L(t)$, as shown in Fig. 10.3a and b respectively. In the first case, we have simply added or overlapped amplitudes of the signal $s_H(t)$ and $s_L(t)$ sample by throughout the entire time region to generate $x_1(t)$. In the second case, we have taken $s_L(t)$ for first half of the time window (from 0 to 1.5 unit of time) and $s_H(t)$ for second half of the time window (from 1.5 to 3 unit of time) to generate $x_2(t)$. As a consequence both the synthesized signal holds identical frequency components, but the location or time region information is missing. From the spectra shown in Fig. 10.3e and f we cannot comment anything about duration of the time regions where the two frequencies are attained. To fulfill this need, *Short Time Fourier Transform (STFT)* came into picture where the entire time scale is divided into small regions (window) and Fourier Transform is executed on each of the windowed signal. We can therefore get a better time-frequency localization from the approach. In light of the present example (Fig. 10.3) we can say, in STFT we will have both the frequencies in all the time windows for first case while we will have high frequency in first half of the time region and low frequency in the second half of the time region for the second case as shown in Fig. 10.3d.

10.2.1 Continuous-Time STFT

In continuous-time STFT, the signal to be transformed is multiplied by a window function which is nonzero for only a short period of time. In other words the function can be called as a rectangular function or a gate function or a single pulse.

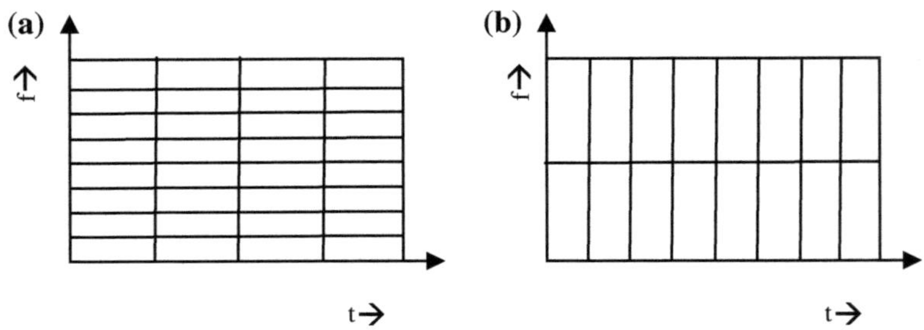

Fig. 10.4 STFT resolutions **a** better frequency resolution (uniform), **b** better time resolution (uniform)

Then 1D Fourier Transform is applied on the resulting windowed signal. As the window is slid along the time axis, it ultimately results in a two-dimensional representation of the signal. Mathematically, this is written as:

$$STFT\{x(t)\} = X(\tau, \omega) = \int_{-\infty}^{\infty} x(t)w(t-\tau)e^{-j\omega t}dt \tag{10.1}$$

where $w(t)$ is the window function, commonly a Hann window or Gaussian "hill" centered around zero, and $x(t)$ is the signal to be transformed. $X(\tau,\omega)$ is essentially the Fourier Transform of $x(t)w(t\text{-}\tau)$, a complex function representing the phase and magnitude of the signal over time and frequency. The time index τ is normally considered to be "*slow*" time and usually not expressed in as high resolution as time t.

10.2.2 Discrete-Time STFT

In the discrete time case, the data to be transformed could be broken up into chunks or frames (which usually overlap each other, to reduce artifacts at the boundary). Each chunk is Fourier transformed, and the complex result is added to a matrix, which records magnitude and phase for each point in time and frequency. This can be expressed as:

$$STFT\{x[n]\} = X(m, \omega) = \sum_{n=-\infty}^{\infty} x[n]w[n-m]e^{-j\omega n} \tag{10.2}$$

In this case, m is discrete and ω is continuous, but in most typical applications the STFT is performed on a computer using the Fast Fourier Transform, so both variables are discrete and quantized.

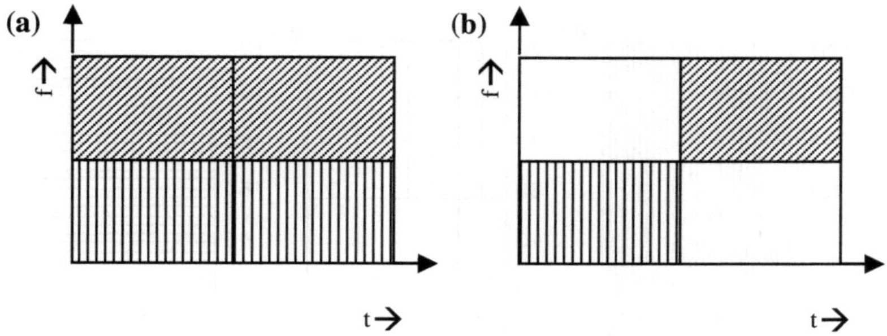

Fig. 10.5 **a** STFT spectrogram of $x_1(t)$(Fig. 10.3c), and **b** STFT spectrogram of $x_2(t)$ (Fig. 10.3d)

10.2.3 Spectrogram

We have already dealt with the time domain representation and frequency domain representation of signals and named them waveform and spectra representations respectively. In multi-resolution analysis, frequently used in STFT and wavelet transform, we need to represent a signal simultaneously in time and frequency domain as understood from the previous discussion. This representation of signal both in time and frequency domain together is called as *spectrogram*. In the Fig. 10.5 two spectrogram of STFT is shown. From spectrogram we can have the time-frequency localized information at a glance. Each rectangle of the spectrogram represents the magnitude (in terms of energy) of the signal at one frequency during a fixed time as follows.

$$spectrogram\{x(t)\} \equiv |X(\tau, \omega)|^2 \tag{10.3}$$

The magnitude is represented by some color intensity codes. In speech processing and analysis it is of great use. When we represent any speech signal in spectrogram, it looks like a 2D arrangement of some small boxes with different colors which is observed as an image and the process is called as "*imaging of speech*".

10.2.4 Limitation

STFT cannot serve the necessity of time frequency or space frequency localization, as in STFT the *sinusoids* and *cosinusoids* are basis functions which are infinitely expanded in time domain. Therefore, in STFT, the effect of windowing hampers the proper definition of the basis and it results in 'incomplete basis functions'.

Although the time and frequency resolution problems are results of a physical phenomenon (the Heisenberg uncertainty principle) and exist regardless of the transform used, it is possible to analyze any signal by using an alternative

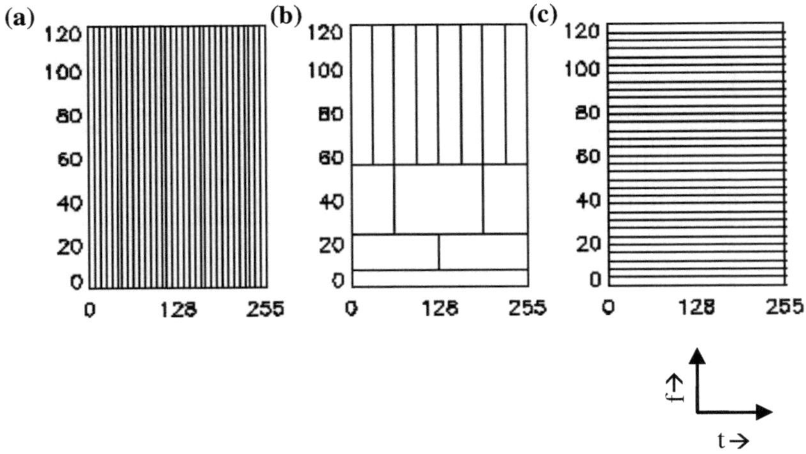

Fig. 10.6 **a** Time resolution in time scale only proposed by Shanon. **b** Time-frequency resolution in multi-resolution analysis (MRA). **c** Frequency resolution in frequency space proposed by Fourier

approach called the multi-resolution analysis (MRA). MRA, as implied by its name, analyzes the signal at different frequencies with different resolutions. Every spectral component is not resolved equally as was the case in the STFT.

MRA is designed to give good time resolution and poor frequency resolution at high frequencies and good frequency resolution and poor time resolution at low frequencies as shown in Fig. 10.6b. This approach makes sense especially when the signal at hand has high frequency components for short durations and low frequency components for long durations. Fortunately, the signals that are encountered in practical applications are often of this type.

10.3 Wavelet Function and Scaling Function

Let's express a real valued function $f(x)$as series summation

$$f(x) = \sum_k \alpha_k \phi_k(x); \tag{10.4}$$

Here k is the integer index of the summation. As k increases, the approximation would be nearer to the original.

$\phi_k(x)$s are now represented using some scaled and shifted functions. We have formed a set which included the required scaled and shifted functions as

$$\{\phi_{r,s}(x)\} \qquad where \begin{cases} r: Scaling\ parameter \\ s: Shift\ parameter \end{cases} \qquad (r,s) \in Z$$

Each of the elements of the set is defined as

$$\phi_{r,s}(x) = 2^{r/2}\phi(2^r x - s) \tag{10.5}$$

For understanding the function given in the above equation,
First of all, without any shifting and scaling let's consider $r = 0, s = 0$

$$\therefore \phi_{0,0}(x) = \phi(x) \tag{10.6}$$

Next, scaling the function by one step, i.e., $r = 1, s = 0$ we get

$$\phi_{1,0}(x) = \sqrt{2}\phi(2x) \tag{10.7}$$

Similarly, scaling the function by one step and shifting by 3 units, (1 unit is 0.5 here), i.e., $r = 1, s = 3$ we get

$$\phi_{1,3}(x) = \sqrt{2}\phi(2x - 3) \tag{10.8}$$

Example 10.2 Let's define a function

$$\phi(x) = \begin{cases} 1; 0 \leq x < 1; \ i.e., x \in [0, 1) \\ 0; \mathit{otherwise} \end{cases}$$

The graphical representation of $\phi_{0,0}(x)$, $\phi_{1,0}(x)$ and $\phi_{1,3}(x)$ are given in the Fig. 10.7

From the Eq. 10.5 and Fig. 10.7, interestingly, we can infer

$$\phi_{0,0}(x) = \frac{1}{\sqrt{2}}\phi_{1,0}(x) + \frac{1}{\sqrt{2}}\phi_{1,1}(x) \tag{10.9}$$

Therefore, the *scaling function* can be represented as

$$\phi(x) = \sum_n h_\phi(n)\sqrt{2}\phi(2x - n) \tag{10.10}$$

From the above discussion, we are now convinced that, the entire square integral real space $(L^2(R))$ can be covered by the aforementioned scaling function just by varying the amounts of shift and scaling, i.e., $\{r, s\}$.

To understand the concept more transparently, let's assume the scaling parameter r is fixed to $r = r_0$. Only by varying shifting parameter s, we cannot cover the entire $L^2(R)$ space. Conversely, by taking $r = r_0 = 0$ we can cover a limited subspace of $L^2(R)$. The subspace, V_0 will necessarily be central, as scaling is 0, as shown in Fig. 10.8.

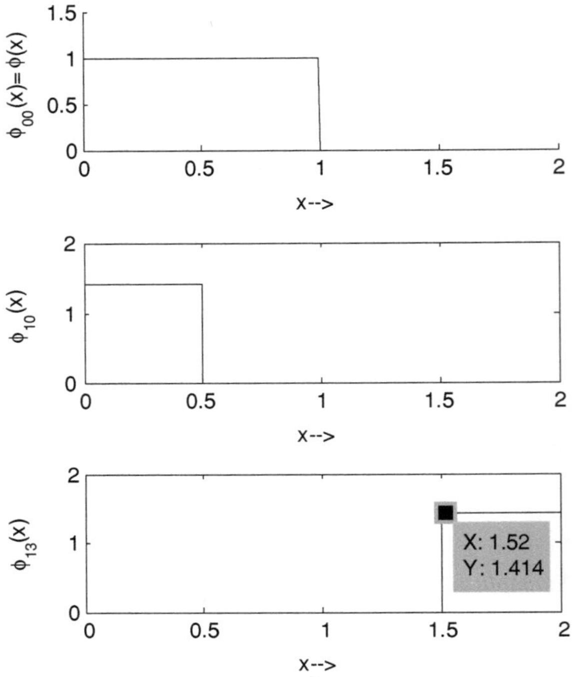

Fig. 10.7 Representation of scaling functions in different scale and shift

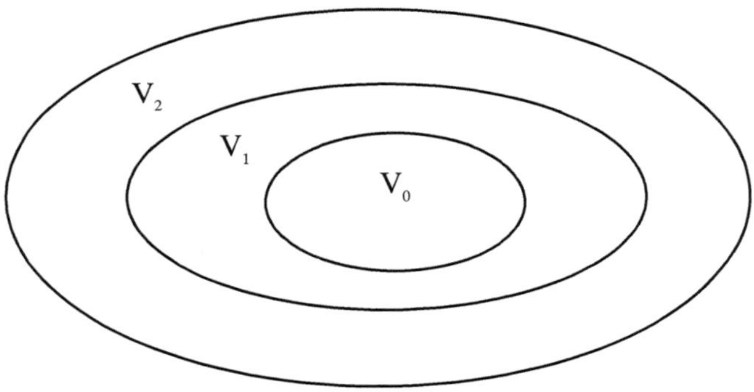

Fig. 10.8 Relation between square integral real subspaces

Next, if we go on increasing the amount of scale, to each unit increase of r, the amplitude of ϕ will be increased by an amount $\sqrt{2}$, $\left(\because 2^{r/2} = \sqrt{2}\right)$ and the width of ϕ will be reduced as explained earlier.

Hence the new subspace obtained by taking $r = 1$, is represented by $V_1 : \{\phi_{1,s}(x)\}$. Similarly, for $r = 2$, the next subspace is $V_2 : \{\phi_{2,s}(x)\}$. Now, from the property of inclusion, as described in Eq. 10.8, we can say $V_0 \subset V_1 \subset V_2$, in general, the relation between the adjacent subspaces (Fig. 10.8) is

$$V_{-\infty} \subset \cdots V_{-1} \subset V_0 \subset V_1 \subset V_2 \subset \cdots \subset V_{\infty} \tag{10.11}$$

Here the subspace coverage argument is defined as

$$\phi(x) = \sum_n h_\phi(n)\sqrt{2}\phi(2x - n) \tag{10.12}$$

The above equation signifies that, each shift parameter is associated with a coefficient, h.
Now, the difference between each pair of adjacent subspaces are defined by another function W, represented as

$$W_i = V_{i+1} - V_i \tag{10.13}$$

$$\text{or, } V_{i+1} = V_i \cup W_i \tag{10.14}$$

From the above equations,

$$V_1 = V_0 \cup W_0 \tag{10.15}$$

$$\text{Similarly, } \quad V_2 = V_1 \cup W_1 = V_0 \cup W_0 \cup W_1 \tag{10.16}$$

Next, a function is defined by difference (Eq. 10.13) of subspaces as follows. As the V_i subspaces show LPF characteristics, the different of subspaces must show HPF characteristics, as HPF here is realized by difference between two LPFs. The expression of difference subspace is very similar to square integral subspace equation as

$$\varphi_{r,s}(x) = 2^{r/2}\varphi(2^r x - s) \tag{10.17}$$

The set formed by difference subspaces W_is is defined as $\{\varphi_{r,s}(x)\}$ and named as ***wavelet function***. As seen from the previous discussion and Eq. 10.13, W_i can be realized by the present subspace V_i and next higher subspace V_{i+1}. Is signifies, the wavelet function $\varphi(x)$ can be realized by weighted scaling functions $\phi(x)$ as presented in the following equation.

$$\varphi(x) = \sum_n h_\varphi(n)\sqrt{2}\phi(2x - n) \tag{10.18}$$

Example 10.3 The Haar scaling function $\phi(x)$ as discussed in the previous example and Haar Wavelet function $\varphi(x)$ are represented as

$$\phi(x) = \begin{cases} 1; 0 \le x < 1; i.e., x \in [0, 1) \\ 0; otherwise \end{cases} \tag{10.19}$$

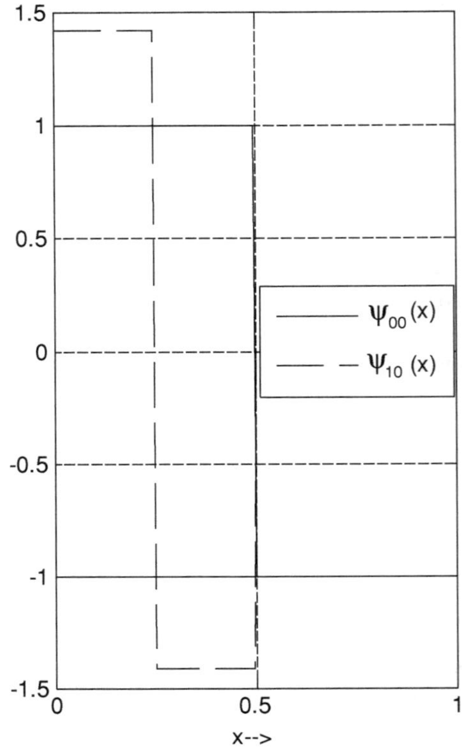

Fig. 10.9 Representation of wavelet functions in different scales

and

$$\varphi(x) = \begin{cases} 1; x \in [0, 0.5) \\ -1; x \in [0.5, 1) \\ 0; otherwise \end{cases} \tag{10.20}$$

respectively (Figs. 10.9, 10.10).

10.4 Wavelet Series

Any signal or function $f(x) \subset L^2(R)$ can be represented in terms scaling functions and wavelet functions. As combining those two functions we can cover the entire square integral real space $L^2(R)$, practically, any continuous function can be realized by those two functions as given in the following equation.

$$f(x) = \sum_s a_{r_0,s}\phi_{r_0,s}(x) + \sum_{r=r_0}^{\infty}\sum_s b_{r,s}\varphi_{r,s}(x) \tag{10.21}$$

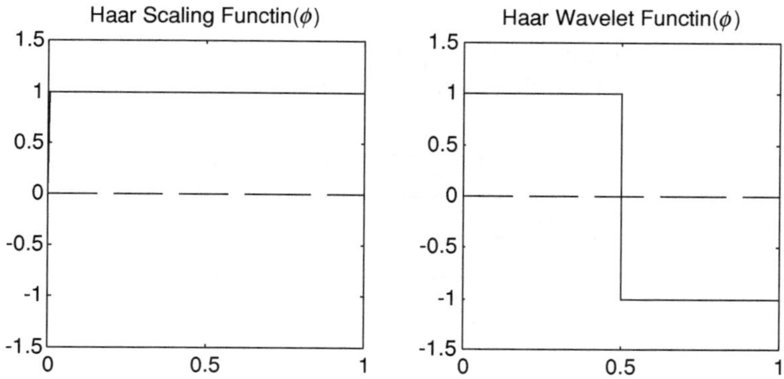

Fig. 10.10 Haar scaling function and wavelet function

The coefficients a and b can be derived from Eq. 10.21

$$a_{r_0,s} = \int f(x)\phi_{r_0,s}(x)dx \tag{10.22}$$

$$b_{r,s} = \int f(x)\varphi_{r,s}(x)dx \tag{10.23}$$

where, $r \geq r_0$

Example 10.4

Expand the following function $f(x)$ in Haar wavelet series

$$f(x) = \begin{cases} x^2; 0 \leq x < 1 \\ 0; elsewhere \end{cases}$$

$$a_{0,0} = \int_0^1 f(x)\phi_{0,0}(x)dx = \int_0^1 x^2 dx = \left.\frac{x^3}{3}\right|_0^1 = \frac{1}{3}$$

$$b_{0,0} = \int_0^1 f(x)\varphi_{0,0}(x)dx = \int_0^{0.5} x^2 dx - \int_{0.5}^1 x^2 dx = -\frac{1}{4}$$

$$b_{1,0} = \int_0^1 f(x)\varphi_{1,0}(x)dx = \int_0^{0.25} x^2\sqrt{2}dx - \int_{0.25}^{0.5} x^2\sqrt{2}dx = -\frac{\sqrt{2}}{32} = K_1$$

$$b_{1,1} = \int_0^1 f(x)\varphi_{1,1}(x)dx = \int_{0.5}^{0.75} x^2\sqrt{2}dx - \int_{0.75}^1 x^2\sqrt{2}dx = -\frac{3\sqrt{2}}{32} = K_2$$

Substituting these values into Eq. 10.21, we get the Haar wavelet series expansion of the given function as (Fig. 10.11)

$$f(x) = \underbrace{\underbrace{\underbrace{\frac{1}{3}\phi_{0,0}(x)}_{V_0} + \underbrace{\left[-\frac{1}{4}\varphi_{0,0}(x)\right]}_{W_0}}_{V_1=V_0\oplus W_0} + \underbrace{\left[-\frac{\sqrt{2}}{32}\varphi_{1,0}(x) - \frac{3\sqrt{2}}{32}\varphi_{1,1}(x)\right]}_{W_1}}_{V_2=V_1\oplus W_1=V_0\oplus W_0\oplus W_1} + \cdots \tag{10.24}$$

10.5 Discrete Wavelet Transform and Multi-Resolution Analysis

The discussions on wavelets and MRA in the previous section were completely in continuous domain. In the present section we shall discuss about the discrete wavelet representation. Let's consider $s(x)$ is sampled uniformly. Total M number of samples are there to represent the signal in discrete domain, as $s(n)$, where $n = 0, 1, 2, 3\ldots(M-1)$. From the discussion of continuous time wavelet series (CTWS), the scaling function coefficient $a_{r_0,s}$ and wavelet function coefficient $b_{r,s}$ were defined in Eqs. 10.22 and 10.23.
In discrete domain, the two coefficients are described similarly as:
Scaling function coefficient

$$W_\phi(j_0, k) = \frac{1}{\sqrt{M}}\sum_n s(n)\phi_{j_0,k}(n) \tag{10.25}$$

Wavelet function coefficient

$$W_\varphi(j, k) = \frac{1}{\sqrt{M}}\sum_n s(n)\varphi_{j,k}(n) \tag{10.26}$$

In this way, a new transformation is obtained. From $s(n)$ we have transformed the representation in terms of $W_\phi(j_0, k)$ and $W_\varphi(j, k)$. The sequence, i.e., discrete time signal $s(n)$ can be expressed in terms of those two coefficients as

$$s(n) = \frac{1}{\sqrt{M}}\sum_n W_\phi(j_0, k)\phi_{j_0,k}(n) + \frac{1}{\sqrt{M}}\sum_n W_\varphi(j, k)\varphi_{j,k}(n) \tag{10.27}$$

The above three equation forms the Discrete Wavelet Transform (DWT) pair. Equations 10.25 and 10.26 represents forward DWT and Eq. 10.27 represents inverse DWT (IDWT).

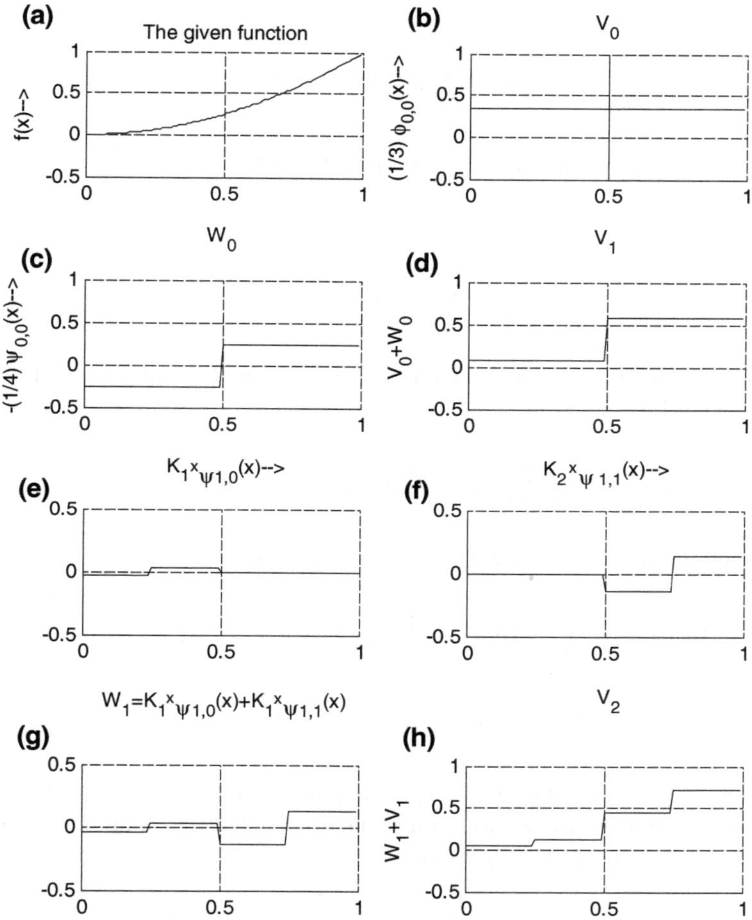

Fig. 10.11 Wavelet series expansion of the function $f(x) = x^2$ using Haar wavelet

From the DWT and IDWT equations, we can define the two functions $\phi_{j_0,k}(n)$ and $\varphi_{j,k}(n)$ as *transformation kernels,* as these two unique functions are responsible for both the operations (forward and inverse DWT).
Let's express the wavelet kernel according to Eq. 10.25 as

$$\varphi_{j,k}(n) = 2^{j/2}\varphi(2^j n - k) \tag{10.28}$$

Putting the value in the expression of wavelet coefficient in Eq. 10.26 we get

$$W_\varphi(j,k) = \frac{1}{\sqrt{M}}\sum_n s(n)2^{j/2}\varphi(2^j n - k) \tag{10.29}$$

Again, we know that

$$\varphi(p) = \sum_{p} h_{\varphi}(p)\sqrt{2}\phi(2n - p) \tag{10.30}$$

$$\therefore \varphi(2^{j}n - k) = \sum_{p} h_{\varphi}(p)\sqrt{2}\phi(2(2^{j}n - k) - p) \tag{10.31}$$

Taking $p = m - 2k$, we get

$$\therefore \varphi(2^{j}n - k) = \sum_{m} h_{\varphi}(m - 2k)\sqrt{2}\phi(2^{j+1}n - m) \tag{10.32}$$

Now, the value of wavelet kernel is considered in the expression of wavelet coefficient of Eq. 10.29

$$W_{\varphi}(j,k) = \frac{1}{\sqrt{M}}\sum_{n} s(n)2^{j/2}\left[\sum_{m} h_{\varphi}(m - 2k)\sqrt{2}\phi(2^{j+1}n - m)\right]$$

Interchanging the order of the above equation, we can rewrite it as

$$W_{\varphi}(j,k) = \sum_{m} h_{\varphi}(m - 2k)\left[\frac{1}{\sqrt{M}}\sum_{n} s(n)2^{j+1/2}\phi(2^{j+1}n - m)\right]$$

$$\therefore W_{\varphi}(j,k) = \sum_{m} h_{\varphi}(m - 2k)W_{\phi}(j + 1, m) \tag{10.33}$$

Equation 10.33 is an interesting expression where we can observe the wavelet coefficient is being expressed as scaling function coefficient. Similarly, the scaling function coefficient can be expressed as

$$W_{\phi}(j,k) = \sum_{m} h_{\phi}(m - 2k)W_{\phi}(j + 1, m) \tag{10.34}$$

In the above two equations, the h functions are the impulse response of filters. Here, h_{ϕ} and h_{φ} are transfer functions of low pass and high pass filters respectively (the reason of being LPF and HPF has already been discussed in the previous section of discussion of continuous wavelet transform). From the Eqs. 10.30 and 10.31 we can infer that, the wavelet and scaling function coefficients can be obtained from scaling function coefficient of higher scale. The input higher scale scaling function can be obtained by convolution with a LPF and a HPF respectively as shown in Fig. 10.12. Another thing to note is that, as the processing is parallel, the effective bandwidth utilization is halved. As we know from the concept of DFT that the number of time domain and frequency domain samples are made forcefully equal, therefore, at the output stage down-sampling (alternate samples are taken) by two i.e., decimation by two is required.

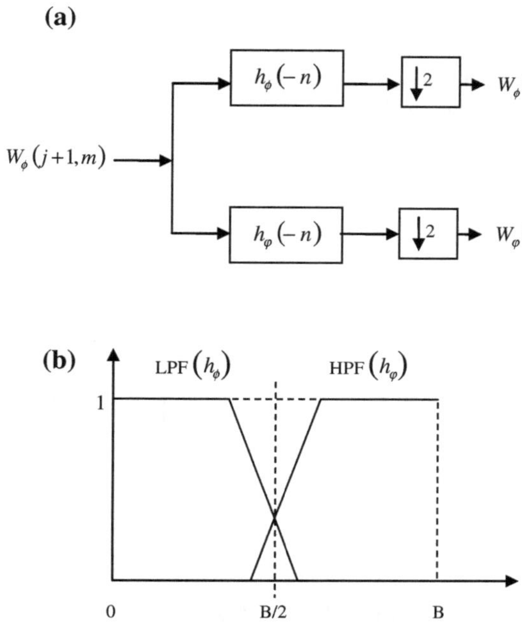

Fig. 10.12 **a** Architecture of obtaining scaling and wavelet coefficient from higher scale functions, **b** efficient bandwidth utilization by HP and LP pair model

10.5.1 Analysis Filter Bank

As observed from the Fig. 10.12, the function W_ϕ is regenerated per iteration of filtering through $h_\phi(-n)$. This hints the computation of multistage DWT coefficients for two or more successive scales by iterating the filter banks. In the Fig. 10.13 we have shown the two stage filter bank for generating the coefficients at the two highest scales of the transform. It is to be noted that the highest scale coefficients are assumed to be samples of the given function itself. Here $s(n) = W_\phi(j+1, m)$, where $j+1$ is the highest scale. The first filter bank in Fig. 10.13a splits the original function into a low pass component correspond to scaling coefficient and a high pass component corresponding to wavelet coefficient. This is clearly understood from the Fig. 10.13b where the scaling space V_{j+1} splits into wavelet subspace W_j and scaling subspace V_j. The spectrum of the original function splits into two half-band components in this way. The second stage of the filter bank in Fig. 10.13a again splits the spectrum of subspace V_j into quarter band subspaces W_{j-1} and V_{j-1}.

This two stage filter bank can easily be extended to any number of scales. If the input function for analysis is sampled well above the Nyquist rate, its samples are good approximations of the scaling coefficients at the sampling resolution and can be used as the starting high-resolution scaling coefficient inputs for the next stage for filtering. Thus no wavelet or detailed coefficients are needed at the sampling scale.

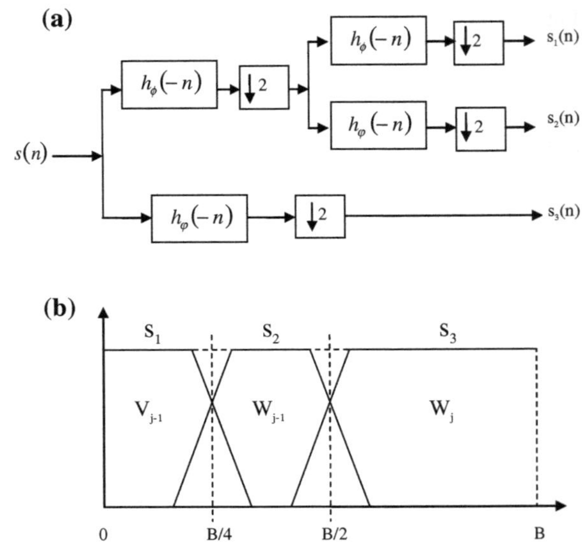

Fig. 10.13 **a** Signal analysis filter banks, **b** transfer characteristics of the hierarchical low pass and high pass filters and bandwidth utilization

10.5.2 Synthesis Filter Bank

In the previous sub-section, we have analyzed a discrete time sequence with the help of forward Discrete Wavelet Transform (DWT). The sequence could be decomposed into different scales of low pass and high pass samples. In the present sub-section we'll go for just the opposite. Here the decomposed components are to be used for realizing the original sequence $s(n)$. As the operation is just the opposite, we would use the Inverse Discrete Wavelet Transform (IDWT) here for sequence synthesis. From the understanding of DWT and IDWT from the expressions of Eqs. 10.33 and 10.34 we can also establish the relations between the synthesis filters (g) and analysis filters (h) as

$$g(n) = h(-n) \tag{10.35}$$

This relation is true both for scaling filter and wavelet filters. As shown in the Fig. 10.14, the component sequences are firstly up-sampled by two. The up-sampled sequences are then passed through stages of low pass (g_ϕ) and high pass (g_φ) filters. The convolved quarter band sequences are added then to form a down-sampled half band low pass sequence. In the next stage, this half band low pass sequence is convolved with the inverse scaling filter after next stage of up-sampling. Finally this scaling filter output is added with the single stage wavelet filter output to retrieve the original sequence $s(n)$.

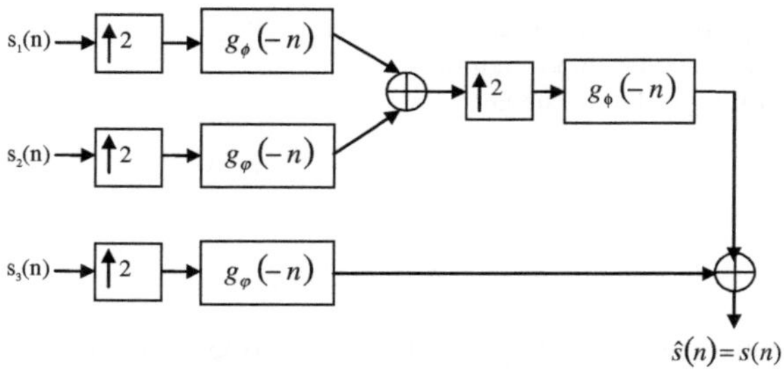

Fig. 10.14 Signal synthesis filter banks: wavelet transform always recovers undistorted signal

10.6 Image Decomposition Using DWT

10.6.1 Concept of 2D Signal Decomposition Using Analysis Filter

As we know from the basic discussion about the concept of spatial frequency in two dimensional signals, we can consider the frequency in the either direction of the 2D signal (like image) in completely uncorrelated manner [3]. Therefore, the applicability of digital wavelet transform in terms of analysis filter bank also can be considered separately in the column and row direction of an image like 2D signal.

As expressed in Eqs. 10.33 and 10.34, in two dimensional signals also the low pass scaling functions and high pass wavelet functions can be defined as follows, considering n_1 and n_2 as two orthogonal directions (like row and column).

$$\phi(n_1,n_2) = \phi(n_1)\phi(n_2) \tag{10.36}$$

$$\varphi^H(n_1,n_2) = \varphi(n_1)\phi(n_2) \tag{10.37}$$

$$\varphi^V(n_1,n_2) = \phi(n_1)\varphi(n_2) \tag{10.38}$$

$$\varphi^D(n_1,n_2) = \varphi(n_1)\varphi(n_2) \tag{10.39}$$

The above four equations cover the complete analysis description of a two dimensional signal like image. In Eq. 10.36, $\phi(n_1,n_2)$ expresses the scaling function where low pass operation (scaling) is done both in the row and column direction. In the Fig. 10.15, the region of the image is designated as 'LL'. In Eq. 10.37, $\varphi^H(n_1,n_2)$ expresses wavelet function which represents column wise (i.e., horizontally) high pass and row wise (vertically) low pass operation. In the Fig. 10.15, the region of the image is designated as 'HL'. The superscript 'H' in the wavelet function signifies the horizontal high pass filtering. In Eq. 10.37, $\varphi^V(n_1,n_2)$ expresses another wavelet function which is column wise (i.e., horizontally) low pass and row wise (vertically) high pass. In the Fig. 10.15, the region of the image is designated as 'LH'. The superscript 'V' in the wavelet function signifies the vertical high pass filtering. Finally, in Eq. 10.39, high pass operation is done in both the directions, therefore as a

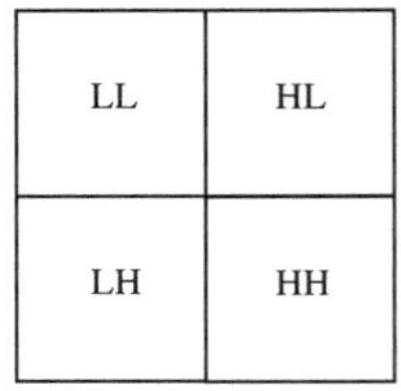

Fig. 10.15 Wavelet based four level decomposition of an image like 2D signal

consequence a diagonal high pass operation is represented by $\varphi^D(n_1, n_2)$. In the Fig. 10.15, the region of the image is designated as 'HH'.

This is to be noted that, as we have considered four different pair of operations, the image area is devided into uniform four parts. To maintain the size of the image, while applying the said operations (LL, HL, LH and HH), we need to under-sample the image. In the next section, we'll discuss how the under-sampled image along with the four level decomposition concepts helps us to make the discrete wavelet transform a reversible one.

10.6.2 DWT on Images

The concept of two dimensional (2D) signal decomposition using stages of analysis filter can directly be applied onto the case of image decomposition. In Eq. 10.36 the 2D scaling function $\phi(n_1, n_2)$and in Eqs. 10.37, 10.38 and 10.39 three 2D wavelet functions, $\varphi^H(n_1, n_2), \varphi^V(n_1, n_2), \varphi^D(n_1, n_2)$ are expressed which are needed for image decomposition in terms of analysis filter banks. These wavelet functions define intensity or gray level variation of image along different directions. $\varphi^H(n_1, n_2), \varphi^V(n_1, n_2)$ and $\varphi^D(n_1, n_2)$ correspond to variation of intensity along column (like horizontal edges), variations along rows (like vertical edges) and variations along diagonal direction of an image. This direction sensitivity is a natural consequence of the separability imposed by the decomposition concept. It does not increase the computational complexity of the two dimensional wavelet transform (Fig. 10.16).

10.6.2.1 DWT

In an image $s(n_1, n_2)$ of size $N_1 \times N_2$, the wavelet forward transformations are given by

$$W_\phi(j_0, k_1, k_2) = \frac{1}{\sqrt{N_1 N_2}} \sum_{n_1=0}^{N_1-1} \sum_{n_2=0}^{N_2-1} s(n_1, n_2)\phi_{j_0,k_1,k_2}(n_1, n_2) \qquad (10.40)$$

$$W_\varphi^i(j, k_1, k_2) = \frac{1}{\sqrt{N_1 N_2}} \sum_{n_1=0}^{N_1-1} \sum_{n_2=0}^{N_2-1} s(n_1, n_2)\varphi_{j_0,k_1,k_2}^i(n_1, n_2); i = \{H, V, D\} \qquad (10.41)$$

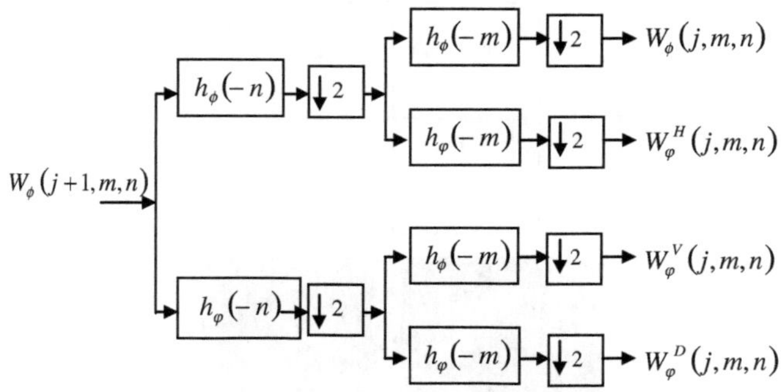

Fig. 10.16 Wavelet based image decomposition: analysis filter bank

Very similar to one dimensional wavelet transform concept, here also, j_0 is the arbitrary starting scale and the $W_\phi(j_0,k_1,k_2)$ coefficients define an approximation of $s(n_1,n_2)$ at scale j_0. The $W_\varphi^i(j,k_1,k_2)$ coefficients on the other hand add horizontal, vertical and diagonal details for scales $j \geq j_0$.

10.6.2.2 IDWT

From the given scaling coefficients $W_\phi(j_0,k_1,k_2)$ and wavelet coefficients $W_\varphi^i(j,k_1,k_2)$ the entire image can be retrieved by inverse discrete wavelet transform in two dimension (IDWT) as given in the following equation.

$$s(n_1,n_2) = \frac{1}{\sqrt{N_1N_2}}\sum_{k_1=0}^{N_1-1}\sum_{k_2=0}^{N_2-1} W_\phi(j_0,k_1,k_2)\phi_{j_0,k_1,k_2}(n_1,n_2)$$
$$+\frac{1}{\sqrt{N_1N_2}}\sum_{i=\{H,V,D\}}\sum_{j=0}^{\infty}\sum_{k_1=0}^{N_1-1}\sum_{k_2=0}^{N_2-1} W_\varphi^i(j,k_1,k_2)\varphi_{j,k_1,k_2}^i(n_1,n_2) \qquad (10.42)$$

Unlike to DCT, the entire signal can be recovered exactly without any errors as we have not applied any type of quantization in any stage.

As discussed, in case of one dimensional signal decomposition through the analysis filter bank, an image $s(n_1,n_2)$ can also be used as the input $W_\phi(j+1,m,n)$. Convolving its rows with $h_\phi(-n)$ and $h_\varphi(-n)$, separately and down-sampling its columns, we get two sub-images whose horizontal resolutions are reduced by a factor of two. The high pass component characterizes the edge like features in vertical orientation. The low pass component characterizes the constant intensity features in vertical orientation. Both the sub-images are again filtered column wise and down-sampled along rows to get four quarter size sub-images $W_\phi(j,m,n)$, $W_\varphi^H(j,m,n)$, $W_\varphi^V(j,m,n)$ and $W_\varphi^D(j,m,n)$ as shown in the Fig. 10.17.

This process of the partitioning one image into sub-images is called as *dyadic partitioning*, where at each complete stage, the image is divided into four

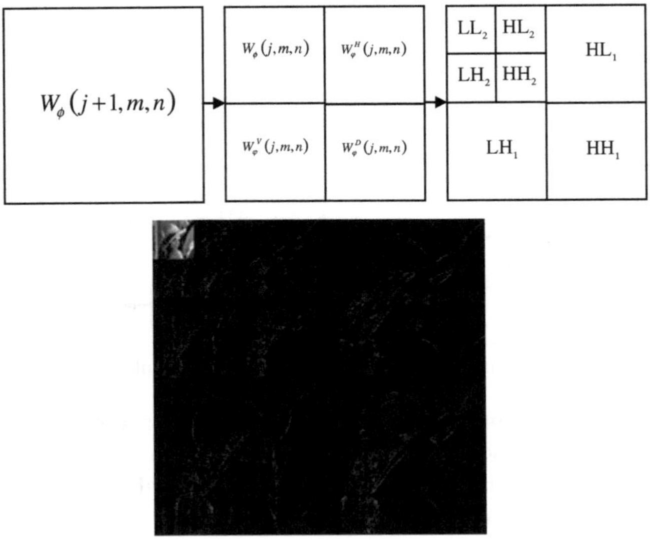

Fig. 10.17 Dyadic decomposition of an image

sub-images with half the row size and half the column size with respect to the image size at the previous scale. The filtering stages shown in the Fig. 10.16 can be iterated by a number of times for greater degree of partitioning. In dyadic partitioning the number of partitions would always be $2^n$for any arbitrary real integer n.

10.7 Image Compression Using DWT: Embedded Zero-Tree Wavelet Encoding

We know from the common understanding of signal representation and signal processing that, to achieve *"compression"*, we need to reduce *"redundancy"* in the signal representation. As an example, say a sinusoid is represented by N number of samples. The sampling frequency maintained is 10 times the frequency of the sinusoid. If each of the samples is represented by 20 bits, the number of bits requires the represent the sinusoid within a defined interval is $20 \times N$. This is a redundancy. We can remove some of the samples intelligently just by ensuring the sampling frequency to be the Nyquist rate, i.e., twice the frequency of the sinusoid. Now the number of samples would be reduced to $N/5$, resulting the number of bits requires the represent the sinusoid within the same defined interval is $20 \times N/5 = 4N$. In this way compression can be achieved. Here we could not have to compromise on the quality of the signal as we can retrieve the original sinusoid from the impulses of alternating polarity simply by low pass filtering. The concept of Continuous Time Fourier Series (CTFS) discussed in the second chapter can support the statement. If the quality of the signal can be restored completely even after the compression, then that compression technique is called as *lossless compression*, i.e., due to compression there is no loss of information. Otherwise the compression is *lossy*.

10.7.1 Relationship Between Decomposed Sub-Bands

In case of wavelet decomposition, we have already found that one coefficient in a coarse resolution (say HL_3) corresponds to four coefficients in the next level of finer resolution (HL_2, LH_2, and HH_2) and corresponds to 16 coefficients in the second next level finer resolution (HL_1, LH_1, and HH_1). This forms a parent–child relationship, considering the coefficient in third level as parent and its corresponding higher resolution coefficients as children. The parent–child relationship between the sub-bands is shown in the Fig. 10.18. Shapiro [4, 5] addressed to this parent–child relationship along with the measure of significance of each coefficient in a customized manner for compression in an elegant way.

The algorithm Embedded Zero Tree Wavelet (EZW) encoding is a multi-pass algorithm. If at one pass it is found that one parent having insignificant value (say 0) and subsequent descendents are also zero, this is forms a Zero Tree (ZT) and the parent will be called as a Zero Tree Root (ZTR). If a zero tree is encountered, there is no need to further encode the entire group of coefficients falling under that tree. Hence, compression is achieved. The zero tree is said to be *embedded* as any time at any pass we can abruptly truncate remaining insignificant coefficients depending upon the bit-budget to achieve higher compression ratio. The beauty of the algorithm will transmit the most significant coefficients in the primary passes and transmit the lower significant coefficients later. Therefore by truncating some of the coefficients any time, we can get reasonably good quality image even at very high compression ratio, whereas this kind of abrupt truncation of DCT coefficients results bad quality image in DCT based image compression.

10.7.2 Successive Approximation Quantization in EZW

In successive approximation (SA) quantization, the threshold of quantization is chosen successively on hierarchy. The approximation (quantization) starts with coarsest and successively refines to the finest. It is equivalent to starting the quantization with most significant magnitude bit and successively including lower order bits up to the least significant bit (LSB). The threshold adaptation at $(j+1)^{th}$ iteration would be half of the threshold at j^{th} iteration as expressed in Eq. 10.40 and Fig. 10.19.

$$T_{j+1} = \frac{T_j}{2} \qquad (10.43)$$

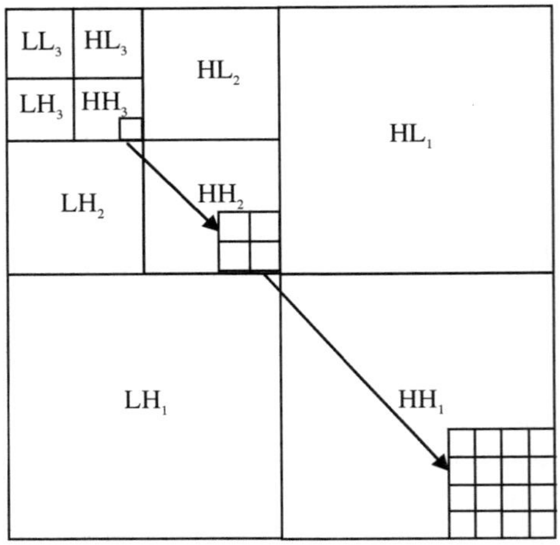

Fig. 10.18 Parent-child relation ship between the wavelet sub-bands

In EZW encoding, the SA quantization concept is used in the following way.

1. Compute the wavelet transform of the image
2. Set a threshold T_0 near the middle of the range of WT coefficient magnitudes.
3. This gives a large "dead zone" that creates of lots of "insignificant values"
(a) These give rise to lots of zero-trees
(b) Zero-trees efficiently handle significance map problem
(c) Send MSB's of significant coefficients
4. Then reduce threshold: $T_{j+1} = \frac{T_j}{2}$
(a) This causes some former insignificant coefficients to become significant
(b) For previously significant: refine by sending next finer bit

10.7.3 EZW Encoding Algorithm

The complete Embedded Zero-tree wavelet (EZW) encoding [4, 5] algorithm deals with some distinct important parameters as discussed here.

- **Sequence of Decreasing Thresholds:**

$$T_0, T_1, T_2, \ldots\ldots T_{N-1}$$

with $T_{j+1} = \frac{T_j}{2}$and |coefficients| $< 2\ T_o$
- **Maintain Two Separate Lists:**
  - *Dominant List:* A list of *coordinates* of DWT coefficients which are not yet found significant.
  - *Subordinate List:* A list of *magnitudes* of coefficients already found to be significant.

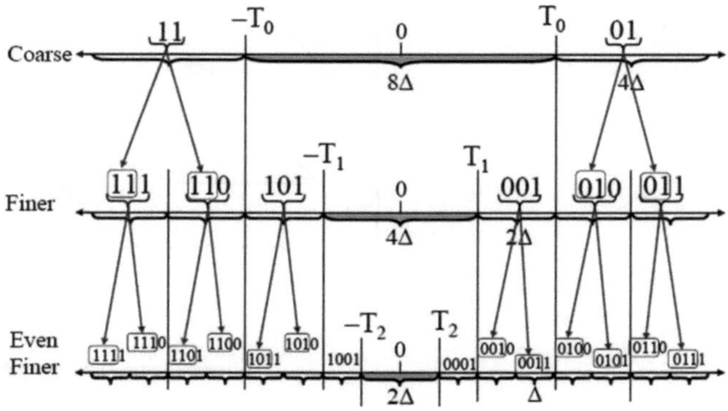

Fig. 10.19 Successive approximation quantization

- **Perform two passes for each threshold** T_j **:**
  - *Dominant (Significance Map) Pass*
    - Coefficients on Dominant List (i.e. currently insignificant) are compared to T_j to know whether this coefficient becomes significant at the new threshold.
    - The resulting significance map as scanned in the Fig. 10.20 is zero-tree encoded and transmitted using four symbols:
      - *Zero-tree Root (ZTR)*
      - *Positive Significant (POS)*
      - *Isolated Zero (IZ),* which is an insignificant coefficient with significant descendents.
      - *Negative Significant (NEG)*
    - For each coefficient that has now become significant (POS or NEG)
      - Put its magnitude on the Subordinate List (making it eligible for future refinement)
      - Remove it from the Dominant List (because it has now been found significant).
    - Provide next lower significant bit on the magnitude of each coefficient on Subordinate List
      - Halve the quantizer cells to get the next finer quantizer.
      - If magnitude of coefficient is in upper half of old cell, provide "1".
      - If magnitude of coefficient is in lower half of old cell, provide "0".
    - Entropy code sequence of refinement bits
  - *Subordinate (Significance Refinement Map) Pass*

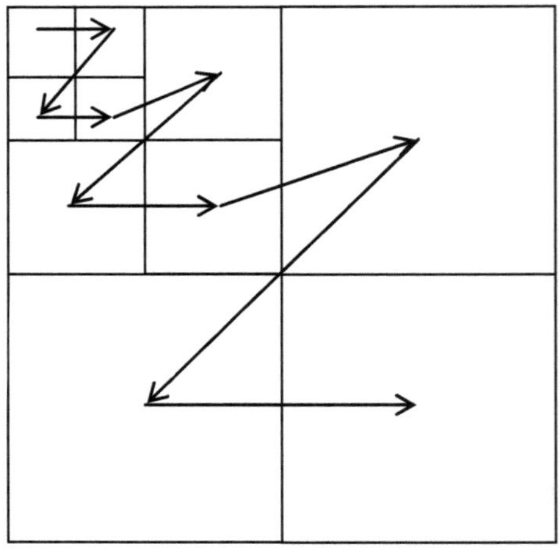

Fig. 10.20 Scanning order of the Sub-bands for Encoding the significance map

- **Repeat with next lower threshold until the bit budget is exhausted.**
- **Encoded stream is an embedded stream**

10.7.4 Image Compression Using EZW: An Example

In the Fig. 10.21, the wavelet coefficients of an 8×8 image are presented. Here DWT is performed for 3-level sub-band dyadic decomposition. According to the algorithm described above, Successive Approximation (SA) quantization is employed to get compressed image representation using EZW coding. The detailed procedure is described as follows (Fig. 10.21):

The top left corner coefficient 63 is the biggest value among all coefficients and represents the DC coefficient. According to the largest coefficient magnitude, the first dominant pass threshold T can be any number between 31.5 and 63. The initial threshold chosen as $T_0 = 32$. The lowest frequency coefficient with magnitude 63 is greater than the threshold and hence is a positive symbol. At the decoder end, this coefficient will be decoded to the centre value of the uncertainty interval [32, 64), which is 48. Next, according to the scan order, the HL_3 coefficient −34 is to be processed. Its magnitude is also greater than 32 and polarity is negative. It will be decoded to −48 in the reconstruction of the image. The coefficient of LH_3 is −31, which is insignificant with respect to the threshold 32. But it has a descendant of magnitude 47 in LH_1. Therefore, −31 is an isolated zero (IZ). Thus, the first dominant pass has completed the process on the finest scale and is going to move to the second scale. At HL_2 sub-band, except 49 is a positive significance, all other three are zero-tree root (ZTR). At LH_2 sub-band, all coefficients are insignificant. However, the coefficient 14

Fig. 10.21 Wavelet coefficients of an 8 × 8 image after 3-level dyadic partitioning

63	-34	49	9	7	13	-12	7
-31	23	14	-12	3	4	6	-1
14	15	3	-12	5	-7	3	9
-9	-7	-14	8	4	-2	3	2
-5	9	-1	47	4	6	-2	2
3	0	-3	3	3	-3	0	4
2	-3	6	-4	3	6	3	5
5	11	5	6	0	3	-4	4

has a significant child. Therefore, 15 is an isolated zero, like −31, and all other three are zero-tree roots. Coefficients in HH_2 sub-band are all zero-tree roots since themselves and their descendants are all insignificant. Then move to the finest scale. Both HL_1 and HH_1 have all zeros. At LH_1 sub-band, except 47 is a positive significant, all others are zeros. This completes the first dominant pass of the sample 8 × 8 image. After the first dominant pass, the first subordinate pass is applied to the significance map. The subordinate pass only looks at the nonzero values and refine them. The uncertainty intervals are refined as [32, 48) and [48, 64). Any significant coefficients greater than 48 will be encoded as symbol "1", and as symbol "0" if lies between 32 and 48. Each symbol "1" will later be decoded as 56, which is the centre of the interval [48, 64), while symbol "0" will later be decoded as 40, the centre of the interval [32, 48). During the second dominant pass, the threshold value is set to be 16 since all the coefficients to be processed are those insignificant ones after the first dominant pass, which are between [0, 32). The procedure is the same as the first dominant pass and results are shown in Table 10.1. The second dominant pass processes all the significant coefficients after the second dominant pass, including those from the first dominant pass. The uncertainty intervals are refined as [16, 24), [24, 32), [32, 40), [40, 48), [48, 56) and [56, 64). In other words, uncertainty intervals starts from the threshold value, and has interval length 16/2 = 8. The SAQ processing continues and can cease at any time.

10.7.5 Experimental Results of Image Compression Using EZW

For any given wavelets coefficients as input with all elements integers, the EZW algorithm encodes the significance map and then the decoder reconstructs the coefficient matrix. The decoder output is proved to be exactly the same as the input coefficients.

Table 10.1 EZW: first and second dominant and sub-ordinate passes

Sub-band	Co-eff	1st dominant pass (T = 32)		1st subordinate pass [56 40]			2nd dominant pass (T = 16)			2nd subordinate pass [60 52 44 36 28 20]
LL3	63	POS	48	1		56		0		60
HL3	−34	NEG	−48	0		40		0		36
LH3	−31	IZ	0		0		NEG		−24	28
HH3	23	ZTR	0		0		POS		24	20
HL2	49	POS	48	1		56		0		52
HL2	9	ZTR	0		0		ZTR		0	0
HL2	14	ZTR	1		0		ZTR		0	0
HL2	−12	ZTR	2		0		ZTR		0	0
LH2	14	ZTR	3		0		ZTR		0	0
LH2	15	IZ	0		0		ZTR		0	0
LH2	−9	ZTR	5		0		ZTR		0	0
LH2	−7	ZTR	6		0		ZTR		0	0
HL1	7	Z	0		0		ZTR		0	0
HL1	13	Z	0		0		ZTR		0	0
HL1	3	Z	0		0		ZTR		0	0
HL1	4	Z	0		0		ZTR		0	0
HL1	−1	Z	0		0		ZTR		0	0
HL1	47	POS	48	1		56		0		44
HL1	−3	Z	0		0		ZTR			0
HL1	3	Z	0		0					0

The EZW coder was applied to the 256 × 256 grey level (8bit/pixel) test image "lena256.bmp", which is shown in Fig. 10.22. The coding results in bit-rate of 0.31 bpp (bits per pixel) in average. Therefore, total number of bits required to represent and transmit the compressed image is $0.31 \times 256 \times 256 = \lceil 20316.16 \rceil = 20317$bits. On the other hand, to represent and/or transmit the original image, we require $8 \times 256 \times 256 = 524288$bits. Therefore, in this example, we have achieved a compression ratio of

$$r = \frac{524288}{524288} = 26 : 1 \tag{10.44}$$

10.7.5.1 The peak signal to noise ratio (PSNR)

PSNR or peak signal to noise ratio is an important measure of the amount of deterioration of a signal (here, image) by compression. To get PSNR performance, first we have to measure the mean squared (MSE) error between the original and compressed image as follows:

$$MSE = \frac{\sum_{r=1}^{M} \sum_{c=1}^{N} (original\ image - compressed\ image\ )^2}{M \times N} \tag{10.45}$$

Fig. 10.22 Performance of EZW based compression (26:1) on Lena's image

$$\text{Total number of gray levels}, Q = 256 \tag{10.46}$$

Therefore,

$$PSNR = 10\log_{10}\frac{Q^2}{MSE} \tag{10.47}$$

In the present experiment the PSNR is found to be 26.21 dB.

10.8 Matlab Programs

10.8.1 Haar Scaling and Wavelet Function

```
% Developed by Apurba Das (02/10/2011)
% Haar
clc;
clear all;
close all;
[phi psi xval]=wavefun('haar',10);
```

```
        xaxis=zeros(size(xval));
        subplot(121);
        plot(xval,phi,'k',xval,xaxis,'--k');
        axis([0 1 -1.5 1.5]);
        axis square;
        title('Haar Scaling Functin(\phi)');
        subplot(122);
        plot(xval,psi,'k',xval,xaxis,'--k');
        axis([0 1 -1.5 1.5]);
        axis square;
        title('Haar Wavelet Functin(\psi)');
```

10.8.2 Wavelet Series Expansion

```
% Wavelet series
% Apurba Das (03-Dec-2011)

clear all;
close all;
clc;

x=0:1/100:1-0.01;
f=x.^2;
subplot(421);
axis([0 1 -0.5 1]);hold on;
plot(x,f,'k');
title('The given function')
ylabel('f(x)-->')
xlabel('(a)');
grid on;
subplot(422);
axis([0 1 -0.5 1]);hold on;
sc_00=(1/3)*ones(1,size(x,2));
plot(x,sc_00,'k');
title('V_0')
ylabel('(1/3) \phi_{0,0}(x)-->')
xlabel('(b)');
grid on;
subplot(423);
axis([0 1 -0.5 1]);hold on;
wlet_00=[ones(1,size(x,2)/2) -ones(1,size(x,2)/2)];
r_wlet_00=-0.25*wlet_00;
```

```
plot(x,r_wlet_00,'k');
title('W_0')
ylabel('-(1/4) \psi_{0,0}(x)-->')
xlabel('(c)');
grid on;
subplot(424);
axis([0 1 -0.5 1]);hold on;
v1=r_wlet_00+sc_00;
plot(x,v1,'k');
title('V_1');
xlabel('(d)');
ylabel('V_0+W_0');
grid;
subplot(425);
axis([0 1 -0.5 0.5]);hold on;
wlet_10=(1/(2^0.5))*[ones(1,size(x,2)/4) -
ones(1,size(x,2)/4) -zeros(1,size(x,2)/2)];
r_wlet_10=-((2^0.5)/32)*wlet_10;
plot(x,r_wlet_10,'k');
title('K_1 * \psi_{1,0}(x)-->')
xlabel('(e)');
grid;
subplot(426);
axis([0 1 -0.5 0.5]);hold on;
r_wlet_11= -(3*(2^0.5)/32)*[zeros(1,size(x,2)/2)
ones(1,size(x,2)/4) -ones(1,size(x,2)/4) ];
plot(x,r_wlet_11,'k');
title('K_2 * \psi_{1,1}(x)-->')
xlabel('(f)');
grid;
subplot(427)
w1=r_wlet_10+r_wlet_11;
axis([0 1 -0.5 0.5]);hold on;
plot(x,w1,'k');
title('W_1=K_1*\psi_{1,0}(x)+K_1*\psi_{1,1}(x)');
xlabel('(g)');
grid on;
subplot(428)
axis([0 1 -0.5 1]);hold on;
w2=w1+v1;
plot(x,w2,'k');
ylabel('W_1+V_1');
title('V_2');
xlabel('(h)');
grid on;
```

10.8.3 Wavelet Decomposition of Image (4 level)

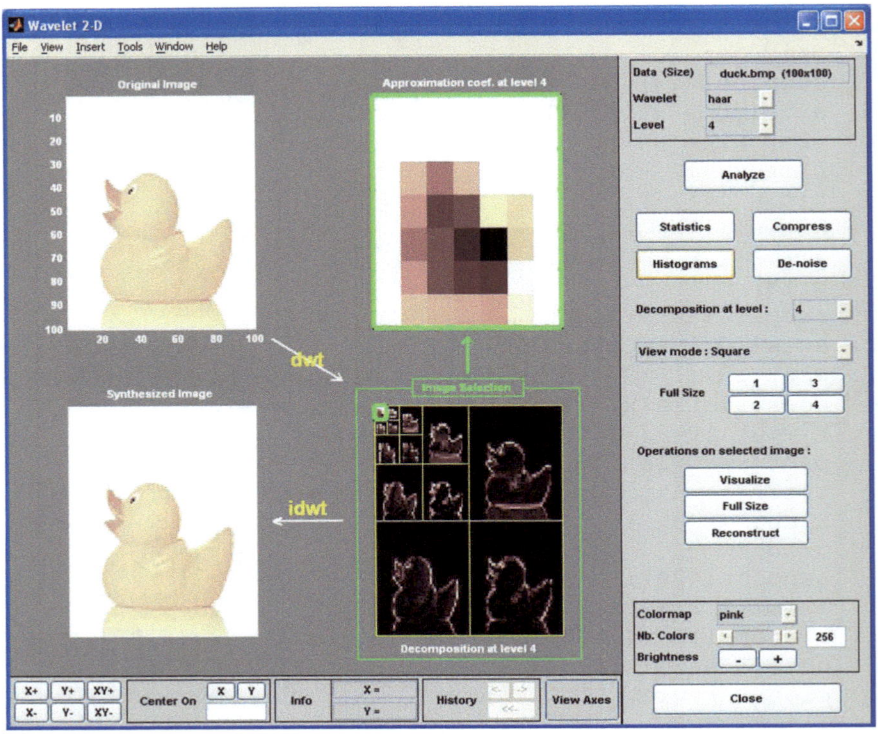

10.8.4 Image Compression by EZW Encoding

```
% Image Compression by EZW encoding
clear all;
close all;
clc;

infilename = 'lena256.bmp';
outfilename = 'lena256_reconstruct.bmp';

fprintf('-----------    Load Image    ----------\n');

img_orig = double(imread(infilename));
```

```
fprintf('done!\n');
fprintf('-----  Wavelet Decomposition   ----\n');
n = size(img_orig, 1);
n_log = log2(n);
level = n_log;  % wavelet decomposition level

type = 'bior4.4';   %wavelet basis
[Lo_D,Hi_D,Lo_R,Hi_R] = wfilters(type);

%img_wavedata: wavelet coefficients of the input image
[img_wavedata,S]=func_DWT(img_orig, level, Lo_D, Hi_D);

fprintf('done!\n');
fprintf('-----------   EZW Encoding   -----------\n');
ezw_encoding_threshold = 50;

[img_enc_significance_map, img_enc_refinement] = ...
func_ezw_enc(img_wavedata, ezw_encoding_threshold);

img_ezw_stream_bit = ...
func_huffman_encode(img_enc_significance_map, ...
img_enc_refinement);

fprintf('done!\n');
fprintf('-----------   EZW Decoding   -----------\n');
treshold = ...
pow2(floor(log2(max(max(abs(img_wavedata))))));

[img_dec_significance_map, img_dec_refinement] = ...
func_huffman_decode(img_ezw_stream_bit);

img_wavedata_dec = func_ezw_dec(n, treshold, ...
img_dec_significance_map, img_dec_refinement);

fprintf('done!\n');
fprintf('--  Inverse Wavelet Decomposition   ------\n');

img_reconstruct = func_InvDWT(img_wavedata_dec, S,
Lo_R, Hi_R, level);

fprintf('done!\n');
fprintf('-----------   Performance   -----------\n');
```

```
imwrite(img_reconstruct, gray(256), outfilename,'bmp');
fprintf('The bitrate is %.2f bpp \n', ...
length(img_ezw_stream_bit)/size(img_orig,1)/ ...
size(img_orig,2), ezw_encoding_threshold);

Q = 255;
MSE = sum(sum((img_reconstruct-...
img_orig).^2))/size(img_orig,1)/size(img_orig,2);
fprintf('The psnr performance is %.2f dB\n', ...
10*log10(Q*Q/MSE));
subplot(211)
imshow(img_orig,[]);
title('Original Image');
subplot(212)
imshow(img_reconstruct,[]);
title('Compressed Image by EZW (26:1)');
```

References

1. Daubechies, I.: Ten lectures on wavelets. CBMS, SIAM **61**, 194–202 (1994)
2. Mallat, S.: A Wavelet Tour of Signal Processing, 2nd edn. Elsevier, Amsterdam (1999)
3. Gonzalez, R.C.: Digital Image Processing, 2nd edn. Prentise Hall of India (2004)
4. Shapiro, J.M.: Embedded image coding using zerotrees of wavelet coefficients. IEEE Trans. Signal Process. **41**, 3445–3462 (1993)
5. Shapiro, J.M.: An embedded wavelet hierarchical image coder. In: Proceedings of IEEE International Conference Acoustics Speech, Signal Processing, San Francisco, CA, Mar 1992

Chapter 11
Steganography: Secret Data Hiding in Multimedia

11.1 Introduction

Steganography is the technique of hiding confidential information within multimedia. People often get confused among steganography and cryptography because both of them are similar in the way that they are used to protect confidential information. The difference between the two is in the appearance in the processed output; the output of steganography operation is not apparently visible but in cryptography the output is scrambled so that it can draw attention. Steganlysis is process to detect of presence of steganography. In this chapter I have discussed different approaches towards implementation of steganography using 'multimedia' file (text, static image, audio and video) and Network IP datagram as cover. Some methods of steganography, both forward (embedding) and reverse (reconstruction) are described in detail. At the end of the chapter we have shown how the capacity of a cover multimedia is measured in terms of security. This is defined as steganography capacity. I have also shown some statistical measure of image parameters for steganalysis and stego-capacity definition.

11.2 Steganography and Steganalysis

Modern steganography uses the opportunity of hiding information into digital multimedia files and also at the network packet level. Hiding information into a media requires following elements

- The cover media (*C*) that will hold the hidden data
- The secret message (*M*), may be plain text, cipher text or any type of data
- The stego function (*Fe*) and its inverse (Fe^{-1})
- An optional stego-key (*K*) or password may be used to hide and unhide the message.

A. Das, *Signal Conditioning*, Signals and Communication Technology,
DOI: 10.1007/978-3-642-28818-0_11, © Springer-Verlag Berlin Heidelberg 2012

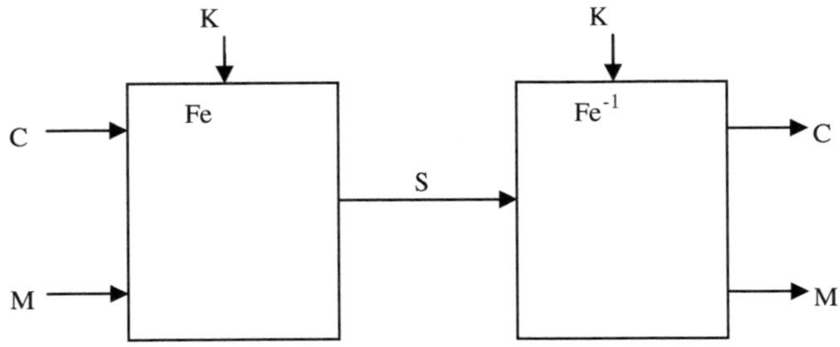

Fig. 11.1 The forward and reverse operations of steganography

The stego function operates over cover media and the message (to be hidden) along with a stego-key (optionally) to produce a stego media (S). The schematic of steganographic operation is shown in Fig. 11.1.

Steganalysis can be defined as the attack of a foreign system. The idea of steganalysis is to catch the stego object in media (channel). As we already understood that the stego and cover media are not different at all with respect to the perception of the human sensory system (vision or auditory), it is even a big challenge to detect whether the media is a normal untampered object or a stego tampered object. Steganalysis can be executed in three serial stages as follows:

1. ***Detection:*** To detect whether the object is tampered or untampered. If the object is stego then the output if this test is 'positive'.
2. ***Extraction:*** If the detection test result is positive and the algorithm of data hiding is also detected, then the tampered or corrupted bits are extracted from the stego media. In this stage we can segregate the cover media and hidden secret data. But the understanding of the hidden data is not yet ensured. Only some corrupted bits in scrambled manner are recovered.
3. ***Reconstruction:*** In this last stage of steganalysis the target English text or target image or other hidden object is perfectly reconstructed from the extracted binary bits. Some algorithms use encryption while hiding the data. In the reconstruction stage, decryption is executed to get the understandable data.

11.3 Plaintext Steganography

Message can directly be embedded into another message. This steganography can be considered as plaintext-into-plaintext (PIP) steganography. PIP can be achieved successfully in two different ways generally. We can propose one algorithm depending upon a pre-defined pattern. Demodulator can retrieve the hidden message from the pattern defined positions of the stego text. Otherwise, we can use non printable keyboard entries to append the desired data. From a look up table available with the demodulator can decode the message and reconstruct the hidden data. Both the techniques have their own advantages and disadvantages. The techniques are described with examples for the ease of understanding. Two probable algorithms for plain text steganography are presented, following these two concepts.

11.3.1 Patterned Position in a Sentence

In this type of steganography, position of each letter into each word is considered as the important feature/parameter. The sender chooses sentences in such a way that he can hide his data into the sentence following some pattern with a prior agreement with the intended receiver that the secret message is hidden within the respective position of subsequent words of the cover text. The demodulation algorithm will know the specific pattern and can extract the hidden letters from the stego sentence to reconstruct the hidden message.

Here I have proposed one algorithm which embed message according to a specific pattern, say Fibonacci series.

11.3.1.1 Algorithm

1. Pick the characters from the text according to Fibonacci series (0, 1, 1, 2, 3, 5, 8, 13……).
2. Index the 1st character as zero, 2nd character as one, and so on.
3. Consider only the alpha-characters (from 'a' to 'z' and from 'A' to 'Z').
4. If the length of the word is less than the index to be considered according to Fibonacci number, concatenate the adjacent words to get the desired length.
5. After getting the desired letter from stego word, jump to next word and start searching for the next letter.
6. Repeat all the steps until the end of sentence is reached.
7. Concatenate the retrieved letters to reconstruct the hidden meaningful data.

11.3.1.2 Example

1. Let us consider a stego plain text "**T**hese s**h**irts a**r**e gre**e**n. The **e**lephant is hap**p**y. Apurba & To**m**"
2. The position pattern is following the Fibonacci series (0, 1, 1, 2, 3, 5, 8, …..).
3. The 1st element of the Fibonacci series is '0'. 0th index of the text signifies the 1st letter of the 1st word (These), i.e., 'T'.
4. The 2nd element of the Fibonacci series is '1'. 1st index of the text signifies the 2nd letter of the 2nd word (shirts), i.e., 'h'.
5. The 3rd element of the Fibonacci series is '1'. 1st index of the text signifies the 2nd letter of the 3rd word (are), i.e., 'r'.

6. The 4th element of the Fibonacci series is '2'. 2nd index of the text signifies the 3rd letter of the 4th word (green), i.e., 'e'.
7. The 5th element of the Fibonacci series is '3'. 3rd index of the text signifies the 4th letter of the 5th word (The). But the word itself has 3 characters, only. Therefore, according to the 4th step of the algorithm, the two adjacent words are concatenated as "Theelephant", removing the 'space'. The 4th letter of the modified word is 'e'.
8. The 6th element of the Fibonacci series is '5'. 5th index of the text signifies the 6th letter of the concatenated word "is happy", that is 'p'.
9. The 7th element of the Fibonacci series is '8'. 8th index of the text signifies the 9th letter of the concatenated word "ApurbaTom", ignoring the special character '&' according to the 3rd step of the algorithm. The extracted letter is therefore 'm'.
10. Finally we reconstruct the hidden text "Three pm" or 3 pm.

11.3.1.3 Drawbacks

1. Difficult to make a stego sentence like this for any input data to be hidden.
2. If the stego sentence does not seem to carry a meaning, it can be a perceptual hint to the steganalysis attack (reverse engineering to retrieve the data from the stego message).
3. It is difficult to incorporate further level of security in terms of encryption, as both the hidden data and stego data must carry some meaning.

11.3.2 Invisible ASCIIs

There are some significant keyboard keys/characters which are very much used in typing but they can be appended to some visible character by ensuring themselves invisible. We can call them invisible ASCIIs. Some keys to be mentioned are as follows:

1. 'Space', whose ASCII value is 20_H
2. 'Tab', whose ASCII value is 09_H
3. 'CR' or Carriage Return, whose ASCII value is $0D_H$
4. 'LF' or line feed or newline, whose ASCII value is $0A_H$.

We can use any password to encrypt the text to be hidden and the cover text can be appended by some invisible ASCIIs. The WHERE and WHAT of the ASCII character is chosen from a pre-defined look up table.

In the example below, I have chosen a cover .txt file file containing only a character 'A'. I have embedded another encrypted alpha-numeric character by a six character length password. If you see the file in notepad, you can still see the same character 'A', as shown in Fig. 11.2c. In hex editor, we can clearly see the appended invisible ASCIIs to embed the secret message with secret password (Fig. 11.2a, b).

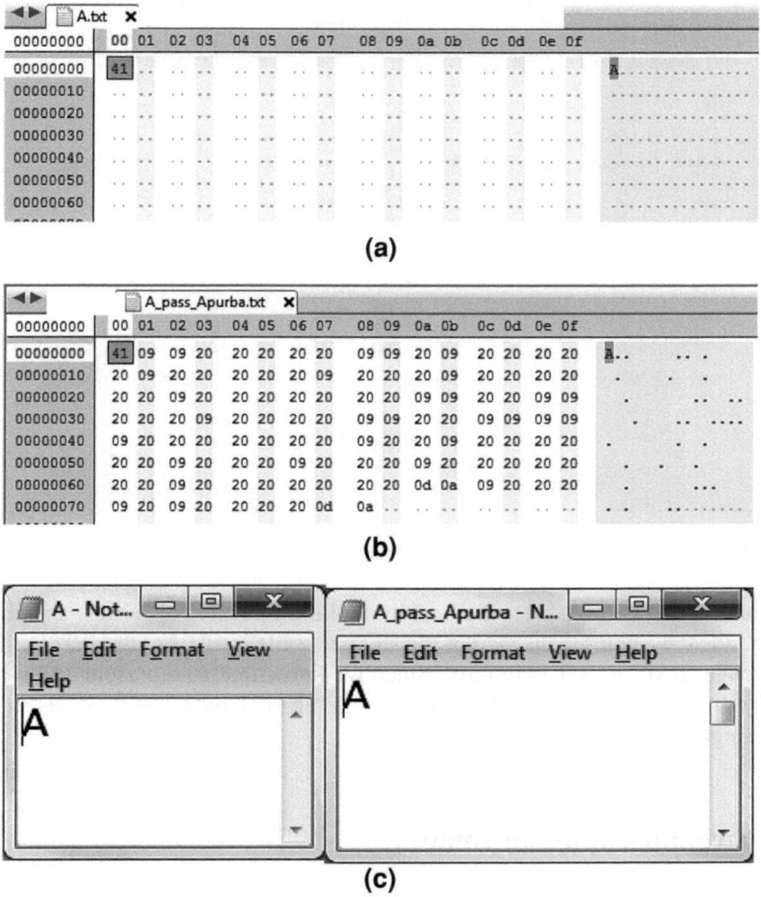

Fig. 11.2 Steganography by invisible ASCII. **a** Cover text in hex editor. **b** Stego text in hex editor. **c** Cover and stego text in notepad

11.3.2.1 Advantages

1. We can embed any message of any size using this method of steganography.
2. Not only text, by designing a suitable LUT (Look Up Table) we can embedded any multimedia secret object into any size of text file.
3. It does not directly affect the cover message. The message borne by the cover is not altered at all.

11.3.2.2 Drawbacks

We know the definition of steganography is statistical not perceptual. The present algorithm of data hiding follows the definition in one sense and hampers it in other. If we look at the previous figure (Fig. 11. 2c), there is no difference between the

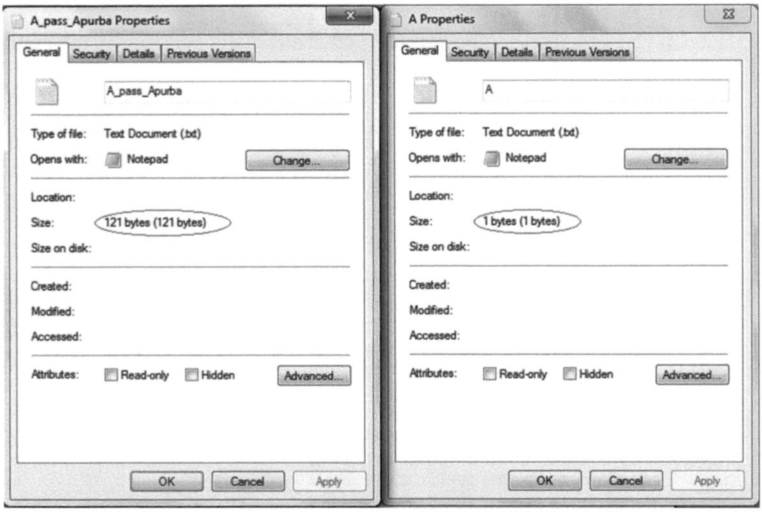

Fig. 11.3 Size mismatch between cover and stego text by invisible ASCII method

cover and the stego text files. But just by looking into the properties, as shown in the Fig. 11.3, we can see the size of the stego text file is 121 bytes whereas the size of the cover text file is 1 byte only. Since we know that to represent 'A' we need only 8 bits of data, looking at the size of stego file (121 bytes), it may arise a doubt to the steganalysis attacker.

11.4 Steganography on Images

The most popular way of doing steganography in multimedia is steganography using digital still image as the cover object. The main reasons of choosing digital image as most suitable cover are its affordable size, perceptual redundancies and availability. When we transfer any secret message embedded stego object through email, these three factors play important role. This steganography technique exploits the weakness of the human visual system (HVS). HVS cannot detect the variation in luminance of color vectors at higher frequency side of the visual spectrum. A picture can be represented by a collection of color pixels. The individual pixels can be represented by their optical characteristics like 'brightness', 'chroma' etc. Each of these characteristics can be digitally expressed in terms of 1s and 0s.

For example: a 24-bit bitmap will have 8 bits, representing each of the three primary color separations (red, green, and blue) at each pixel. If we consider just one of the separations, there will be 2^8 different values of each separation. The difference between 11111111 (255 in decimal) and 11111110 (254 in decimal) in the value for red green or blue intensity is likely to be undetectable by the human eye. Hence, if the terminal recipient of the data is nothing but human visual system

Fig. 11.4 Cover and stego image in text within image LSB steganography

(HVS) then the Least Significant Bit (LSB) can be used for hiding the secret data. The technique is applicable to any image format like BMP, JPEG, JPEG-2000, GIF, etc. The uncompressed 24 bit bitmap (BMP) image can be used for LSB steganography, JPEG and JPEG-2000 can be used for DCT and DWT based steganography respectively, whereas GIF covers can be used for palette based steganography. All these three concepts are discussed in the present section.

11.4.1 LSB Steganography

LSB of any media can be considered as virgin bit plane. If the terminal recipient of the data is nothing but human visual system (HVS) then the Least Significant Bit (LSB) can easily be used for hiding the secret data [1]. As we know that the significance of LSB is $\frac{1}{2^8} = \frac{1}{256}$ with respect to the MSB, by altering the LSB we are exploiting the weakness of human visual system for secret data hiding. The two methods are as follows.

11.4.1.1 Text Within Image

In this method binary equivalent of the message (to be hidden) is distributed among the LSBs of each pixel. For example, if we have to hide the character ‘A’, the binary bit stream representation of the ASCII of ‘A’, i.e., 01000001 can be embedded into the LSB of an 8-bit BMP image, or we can take any of the three planes (red or green or blue) of a 24 bit BMP image to embed the text.

I have included two Matlab codes (www.extras.springer.com) for LSB steganography in text within image framework. Here I have used the red LSB plane for hiding secret data. As we understand, to embed one character (i.e., 8 bits), we need 8 pixels, we have handled the exception by checking the size of the image column at least higher than 8 times the number of characters to be hidden. The complete code, result and intermediate data are as follows (Figs. 11.4, 11.5, 11.6).

MATLAB code for Text within Image LSB steganography—Transmitter

```
% Steg
% Prog. By Apurba Das (10/09/2010)
%%
clc;
clear all;
close all;
a=imread('tajmahal.bmp');
subplot(121)
imshow(a);
title('Original');
[ro co v]=size(a);
% b=rgb2gray(a);
st=input('Enter text to be hidden into the image : ');
t=st-1+1;
[z sz]=size(t);
if (co/8)<(sz+1)
    disp('Reduce length of the hidden text.....Run the
program again');
    break;
end
t_bin=dec2bin(t,8);
r=a(:,:,1);
g=a(:,:,2);
b=a(:,:,3);

% Go to middle row
new=zeros(1,8*(sz+1));
new(1:8)=dec2bin(sz,8)-48; % subtract 48 to convert
from string to int.
j=1;
for i=1:sz
    new(8+j:j+15)=t_bin(i,1:8)-48;
    j=j+8;
end
for i=1:8*(sz+1)
    if(new(i)==1)
        if rem(r(ro/2,i),2)==0
            r(ro/2,i)=min((r(ro/2,i)+1),255);
        end
    else
        if rem(r(ro/2,i),2)==1
            r(ro/2,i)=max((r(ro/2,i)-1),0);
        end
    end
end
a(:,:,1)=r;
subplot(122)
imshow(a);
title('Tampered');
imwrite(a,'tamp.bmp','bmp');
```

MATLAB code for Text within Image LSB steganography—Receiver

```
% Unsteg
% Prog. By Apurba Das (10/09/2010)
%%

clc;
clear all;
close all;

%Unsteg
%%
im=imread(uigetfile('*.bmp'));
r1=im(:,:,1);
[ro co]=size(r1);
ss=int2str(rem(r1(ro/2,1:8),2));
siz=bin2dec(ss);
j=1;
for i=1:siz
    bb=(int2str(rem(r1(ro/2,8+j:15+j),2)));
    B(i)=bin2dec(bb);
    j=j+8;
end
int2str(bb)
disp('The hidden text is : ')
char(B)
```

11.4.1.2 Image Within Image

In the same way, we can embed a complete image into another larger image by using the LSB plane of the cover. Let's consider two 24 bmp images. Both the secret image and the cover image are having three separations like red, green and blue. We can use only 16 dedicated pixels of the red plane for hiding the width and height of the secret image. These two parameters would be useful while recovering the secret image from stego. The red, green and blue secret image data can be hidden into the red, green and blue plane LSBs of the cover image respectively. In Fig. 11.7 the cover and stego image is shown. We do not fing any perceptual difference among them. But applying the extraction and reconstruction algorithm, we can successfully recover the secret image (the gun). By this was the 'gun' could be hidden into the 'rose bouquet' without any perceptual differentiation.

11.4.2 DCT and DWT Based Steganography

Especially in image the low frequency component are dominant over the high frequency components. Both in the case of DCT (JPEG) and DWT (JPEG-2000) the concept of insignificant coefficients are there. In digital image, all the

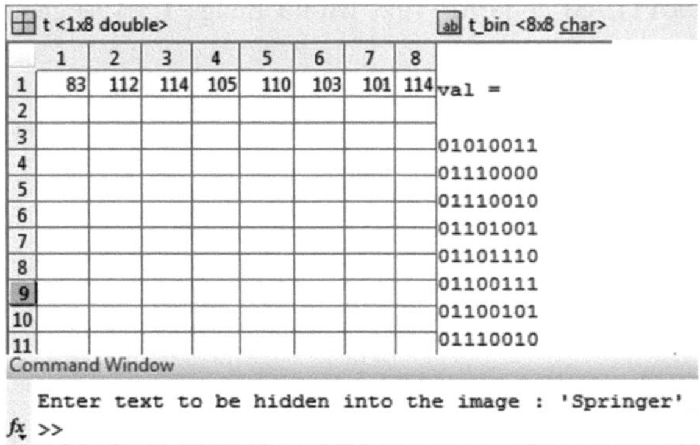

Fig. 11.5 User interaction and intermediate operations on the secret data 'Springer'

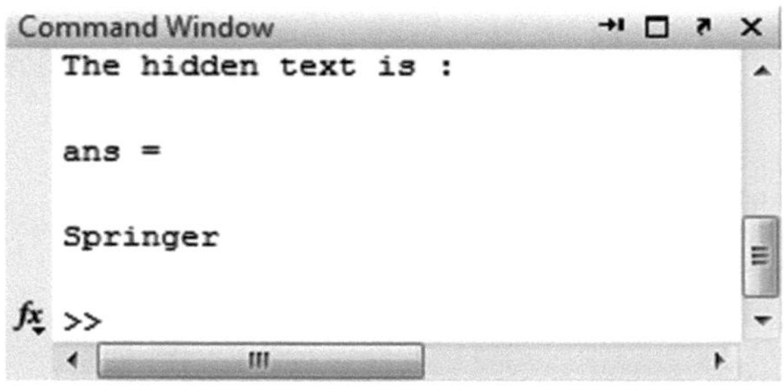

Fig. 11.6 Reconstruction of the embedded data from the stego image

components are encoded in binary bit stream. We can easily use the LSBs of the insignificant DCT or DWT components to embed secret information. One threshold can be set by the designer of the algorithm to fix the extent of insignificance of the transformed coefficients.

DCT based steganography technique follows the following steps. DWT based steganography is mostly similar to this technique.

1. The Image is broken into data units each of them consists of 8 × 8 block of pixels.
2. Working from top-left to bottom-right of the cover image, DCT is applied to each block.
3. After applying DCT, one DCT Coefficient is generated for each block in data unit.
4. Each DCT coefficient is then quantized against a reference quantization table.
5. The LSB of binary equivalent the quantized DCT coefficient is replaced by a bit from secret message.
6. Encoding is then applied to each modified quantized DCT coefficient to produce compressed Stego Image.

Fig. 11.7 Cover and stego image in image within image LSB steganography: the gun image is embedded into the stego image (not in scale)

11.4.3 Palette Based Steganography and PoV

Some compressed image formats like GIF, uses the concept of palette in image description and representation. The concept of palette based image representation has quite resembles with the act of painting. In the image palette, a number of colors are defined in an indexed manner. The colors are non-necessarily being primary. It can hold multiple composite colors. Each of the pixels only has to refer the suitable palette index which essentially gives a good amount of compression. It is not difficult to understand that if the number of color variations increases, the number of palette entries also increases, reducing the compression ration.

Palette image based steganography uses a nice technique of PoV. If we just sort the element colors residing into the palette with respect to their perceptual dominance defined by the following equation, it is obvious that the similar colors reside closer into the sorted palette. Each pair of close colors in the sorted palette is called pair of values (PoV) [2].

$$\begin{aligned} elementPalette(i) =& (-256)^3 + red(i) \times 256^2 \\ &+ green(i) \times 256^1 + blue(i) \times 256^0 \end{aligned} \tag{11.1}$$

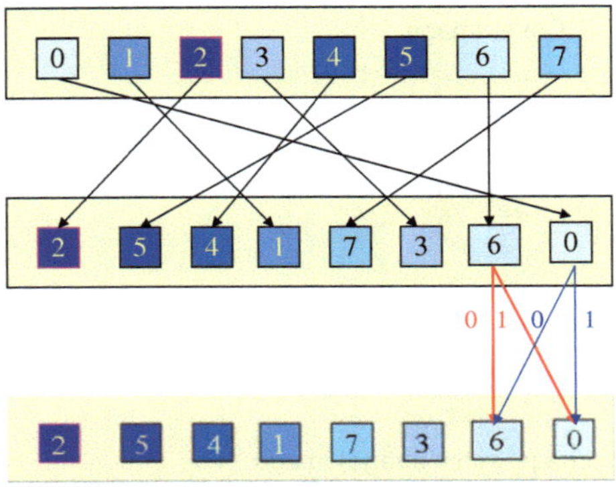

Fig. 11.8 Process of data embedding into pair of values (PoV)

The idea of secret data hiding is, we can embed one bit (either 1 or 0) into the sorted palette index not in the unsorted or original palette index. As shown in Fig. 11.8, when bit 1 is to be embedded into any even index palette entry, it jumps to the next palette entry which is its PoV. If zero is to be embedded into even indexed palette entry, the index does not change. For the odd palette entries does just the opposite. Therefore data hiding only affect the corresponding PoV. It is easily understandable that, the color pallet is sorted so that the jump between any two successive colors is relatively small.
The properties of PoV based steganography are as follows.

- After sorting the color palette, each consecutive pair naturally would not have a large color deviation. These pairs are called as Pair of Values (PoV).
- The significance of forming PoV is prevention of utmost visual attack.
- After embedding message into an image by PoV based steganography algorithms, frequency of each member of each particular PoV closes to their statistical mean.

The last property is really significant from the steganalysis point of view. Just by statistical analysis one attacker can mark one image suspect to be a stego image. As the PoV is the close region of bit exchange, it is quite obvious that for good amount of pay-load (size of hidden data with respect to size of cover) the bit exchange will normalize the height of each member of each PoV to their corresponding statistical mean. This PoV based statistical steganalysis can generate doubt to the image to be stego (Fig. 11.9).

11.5 Audio and Video Steganography

In audio steganography, secret message is embedded into digitized audio signal which result slight altering of binary sequence of the corresponding audio file. But as the alteration is done on the redundant bit level, perceptually the audio becomes unchanged to the listeners. Some of the techniques are described here.

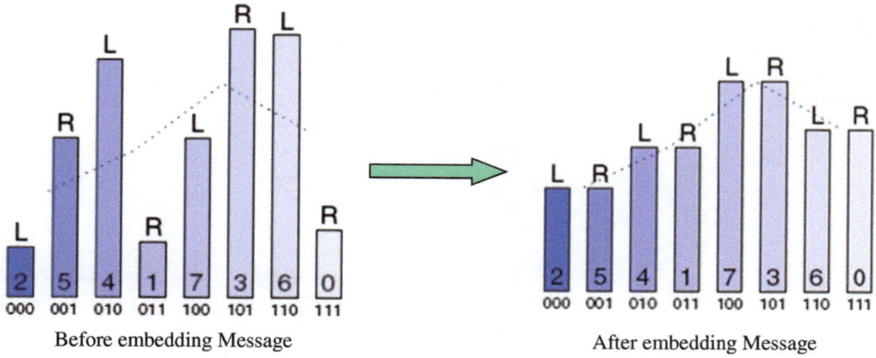

Fig. 11.9 Palette based image steganography: histogram of the sorted color palette

11.5.1 LSB Coding

Least Significant Bit (LSB) is proved to be the perceptually insignificant bit plane. Here mostly 16 bit quantization is employed as indicated in Fig. 11.10, as sample width. The weight of LSB is here 2^{-16} with respect to the MSB. It means, by changing LSB from 0 to 1 or 1 to 0 we are actually changing the sample height by an amount of $\frac{1}{2^{16}} = \frac{1}{65{,}536}$.

Sampling (in the example as shown in Fig. 11.10, 256 samples have been considered) followed by quantization [3] converts one analog audio signal to digital binary bit stream.

For example if we want to hide the letter 'A' to an digitized audio file where each sample is represented with 16 bits, then LSB of 8 consecutive samples (each of 16 bit size) is replaced with each bit of binary equivalent of the letter 'A'.

In Table 11.1, we can see the process of embedding of the secret data 'A'. Here the ASCII value of 'A', i.e., 65 is encoded digitally as '01000001'. We have just embedded these 8 bits in the LSBs of eight consecutive samples of the audio bit stream.

11.5.2 Spread Spectrum Technique

Direct Sequence is the best known Spread Spectrum Technique which can be employed for steganography in audio and video domain. Here, the secret data signal (with bit duration T_B) is multiplied by a Pseudo Random Noise Code (PN code) with chip duration T_C, where $T_C \ll T_B$.

A PN code is a sequence of chips valued -1 and 1 (polar) and has noise-like properties [3]. This results in low cross-correlation values among the codes and the difficulty to jam or detect a data message.

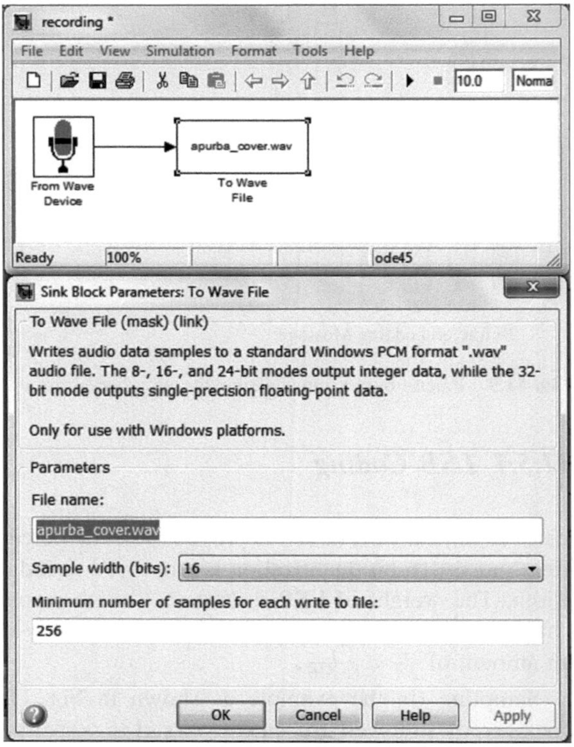

Fig. 11.10 Recording an audio (wav) signal live in 256 sampling rate and 16 bit quantization

Table 11.1 Audio Steganography using LSB embedding technique

Cover audio (8 consecutive bit stream)	'A'	Stego audio (8 consecutive bit stream)
1001 0100 0111 1101	0	1001 0100 0111 110**0**
1100 0010 1111 0100	1	1100 0010 1111 010**1**
1100 1100 0011 1000	0	1100 1100 0011 100**0**
0100 0101 1011 1100	0	0100 0101 1011 110**0**
1011 1101 1000 0111	0	1011 1101 1000 011**0**
1011 0110 0110 0110	0	1011 0110 0110 011**0**
0110 0010 0111 0101	0	0110 0010 0111 010**0**
0110 0101 1000 0110	1	0110 0101 1000 011**1**

As shown in the Fig. 11.11, the data d(t) is multiplied by the high frequency chip sequence (PN code) c(t). It results in bandwidth spreading.

As, $c^2(t) = 1$, if, the spread sequence is again multiplied by c(t), the output of the receiver would be $d(t)\ c^2(t) = d(t)$. The chips are actually chopping the bit sequence as shown in the Fig. 11.12.

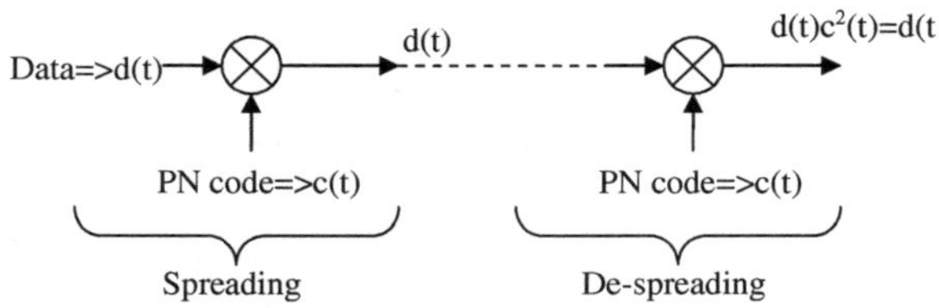

Fig. 11.11 Direct sequence spread spectrum

As the bandwidth of the data signal is now increased, as the frequency of PN sequence is much higher to the data sequence, the power contents however remain unchanged, with the result that the power spectral density lowers.

Here, the spreading is done by multiplying the digital data with PN chip sequence. This is very much analogous to Amplitude Modulation technique as (1) the modulation is done by simple multiplication and (2) the amplitude of the data is being modified/modulated, both of which are characteristics of Amplitude Modulation (AM) as discussed in Chap. 4.

In the receiver, the received signal is multiplied again by the same (synchronized) PN code. Since the code existed of +1s and −1s, this operation completely removes the code from the signal and the original secret data-signal is recovered (Fig. 11.12). Another observation is that the de-spread operation is the same as the spread operation. If the PN code is different in terms of one bit even, the secret data cannot be recovered. This PN code can be considered as secured password in this application and password is must for data hiding and recovery.

11.5.3 Echo Hiding

In the method of echo hiding, the secret message/data is embedded/hidden into cover audio signal as an echo. Three parameters of the echo of the cover signal namely amplitude, decay rate and offset from original signal are varied to represent encoded secret binary message. They are set below to the threshold of Human Auditory System (HAS) so that the incremental change in echo can't be easily perceived.

Video files are generally consists of images and sounds, so most of the relevant techniques for hiding data into images and audio are also applicable to video media. In the case of video steganography sender sends the secret message to the recipient using a video sequence as cover media. In video there are two types of compression employed, namely inter-frame and intra-frame compression. Optional secret key 'K' can here be used during embedding the secret message to the cover media to produce 'stego-video'. After that the stego-video is communicated over public channel to the receiver. At the receiving end, receiver uses the secret key along with the extracting algorithm to extract the secret message from the stego-object. The original cover video consists of frames (I frame, P frame and B frame)

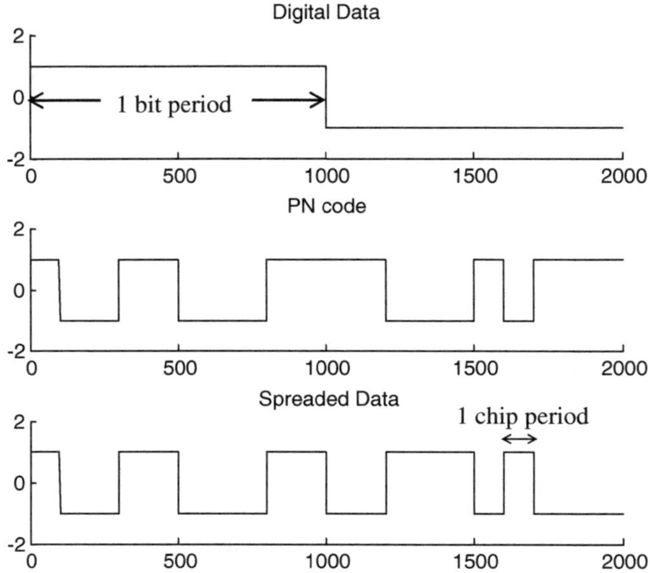

Fig. 11.12 Direct-sequence spreading by bit chopping

are represented by $\Gamma_k(m,n)$ where $1 \leq k \leq N$. 'N' is the total number of frames and m, n are the row and column indices of the pixels at each frame, respectively. The binary secret message denoted by $M_k(m,n)$ is embedded into the cover video media by modulating it into a signal. $M_k(m,n)$ is defined over the same domain as the cover $\Gamma_k(m,n)$. The stego-video signal is represented by the equation

$$S_k(m,n) = \Gamma_k(m,n) + \alpha_k(m,n)M_k(m,n); \quad k = 1,2,3,\cdots \tag{11.2}$$

Where, $\alpha_k(m,n)$ is a scaling factor.

For simplicity $\alpha_k(m,n)$ can be considered to be constant over all the pixels and frames. So the equation becomes:

$$S_k(m,n) = \Gamma_k(m,n) + \alpha \times M_k(m,n); \quad k = 1,2,3,\cdots \tag{11.3}$$

11.6 IP Datagram Steganography

This is another approach of steganography, which employs hiding data in the network datagram level in a TCP/IP based network like Internet [4, 5]. Network Covert Channel is the synonym of network steganography. Overall goal of this approach to make the stego datagram is undetectable by Network watchers like sniffer, Intrusion Detection System (IDS) etc.

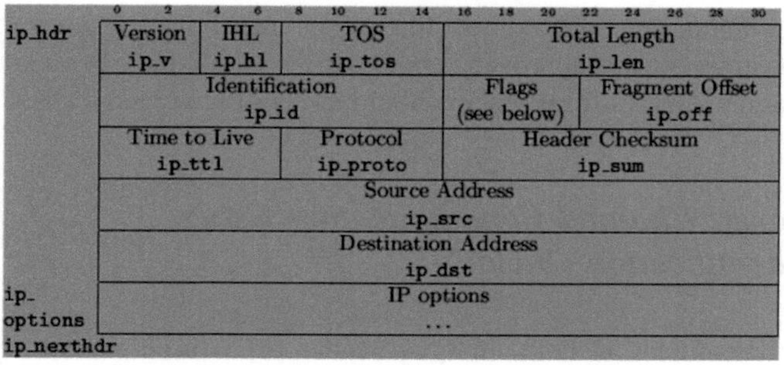

Fig. 11.13 IPv4 header

Table 11.2 IP datagram steganography in flag field

Datagram	3 Bit flag field	13 Bit fragment offset	Action
1	010	000...0000	Datagram 1 covertly communicating '1'.
2	000	000....0000	Datagram 2 covertly communicating '0'.

In this approach information to be hidden is placed in the IP header of a TCP/IP datagram. Some of the fields of IP header and TCP header in an IPv4 network are chosen for data hiding.

First we will demonstrate how 'Flags' and 'Identification' field of Ipv4 header can be exploited by this methodology (Fig. 11.13).

11.6.1 Covert Channel Communication Using 'Flags'

The size of Flag field is 3 bit. There are three flags denoted by each bit. First bit is reserved. Second and third one denoted by DF (Don't fragment) and MF (More Fragment) respectively. An un-fragmented datagram has all zero fragmentation information (i.e. MF = 0 and 13-bit Fragment Offset = 0) which gives rise to a redundancy condition, i.e. DF (Do not Fragment) can carry either "0" or "1" subject to the knowledge of the maximum size of the datagram.

Now if sender and recipient both have a prior knowledge of Maximum Transfer Unit (MTU) of their network then they can covertly communicate with each other using DF flag bit of IP header. Datagram length should be less than path MTU otherwise packet will be fragmented and this method will not work. Table 11.2 shows the how the sender communicates 1 and 0 to the recipient by using DF flag bit.

Table 11.2 shows an example of covert communication since there is no way to the network monitoring devices like IDS or sniffer to detect the communication because cover datagram is a normal datagram. As the payload is untouched, there is no way an IDS or any other content filtering device could recognize this activity. In major constraint of this approach is both parties should have prior knowledge of path MTU and datagram from sender should not be fragmented further in the way.

11.6.2 Covert Channel Communication Using 'Identification' Field

The '16-bit identification field' in Ipv4 header [6] is used to identify the fragmented packed of an IP datagram. If there is no fragmentation of datagram, then this Identification field can be used to embed sender specified information.

11.6.3 Covert Channel Communication Using ISN (Initial Sequence Number) Field

ISN (Initial Sequence Number) in TCP header [7] is another candidate for cover media for network steganography. Initial sequence number is a 32-bit digit generated during three-way TCP/IP handshaking between client and server, which is as follows.

1. Client sends a TCP/IP packet with SYN flag is ON. This is segment 1 where client specifies the port number of the server that it wants to connect to, and the client's ISN.
2. Sever responds with a TCP/IP packet with SYN flag is ON and also containing the server's ISN. Server also acknowledges the client's SYN by ACK flag is ON in this packet with acknowledgement number which is client's ISN + 1. This is segment 2.
3. The client must acknowledge this server's SYN packet by sending a packet with ACK flag is ON with acknowledgement number, which is server's ISN + 1. This is segment 3.

The large 32-bit address space of the Sequence Number field can be used for covert channel. The sender will craft a TCP/IP packet where the secret binary message can be embedded over the Sequence Number field and the passively listening receiving party will then extract the data. Source Port and Checksum in UDP header Code field in ICMP header is also good candidate for Cover item for Network Steganography (Fig. 11.14).

11.7 Steganography Capacity: A Measure of Security

Steganography capacity of a particular cover media is defined by the amount of secret data the cover can allow to be embedded into itself without significant perceptual change. As we know that the definition of steganography is statistical

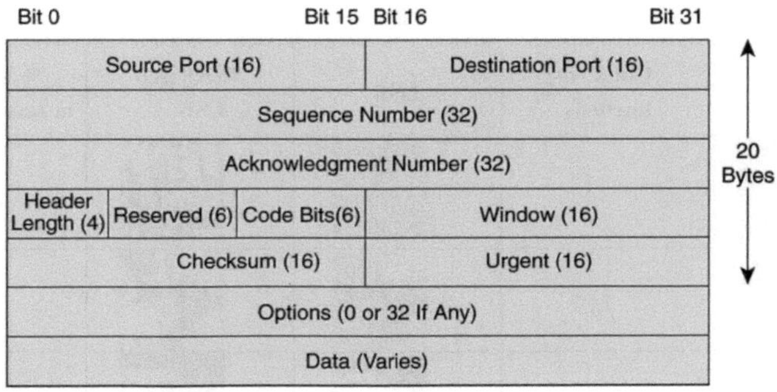

Fig. 11.14 TCPv4 header

not perceptual [8], determining steganography capacity is an important research topic.

As per Cachin's [9] definition, a steganography scheme is said to be ε-secure if the Kullback–Leibler(K–L) divergence between the stego and the cover object is at most ε. In 2009 we (Mazumdar, Das and Pal) have included the statistical properties [10] of the cover image into the Cachin's measure of steganography capacity which is more relevant to the real life data.

A new steganalysis algorithm in which the LSB plane is modelled as a 2D Ising lattice, and Gibbs Markov statistical distribution is considered to be the a-priory distribution over the lattice. In the framework of Ising model the local randomness in the LSB plane is studied and its influence on the steganography capacity is computed.

Cachin defines a steganographic method to be ε-secure ($\varepsilon \geq 0$) if the relative entropy between the stago and the cover probability distribution (P_C and P_S respectively) is at most ε, i.e., the Kullback–Leibler (K–L) divergence

$$D(P_C \| P_S) = \int_{\Omega} P_C \log \frac{P_C}{P_S} \leq \varepsilon \tag{11.4}$$

where, $\Omega \subset^{\phi} R^d$. A stego technique is said to be perfectly secure if $\varepsilon = 0$, while Cachin opined that the perfectly secure algorithm exists [9]. Chandramouli et. al [11] explained that there are several shortcoming to the definition. According to them, real-life cover images have a rich statistical structure in terms of correlation, higher-order dependence, etc. We tried to quantify these effects through our MRF model of the LSB plane [10].

Let, P_θ and P_γ be the joint probability distribution function of the non-stego and stego image respectively. Both the two pdfs follow the Gibbs distribution with the respective potential functions u_θ and v_γ.

$$P_\theta(x) = \frac{e^{-u_\theta(x)}}{z_\theta}, \quad \text{where} \quad Z_\theta = \int e^{-u_\theta(x)} dx \tag{11.5}$$

Table 11.3 Reconstruction vs kurtosis

Image	Coeff^t of kurtosis	LSB	MRF gen LSB	% of error in reconstruction
	1			26.9%
	1.07			14.77%
	3.9787			10.54%
	6.8263	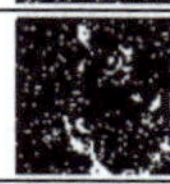	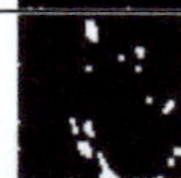	6.52%
	8.1015	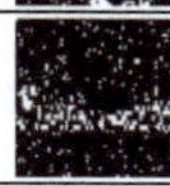	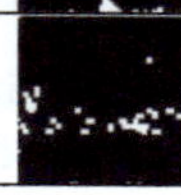	6.18%

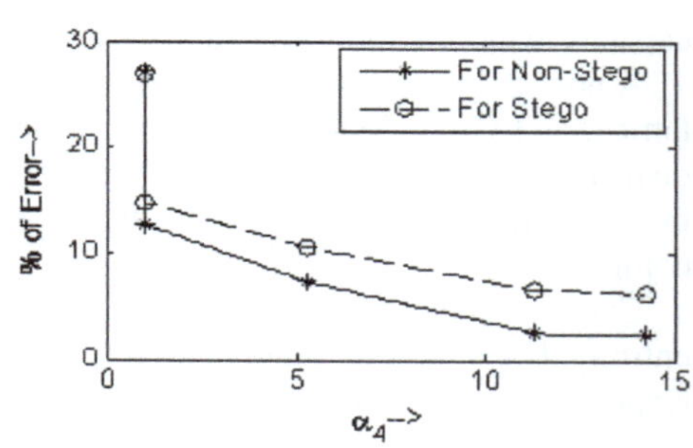

Fig. 11.15 Dependency of reconstruction error on coefficient of kurtosis

$$P_\gamma(x) = \frac{e^{-v_\gamma(x)}}{z_\gamma}, \quad \text{where} \quad Z_\gamma = \int e^{-v_\gamma(x)} dx \tag{11.6}$$

Noting the expectation with respect to pdf P_θ as $E_\theta\,[\,]$ (similarly $E_\gamma\,[\,]$ for P_γ) one gets:

$$D(P_C \| P_S) = E_\theta\left[\log \frac{P_\theta}{P_\gamma}\right] = \log \frac{Z_\gamma}{Z_\theta} - E_\theta[u_\theta - v_\gamma] \tag{11.7}$$

We compute the K–L divergence for many natural images. We have experimentally proved that K–L divergence cannot be the sole determining parameter of steganography capacity. We have presented some pair of images (cover and stego) with identical K–L divergence but of different level of perceptual differentiation.

We have also shown that the extraction of hidden bits is larger where the image coefficient of kurtosis is larger as shown in Table 11.3 and Fig. 11.15.

The problem of parameter estimation is an aspect of the research related to the modeling of any image using Markov Random Field (MRF). We have used an EM (Expectation Maximization) algorithm first proposed by Besag et. al [12] in restoring predicted cover LSB image from stego LSB, considered as degraded binary MRF images. The algorithm is essentially consists of two major steps. The LSB plane is essentially a binary lattice. By considering this binary LSB bit plane as Ising lattice, we have applied two dimensional MRF for parameter estimation in iterative manner. We have found that the high kurtosis LSB images can easily detect the tampered bits where as low kurtosis (random) LSB images, are having more steganography capacity. It is difficult to extract or predict the tampered bits statistically.

References

1. Dunbar, B.: Steganographic Techniques and their Use in an Open-Systems Environment. The Information Security Reading Room, SANS Institute (2002)
2. Stanley, C.A.: Pairs of values and the chi-squared attack. Master's Thesis, Department of Mathematics, Iowa State University (2005)
3. Das, A.: Digital Communication—Principles and System Modelling. Springer, (2010)
4. Ahsan, K., Kundur, D.: Practical data hiding in TCP/IP. In: Proceedings of Workshop on Multimedia and Security at ACM Multimedia, Juan-les-Pins, France, 6 Dec 2002
5. Provos, N., Honeyman, P.: Hide and seek: an introduction to steganography. IEEE Secur. Privacy Mag. **1**(3), 32–44 (2003)
6. Johnson, N.F., Jajodia, S.: Exploring steganography: seeing the unseen. IEEE Computer **31**(2), 26–34 (1998). (George Mason University)
7. RFC793—Transmission control protocol, September (1981)
8. Mazumdar, D., Mitra, S., Dhali, S., Pal, S.K.: A chosen plaintext steganalysis of hide 4PGP v 2.0. In: Pal, S.K., Bandyopadhyay, S., Biswas, S. (eds.) PReMI 2005. LNCS, vol. 3776, pp. 459–464. Springer, Heidelberg (2005)
9. Cachin, C.: An information-theoretic model for steganography. Inf. Comput. **1**, 41–56 (2004)
10. Mazumdar, D., Das, A., Pal, S.K.: MRF based LSB steganalysis–a new measure of steganography capacity. In: PREMI'2009, LNCS-Springer, pp. 420–425
11. Chandramouli, R., Kharrazzi, M., Memon, N.: Image steganography and steganalysis: concepts and practice. In: Lecture Notes in Computer Science, vol. 2939, Springer-Verlag (2004)
12. Besag, J.E.: Spetial interaction and the statistical analysis of lattice system. J. Roy. Statis. Soc. B **36**, 192–336 (1974)

Appendix
Frequently Used MATLAB Functions

plot()

Linear 2-D plot

Syntax

plot(Y)
plot(X1,Y1,...)
plot(X1,Y1,LineSpec,...)
plot(...,'PropertyName',PropertyValue,...)
plot(axes_handle,...)
h = plot(...)
hlines = plot('v6',...)

Description

plot(Y) plots the columns of Y versus their index if Y is a real number. If Y is complex, plot(Y) is equivalent to plot(real(Y),imag(Y)). In all other uses of plot, the imaginary component is ignored.
plot(X1,Y1,...) plots all lines defined by Xn versus Yn pairs. If only Xn or Yn is a matrix, the vector is plotted versus the rows or columns of the matrix, depending on whether the vector's row or column dimension matches the matrix.
plot(X1,Y1,LineSpec,...) plots all lines defined by the Xn,Yn,LineSpec triples, where LineSpec is a line specification that determines line type, marker symbol, and color of the plotted lines. You can mix Xn,Yn,LineSpec triples with Xn,Yn pairs: plot(X1,Y1,X2,Y2,LineSpec,X3,Y3)

imshow()

Display image

A. Das, *Signal Conditioning*, Signals and Communication Technology,
DOI: 10.1007/978-3-642-28818-0, © Springer-Verlag Berlin Heidelberg 2012

Syntax

imshow
imshow(I)
imshow(I,[low high])
imshow(RGB)
imshow(BW)
imshow(X,map)
imshow(filename)
himage = imshow(...)
imshow(...,param1,val1,param2,val2)

Description

imshow displays the file chooser dialog box so you can select an image file interactively.
imshow(I) displays the intensity image I.
imshow(I,[low high]) displays I as a grayscale intensity image, specifying the display range for I. The value low (and any value less than low) displays as black; the value high (and any value greater than high) displays as white. Values in between are displayed as intermediate shades of gray, using the default number of gray levels. If you use an empty matrix ([]) for [low high], imshow uses [min(I(:)) max(I(:))]; that is, the minimum value in I is displayed as black, and the maximum value is displayed as white.
imshow(RGB) displays the truecolor image RGB.
imshow(BW) displays the binary image BW. imshow displays pixels with the value 0 (zero) as black and pixels with the value 1 as white.
imshow(X,map) displays the indexed image X with the colormap map. A color map matrix may have any number of rows, but it must have exactly 3 columns. Each row is interpreted as a color, with the first element specifying the intensity of red light, the second green, and the third blue. Color intensity can be specified on the interval 0.0 to 1.0.
imshow(filename) displays the image stored in the graphics file filename. The file must contain an image that can be read by imread or dicomread. imshow calls imread or dicomread to read the image from the file, but does not store the image data in the MATLAB workspace. If the file contains multiple images, the first one will be displayed. The file must be in the current directory or on the MATLAB path.
himage = imshow(...) returns the handle to the image object created by imshow

drawnow()

Complete pending drawing events

Syntax

drawnow

Description

drawnow flushes the event queue and updates the figure window.

Remarks

Other events that cause MATLAB to flush the event queue and draw the figure windows include Returning to the MATLAB prompt A pause statement A waitforbuttonpress statement A waitfor statement A getframe statement A figure statement

Examples

Executing the statements x = -pi:pi/20:pi;
plot(x,cos(x))
drawnow
title('A Short Title')
grid on
as an M-file updates the current figure after executing the drawnow function and after executing the final statement.

stairs()

Stairstep plots display data as the leading edges of a constant interval (i.e., zero-order hold state). This type of plot holds the data at a constant y-value for all values between x(i) and x(i + 1), where i is the index into the x data. This type of plot is useful for drawing time-history plots of digitally sampled data systems.

Example—Stairstep Plot of a Function

Define a function f that varies over time, alpha = 0.01;
beta = 0.5;
t = 0:10;
f = exp(-alpha*t).*sin(beta*t);
Use stairs to display the function as a stairstep plot and a linearly interpolated function. stairs(t,f)
hold on
plot(t,f,'–*')
hold off

int2str()

Integer to string conversion

Syntax

str = int2str(N)

Description

str = int2str(N) converts an integer to a string with integer format. The input N can be a single integer or a vector or matrix of integers. Noninteger inputs are rounded before conversion.

Examples

int2str(2 + 3) is the string '5'.
One way to label a plot is title(['case number' int2str(n)])
For matrix or vector inputs, int2str returns a string matrix: int2str(eye(3))
ans =

1 0 0
0 1 0
0 0 1

conv()

Cell arrays can replace comma-separated lists of MATLAB variables in Function input lists Function output lists Display operations Array constructions (square brackets and curly braces)
If you use the colon to index multiple cells in conjunction with the curly brace notation, MATLAB treats the contents of each cell as a separate variable. For example, assume you have a cell array T where each cell contains a separate vector. The expression T{1:5} is equivalent to a comma-separated list of the vectors in the first five cells of T.
Consider the cell array C: C(1) = {[1 2 3]};
C(2) = {[1 0 1]};
C(3) = {1:10};
C(4) = {[9 8 7]};
C(5) = {3};
To convolve the vectors in C(1) and C(2) using conv, d = conv(C{1:2})
d =
1 2 4 2 3

conv2()

Two dimensional convolution

C = conv2(A, B) performs the 2-D convolution of matrices A and B.
If [ma,na] = size(A), [mb,nb] = size(B), and [mc,nc] = size(C), then
mc = max([ma + mb − 1,ma,mb]) and nc = max([na + nb − 1,na,nb]).

C = conv2(H1, H2, A) convolves A first with the vector H1 along the rows and then with the vector H2 along the columns. If n1 = length(H1) and n2 = length(H2), then mc = max([ma + n1 − 1,ma,n1]) and nc = max([na + n2 − 1,na,n2]).
C = conv2(..., SHAPE) returns a subsection of the 2-D convolution with size specified by SHAPE:
'full'—(default) returns the full 2-D convolution,
'same'—returns the central part of the convolution

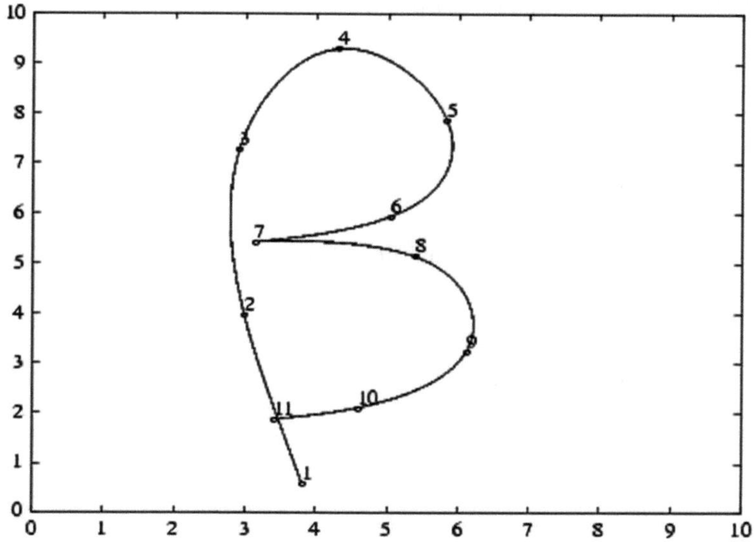

Fig. A.1 Plotting an interpolated curve

that is the same size as A.
'valid'—returns only those parts of the convolution
that are computed without the zero-padded edges.
size(C) = max([ma-max(0,mb − 1),na-max(0,nb − 1)],0).

ginput()

Interactive Plotting
The ginput function enables you to use the mouse or the arrow keys to select points to plot. ginput returns the coordinates of the pointer's position, either the current position or the position when a mouse button or key is pressed. See the ginput function for more information.
Example—Selecting Plotting Points from the Screen (Fig. A.1)

This example illustrates the use of ginput with the spline function to create a curve by interpolating in two dimensions.
First, select a sequence of points, [x,y], in the plane with ginput. Then pass two one-dimensional splines through the points, evaluating them with a spacing one-tenth of the original spacing. axis([0 10 0 10])
hold on
% Initially, the list of points is empty.
xy = [];
n = 0;
% Loop, picking up the points.
disp('Left mouse button picks points.')

```
disp('Right mouse button picks last point.')
but = 1;
while but == 1
   [xi,yi,but] = ginput(1);
   plot(xi,yi,'ro')
   n = n + 1;
   xy(:,n) = [xi;yi];
end
% Interpolate with a spline curve and finer spacing.
t = 1:n;
ts = 1: 0.1: n;
xys = spline(t,xy,ts);
% Plot the interpolated curve.
plot(xys(1,:),xys(2,:),'b-');
hold off
```

bitget()

bitget Get bit.
C = bitget(A,BIT) returns the value of the bit at position BIT in A. A must be an unsigned integer or an array of unsigned integers, and BIT must be a number between 1 and the number of bits in the unsigned integer class of A e.g., 32 for UINT32s.

Example

Prove that INTMAX sets all the bits to 1:

```
a = intmax('uint8')
if all(bitget(a,1:8)), disp('All the bits have value 1.'), end
```

bitset()

bitset Set bit.
C = bitset(A,BIT) sets bit position BIT in A to 1 (on). A must be an unsigned integer or an array of unsigned integers, and BIT must be a number between 1 and the length in bits of the unsigned integer class of A, e.g., 32 for UINT32s.
C = bitset(A,BIT,V) sets the bit at position BIT to the value V.
V must be either 0 or 1.

Example

Repeatedly subtract powers of 2 from the largest UINT32 value:

```
a = intmax('uint32')
for i = 1:32, a = bitset(a,32 - i + 1,0), end
```

dec2bin()

Convert decimal to binary number in string

Syntax

str = dec2bin(d)
str = dec2bin(d,n)

Description

str = dec2bin(d) binary representation of d as a string. d must be a nonnegative integer smaller than 2^{52}.
str = dec2bin(d,n) produces a binary representation with at least n bits.
The output of dec2bin is independent of the endian settings of the computer you are using.

Examples

Decimal 23 converts to binary 010111:
dec2bin(23)
ans =
10111

fft2()

2-D fast Fourier transform

Syntax

Y = fft2(X)
Y = fft2(X,m,n)

Description

Y = fft2(X) returns the two-dimensional discrete Fourier transform (DFT) of X, computed with a fast Fourier transform (FFT) algorithm. The result Y is the same size as X.

Y = fft2(X,m,n) truncates X, or pads X with zeros to create an m-by-n array before doing the transform. The result is m-by-n.
Algorithms

fft2(X) can be simply computed as
fft(fft(X).').'
This computes the one-dimensional DFT of each column X, then of each row of the result. The execution time for fft depends on the length of the transform. It is fastest for powers of two. It is almost as fast for lengths that have only small prime factors. It is typically several times slower for lengths that are prime or which have large prime factors.

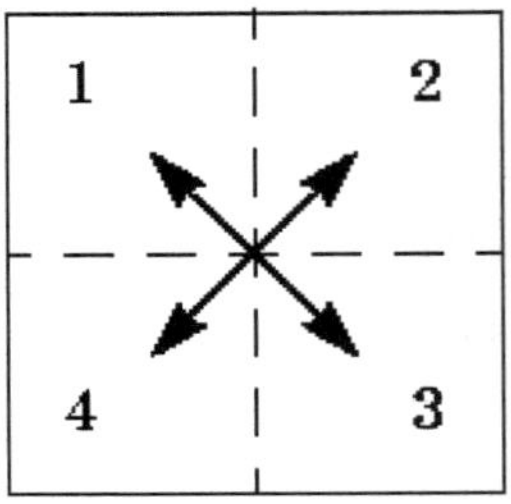

Fig. A.2 Quadrant swapping in fftshift()

fftshift()

Shift zero-frequency component to center of spectrum

Syntax

Y = fftshift(X)
Y = fftshift(X,dim)

Description

Y = fftshift(X) rearranges the outputs of fft, fft2, and fftn by moving the zero-frequency component to the center of the array. It is useful for visualizing a Fourier transform with the zero-frequency component in the middle of the spectrum.

For vectors, fftshift(X) swaps the left and right halves of X. For matrices, fftshift(X) swaps the first quadrant with the third and the second quadrant with the fourth (Fig. A.2).

For higher-dimensional arrays, fftshift(X) swaps "half-spaces" of X along each dimension.

Y = fftshift(X,dim) applies the fftshift operation along the dimension dim (Fig. A.3).

Note ifftshift will undo the results of fftshift. If the matrix X contains an odd number of elements, ifftshift(fftshift(X)) must be done to obtain the original X. Simply performing fftshift(X) twice will not produce X.

Examples

For any matrix X
Y = fft2(X)
has Y(1,1) = sum(sum(X)); the zero-frequency component of the signal is in the upper-left corner of the two-dimensional FFT. For
Z = fftshift(Y)

this zero-frequency component is near the center of the matrix.

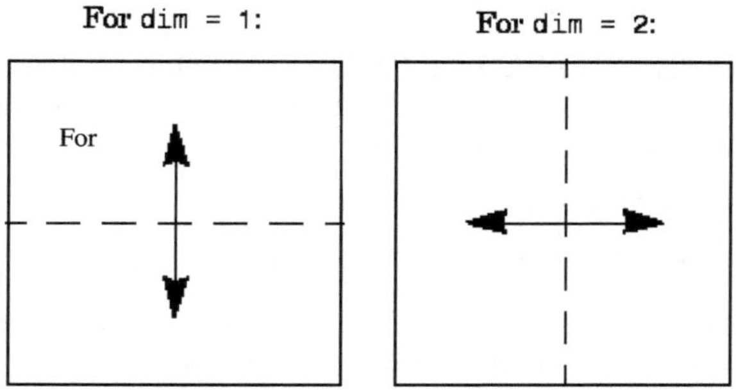

Fig. A.3 fftshift() for single and two dimensional operations

The difference between fftshift and ifftshift is important for input sequences of odd-length.

```
N = 5;
X = 0:N-1;
Y = fftshift(fftshift(X));
Z = ifftshift(fftshift(X));
```

Notice that Z is a correct replica of X, but Y is not.

```
isequal(X,Y),isequal(X,Z)
ans =
0
ans =
1
```

wavefun()

Wavelet and scaling functions

Syntax

[PHI,PSI,XVAL] = wavefun('wname',ITER)
[PHI1,PSI1,PHI2,PSI2,XVAL] = wavefun('wname',ITER)
[PSI,XVAL] = wavefun('wname',ITER)
[...] = wavefun('wname',A,B)

Description

The function wavefun returns approximations of the wavelet function 'wname' and the associated scaling function, if it exists. The positive integer ITER determines the number of iterations computed; thus, the refinement of the approximations.
For an orthogonal wavelet:

[PHI,PSI,XVAL] = wavefun('wname',ITER) returns the scaling and wavelet functions on the points grid XVAL.

For a biorthogonal wavelet:

[PHI1,PSI1,PHI2,PSI2,XVAL] = wavefun('wname',ITER) returns the scaling and wavelet functions both for decomposition (PHI1,PSI1) and for reconstruction (PHI2,PSI2).

For a Meyer wavelet:

[PHI,PSI,XVAL] = wavefun('wname',ITER)

For a wavelet without scaling function (e.g., Morlet, Mexican Hat, Gaussian derivatives wavelets or complex wavelets):

[PSI,XVAL] = wavefun('wname',ITER)

[...] = wavefun(wname,A,B), where A and B are positive integers, is equivalent to [...] = wavefun('wname',max(A,B)), and draws plots.

When A is set equal to the special value 0,

[...] = wavefun('wname',0) is equivalent to

[...] = wavefun('wname',8,0).

[...] = wavefun('wname') is equivalent to

[...] = wavefun('wname',8).

The output arguments are optional.

Example

On the following graph, 10 piecewise linear approximations of the sym4 wavelet obtained after each iteration of the cascade algorithm are shown (Fig. A.4).

```
% Set number of iterations and wavelet name.
iter = 10;
wav = 'sym4';
% Compute approximations of the wavelet function using
the
% cascade algorithm.
for i = 1:iter
    [phi,psi,xval] = wavefun(wav,i);
    plot(xval,psi);
    hold on
end
title(['Approximations of the wavelet ',wav, ...
       ' for 1 to ',num2str(iter),' iterations']);
hold off
```

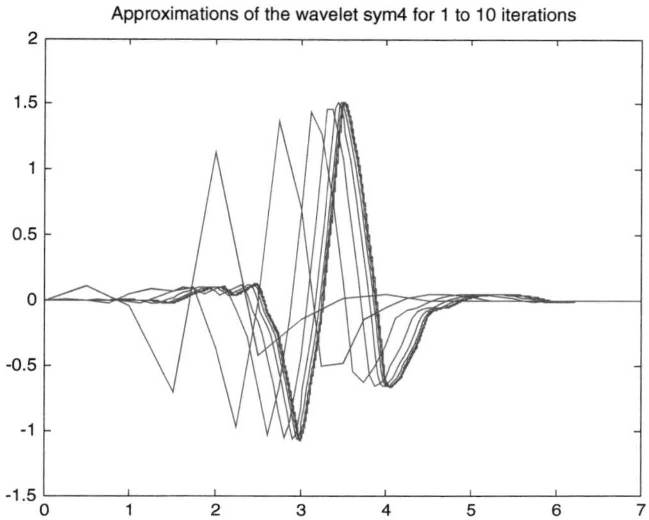

Fig. A.4 Approximation of the wavelet sym4 for first 10 iterations

Fourier Synthesizer GUI

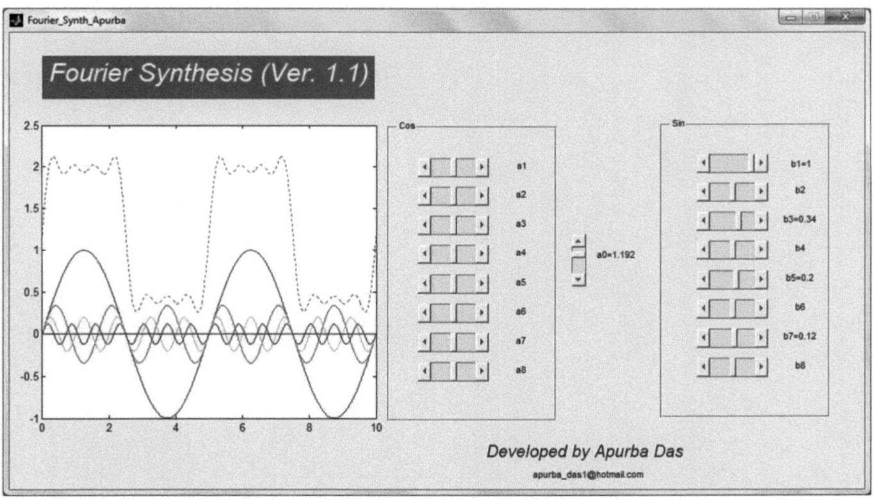

Reference

http://www.mathworks.com

Index

A. Das, *Signal Conditioning*, Signals and Communication Technology,
DOI: 10.1007/978-3-642-28818-0, © Springer-Verlag Berlin Heidelberg 2012